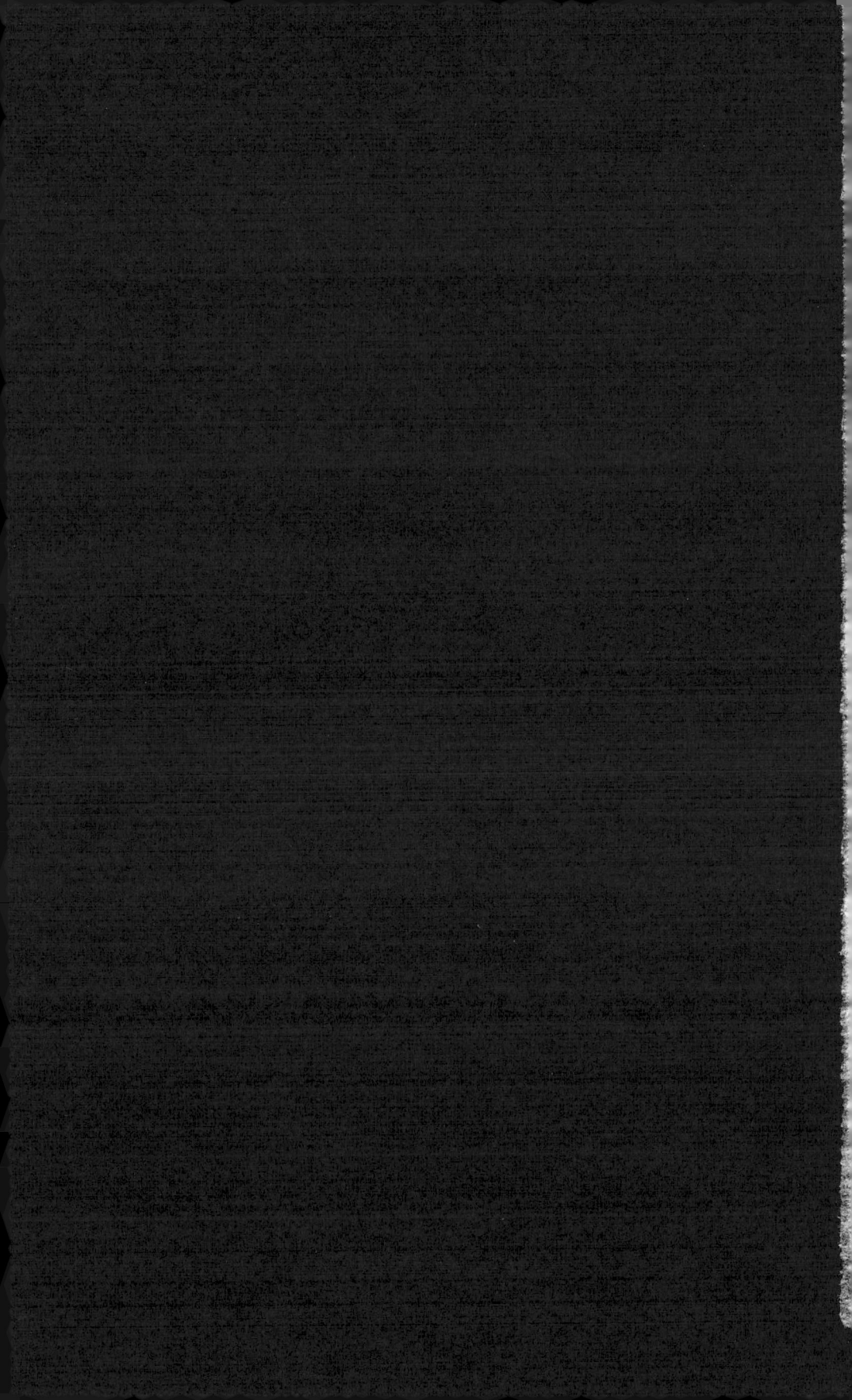

THE BOUNDLESS DEEP

Also by Richard Holmes

One for Sorrow (poems)
Shelley: The Pursuit
Shelley on Love (editor)
Gautier: My Fantoms (translator)
Nerval: The Chimeras (with Peter Jay)
Mary Wollstonecraft and William Godwin: A Short Residence
in Sweden and Memoirs (editor)
De Feministe en de Filosoof
Dr Johnson & Mr Savage
Coleridge: Early Visions
Coleridge: Darker Reflections
Coleridge: Selected Poems (editor)
Footsteps: Adventures of a Romantic Biographer
Sidetracks: Explorations of a Romantic Biographer
Insights: The Romantic Poets and Their Circle
Classic Biographies (series editor)
The Age of Wonder: How the Romantic Generation Discovered
the Beauty and Terror of Science
Falling Upwards: How We Took to the Air
This Long Pursuit: Reflections of a Romantic Biographer

THE BOUNDLESS DEEP

Young Tennyson, Science and the Crisis of Belief

Richard Holmes

WILLIAM
COLLINS

William Collins
An imprint of HarperCollins*Publishers*
1 London Bridge Street
London SE1 9GF

WilliamCollinsBooks.com

HarperCollins*Publishers*
Macken House
39/40 Mayor Street Upper
Dublin 1, D01 C9W8

First published in Great Britain in 2025 by William Collins

1

A catalogue record for this book is available from the British Library

ISBN 978-0-00-738693-2

Set in Adobe Caslon Pro by Six Red Marbles UK, Thetford, Norfolk

Printed and bound in the UK [using 100% renewable electricity at CPI Group (UK) Ltd]

To Rose Tremain, with love

When that which drew from out the boundless deep
Turns again home
– Alfred Tennyson

Contents

I

Old Monster

Was Tennyson ever young? It might seem unlikely. For generations he has been enshrined in the national memory as an ancient Victorian bard with a tremendous beard. That beard tells its own story. It began to flourish between the time Tennyson was appointed Poet Laureate and married in 1850, and the year he published his most famous poem 'The Charge of the Light Brigade' to huge popular acclaim in December 1854.

The defining bearded image appears in the early photographs by Julia Margaret Cameron in the 1860s, and was finally consecrated by a statue in the ante-chapel of Tennyson's alma mater, Trinity College, Cambridge. The entrance chamber, just off Trinity Great Court, is dominated by six massive and imposing monuments; each figure is slightly larger than life-size, carved in white marble, now turned a somnolent grey. Two are previous Masters of the College, both distinguished scientists in their own field. But the remaining four are each symbolic of wider national genius. The first is Sir Isaac Newton, the father of British science. The second is Sir Francis Bacon, the avatar of modern British philosophy. The third is Lord Macaulay, godfather of British narrative history. And the fourth is Alfred Lord Tennyson, Poet Laureate of the British Empire.

The statue was carved by Hamo Thornycroft RA to commemorate Tennyson's solemn centenary in 1909. The poet holds a large folio book on his knee, surely his enormous elegy *In Memoriam* (131 poems and an epilogue) or perhaps his *Idylls of the King* (an Arthurian romance in a cycle of twelve narrative poems) ready for yet another drawing room recital, a favourite occupation of Tennyson's in later years. Today his beard still overflows in stony prophetic locks, scrolling down beneath his chin and onto his mighty bardic chest. A small stone ash tray for his perpetual tobacco pipe lies hidden beneath his chair. There he sits enthroned, monumental and Victorian; apparently immovable, and probably extinct.

Visitors to Trinity today, especially young ones, tend to walk past the Tennyson monument without a second glance. Why should they? His mighty British Empire is mostly lost and derided, and so too is most of his later Laureate poetry, especially the neo-Gothic soap opera of his Arthurian tales, twelve thousand lines of sonorous blank verse relentlessly composed and published over thirty years, already an anachronism in the great age of the realistic Victorian novel, and dedicated (posthumously, as it were) to Queen Victoria's royal consort, the recently deceased Prince Albert. Does anyone even know how to *pronounce* the word 'Idylls', or take seriously the 'chivalric' code of knightly gallantry, or the 'historic' ruins of Tintagel and Camelot?

As his great friend Edward FitzGerald once remarked of the famous Laureate photographs: 'As to the bearded Daguerreotypes, they are very fine, I dare say: but I like the old beardless Alfred Tennyson.'

It is strange to think that some eighty years before this statue was installed, a young and anxious Tennyson, awkward and beardless and very tall, was settling into bare student lodgings in bustling Trumpington Street, well outside the stately calm of Trinity Great Court, writing one of his greatest and most unsettling poems, about an apparently extinct monster, 'The Kraken'.

Below the thunders of the upper deep;
Far, far beneath in the abysmal sea,
His ancient, dreamless, uninvaded sleep
The Kraken sleepeth: faintest sunlights flee
About his shadowy sides: above him swell
Huge sponges of millennial growth and height;
And far away into the sickly light,
From many a wondrous grot and secret cell
Unnumbered and enormous polypi
Winnow with giant arms the slumbering green.
There hath he lain for ages and will lie
Battening upon huge sea-worms in his sleep,
Until the latter fire shall heat the deep;
Then once by man and angels to be seen,
In roaring he shall rise and on the surface die.

The poetry of young Tennyson, before the beard made him a Victorian, continuously displays this kind of unsuspected and suddenly modern magic. It emerges from those essentially 'vagrant years' 1829–1849, *before* he married and became Poet Laureate, and was involved with the small group of friends, especially James Spedding, Edmund Lushington, and above all Edward FitzGerald, who tried to support and sustain him; and the women with whom he was variously in love: Sophie Rawnsley, Rosa Baring and Emily Sellwood. And perhaps one imaginary one, known simply as Maud.

It is sometimes thought that Tennyson's whole existence was essentially elegiac and melancholy, given up to the lifelong mourning for just one lost undergraduate friend, the brilliant golden boy Arthur Hallam, suddenly and tragically dead, aged just twenty-two. This, it seems, was a particular extinction from which the poet Tennyson never recovered. As W.H. Auden summarised it in a single couplet:

Black Tennyson whose talents were
For an articulate despair.

Yet Alfred Tennyson's early life also reveals a quite different kind of articulate hope: strange, energetic, independent-minded, and subtly engaged with contemporary questions. He was driven by a number of major ideas and intellectual obsessions almost forgotten or subdued in later years, or at least overwhelmed by the Victorian legend of the grand established Laureate. During this turbulent early period Tennyson's thought and poetry were fired and animated by the new science and the new scepticism; by ideas of geology and deep time; by the vastness, beauty and terror of the new cosmology; and the challenge of social revolution and women's education. But beneath all these Tennyson was driven by the emergence – or eruption – of three new and fundamentally transformative ideas from science: that of biological evolution, a godless universe, and planetary extinction.

Their impact – not only intellectual, but also emotional and spiritual – challenged Tennyson's whole imaginative world, and those around him. They brought him into contact with the life and scientific work of William Whewell (originally his university tutor), the astronomer John Herschel, the geologist Charles Lyell, the mathematician Mary Somerville, the computer pioneer Charles Babbage, and the brilliant science populariser Robert Chambers. He also shared his visions and anxieties with contemporary writers and social commentators like Thomas Carlyle and Charles Dickens, and poets like Elizabeth Barrett Browning and Edgar Allan Poe.

They led him to grapple with his own problems of destiny and identity, the threat of suicide and depression, the struggle between love and loneliness, between intellectual hope and spiritual despair. This was an Inner Kraken with which the young Tennyson also had to do battle, a Kraken of the mind. Hidden grief and extrovert delight always fought within Tennyson, and shaped the poetry of these early years.

'The Kraken' (1830) was Tennyson's first great poem, written at the age of twenty while an undergraduate at Cambridge. It is a mysterious combination of ancient folklore and modern marine science, summoning a powerful, menacing and even repulsive image, and bringing a new tone and edge to nineteenth-century poetry. It is a fifteen-line sonnet. Perhaps it was originally inspired by his lonely wanderings along the wild North Sea beaches of Lincolnshire, on the bleak dunes between Mablethorpe and Scarborough, during his restless childhood. He thought of these dunes, between the Wolds and the sea, as 'the spine-bone of the world'. He was always aware of the dark undertow of things.

The poem is composed of a fantastic mixture of Norse mythology, eighteenth-century zoology, nineteenth-century science fiction and the biblical Book of Revelations. It produced a great stir among his contemporaries at Cambridge, and foreshadowed a powerful element in the next two decades of Tennyson's life and writing. It seemed to be a projected image of unconscious forces, buried life and energy, and even psychic eruption. In what sense was the Kraken really extinct?

Certainly the poem itself almost disappeared. First published by Tennyson in 1832, it was then suppressed from all subsequent recollections for forty years, until it resurfaced, still largely hidden away, in the huge Imperial Library edition of the *Collected Works* in 1872.

The poem has deep tangled roots, both literary and scientific. The Kraken itself had become a subject of popular (half-amused) speculation in the 1820s, not unlike the Loch Ness monster a century later in the 1930s. Such sea monster stories perhaps came originally from the chapter in Revelations 13, which bears witness: 'And I stood upon the sands of the sea, and saw a beast rise up out of the sea.' But there is also the more violent prophecy from the Book of Isaiah chapter 27, verse 1: 'In that day the LORD ... will punish Leviathan the fleeing serpent, Leviathan the twisting serpent, and he will slay the dragon that is under the sea.'

Tennyson's own researches would find a nicely obscure reference quoted in the family copy of the *Biographie Universelle* of 1823. He noted: 'See the account which Erik Pontoppidan, the Norwegian bishop, born 1698, gives of the fabulous sea-monster – the Kraken'. Pontoppidan describes the creature as variously an enormous cephalopod, a giant crab or a huge starfish; it is so big that when on the surface fishermen have mistaken it for a small island, and fatally anchored alongside it. When moving underwater, it was capable of producing a suction whirlpool like the notorious Norwegian

Maelstrom which was equally lethal (and which would later fascinate Edgar Allan Poe).

Pontoppidan gives the following description of the creature's powers in his *Natural History of Norway*: 'It is said that if the Creature's arms were to lay hold of the largest man-of-war, they would pull it down to the bottom.' All these images of darkness, engulfment and capture would find their many echoes and equivalents in Tennyson's maturing imagination.

As a speculative subject, the Kraken was one of the intriguing topics that Coleridge had broached with the young John Keats, during their famous encounter on Hampstead Heath, a decade before in April 1819. These two Romantic poets of the previous generation, alongside Shelley, were the most important to young Tennyson. It is easy to forget how nearly they all overlapped in time, and how alive their work still seemed to him. Keats had died less than a decade before in 1821; Shelley had drowned in 1822; Byron, extinguished by fever at Missolonghi in 1824. Even Coleridge, though not yet dead, was already a largely mythological presence (though visitable under the right sage-like conditions) at Highgate until 1834; while Wordsworth hung on in the Lake District, like some mighty low-lying luminous mist, doomed to condense into the Laureateship after Robert Southey. So the Romantics were a powerful continuing presence, a poetic generation that never grew old. The very fact of their ageless youth haunted Tennyson's own youth, the glamorous ghosts at his troubled feast.

When Keats met Coleridge walking on London's Hampstead Heath in the spring of 1819 (Tennyson then being a boy of ten in remote Lincolnshire), the Kraken quickly raised its head. Keats never forgot the topics Coleridge mentioned, almost like possible ideas for future poems. Coleridge, as usual, dominated the talk during their slow two-mile walk up towards Highgate, 'at his alderman after-dinner pace'. Keats enthusiastically records the kaleidoscope of potential ideas in one of his letters: 'In these two miles he broached a thousand things. Let me see if I can give you a list – Nightingales, Poetry, on Poetical sensation, Metaphysics, Different genera and species of Dream, Nightmares ... Second Consciousness, Monsters, the Kraken ...'

Tennyson soon knew much of the Romantics' poetry by heart.

Stylistically his own 'Kraken' poem already has echoes of Keatsian phrases like its 'wondrous grot and secret cell', as well as hints of Coleridge's strange opium-dream pathology: 'battening upon huge sea-worms in his sleep'. Tennyson also appears to draw variously on more immediate scientific sources. In 1818 the Scottish journalist James Wilson contributed a long speculative article on the biological nature of the Kraken to *Blackwood's Edinburgh Magazine*.

There were also echoes of the mathematician Mary Somerville's account of hearing of the Kraken legend when a young girl growing up in Scotland 'like a savage' and roaming the remote beach at Burntisland on the Firth of Forth and talking with the fishermen. She was fascinated to hear of it, that enormous deep-sea monster (a 'gigantic flat-fish' or perhaps a 'sea serpent' or a giant squid), rumoured to lurk off the coast of Norway in the icy waters of the North Sea, and sometimes surfacing to devour unsuspecting mariners. 'It was so enormous that when it came to the surface, covered with tangles and sand, it was supposed to be an island, till on occasion, part of a ship's crew landed on it and found out their mistake.' For her it symbolised the mysteries and superstitions of nature that science, and especially mathematics, was trying to rationalise.

So Tennyson's 'Kraken' is in some ways the first modern science fiction poem, and can be compared to the first modern science fiction novel of monstrosity, Mary Shelley's *Frankenstein*, which became popular in its second 1832 printing. It also curiously anticipates the world of Poe's horror stories; and Jules Verne's *Twenty Thousand Leagues under the Sea*. The Kraken surfaces menacingly in the form of 'The Squid' in chapter 59 of Melville's *Moby Dick* (1850), 'curling and twisting like a nest of anacondas'; and hence eventually huge and ravenous in the Steven Spielberg film *Jaws* (1975). Although it dies in the last line of the poem, Tennyson's monster was in fact bursting with future life.

It had an unusual curriculum vitae. The Kraken as an enormous, unnamed multi-tentacled cephalopod was first mentioned scientifically in Pliny's *Naturalis Historia* in AD 77 (shortly before Pliny witnessed another sort of monster, the eruption of Vesuvius). It was first formally catalogued in Carl Linnaeus's *Systema Naturae* (1735).

Yet Linnaeus withdrew it from later editions, when he clarified that
he had 'never actually seen one' himself. Scientific fascination with
undersea life nevertheless grew at the end of the eighteenth century,
especially with the journals, logs and letters written during the many
naval expeditions of the Napoleonic Wars.

The hidden monster, well established as a Romantic literary idea,
was soon to be transformed into a specific mid-Victorian fascination
with the revelations of marine zoology. A host of bizarre undersea
creatures, previously unknown or feared or simply ignored, now
revealed their characteristics. These ranged from the sensitivities of
the humble sea-anemone to the mysterious intelligence of the secretive octopus (which seemed to promise, if not authenticate, the sinister
possibilities of the Kraken). It would lead to the outdoor cult of the
seashore rockpool, and the indoor hobby of aquarium-keeping.

Both activities were popularised by P.H. Gosse and Charles Kingsley in many beautifully illustrated works of natural history in the 1860s.
It was Philip Henry Gosse who early championed the aquarium. After
decades of seashore research in Jamaica and in Devon, with many papers
sent to the Royal Zoological Society, and an animated exchange of letters with Kingsley, Gosse would publish his landmark marine study, *The
Aquarium: an Unveiling of the Wonders of the Deep Sea*, in 1854, shortly
followed by the more practical *A Handbook to the Marine Aquarium*.

This new invention of an enclosed, glass 'water museum', convenient for home study but also for philosophical reflections, provided a
method of descending imaginatively into a completely alien submarine world. The aquarium was in effect a kind of homely glass
bathascope. Gosse's revelation of a previously hidden aspect of the
natural universe was widely praised, as for example by the *Literary
Gazette*: 'Mr Gosse has ... dived into the bejewelled palaces which
Old Neptune has so long kept reluctantly under lock and key, and we
find their treasures set before us with a freshness and fidelity which
afford welcome and instructive lessons to naturalists of all ages.'

The new opportunity of examining sea creatures in great detail, with
the use of portable microscopes and exquisite laboratory drawings and
dissections, was both fascinating and repelling. The impression of such

green alien depth, yet with magnified aquarium clarity, is central to the descriptive lines of Tennyson's poem, with their evocative mixture of Gothic and scientific terms:

> And far away into the sickly light,
> From many a wondrous grot and secret cell
> Unnumbered and enormous polypi
> Winnow with giant arms the slumbering green.

The Kraken theme becomes a universal expression of certain Victorian fascinations and fears, like the contemporary discovery of dinosaurs. They were all monsters of Deep Time, that may – or may not – have escaped extinction. In one version the Kraken is a sort of submarine dinosaur. In another it is a monstrous octopus or appalling squid. When it finally surfaces, it will terrify mankind, but die in convulsions and fury. So the Kraken becomes the symbol of a monstrous idea, which will eventually erupt among humanity: like the idea of relentless evolution; or a universe without a divine creator; or the inevitable extinction of every species on planet earth, including the human race itself.

But it also gradually becomes clear that it is something more intimate: Tennyson's Inner Kraken. It is some dark hidden shape in his mind or personality, in his memory or unconscious, or even perhaps in his soul, which will steadily rise and come forth as he grows older. Perhaps the most immediate of these monsters is Scepticism or Doubt itself, as Tennyson would later put it in No. 124 *In Memoriam*, a vision or a voice which itself rises out of the sea, out of the waves, out of the 'Godless deep'.

> I found Him not in world or sun,
> Or eagle's wing, or insect's eye;
> Nor through the questions men may try,
> The petty cobwebs we have spun:
>
> If e'er when faith had fallen asleep,
> I heard a voice 'believe no more'

>And heard an ever-breaking shore
>That tumbled in the Godless deep;

Victorian critics would only slowly recognise the significance of the Kraken. It would be picked out by the young philosopher of liberty, John Stuart Mill, in 1835. But it was not until 1887 that the militant free-thinker and Darwinian J.M. Robertson could write: 'In the short piece on "The Kraken" we have the [promise] of a fresh kind of achievement in our literature, that weaving of the ideas or the fancies of science into harmonious poetry without the loss of the scientific outline, in respect of which Tennyson stands apart from those poets, like Shelley, who have paraphrased such ideas into allegories ...' It was just such ideas, such monsters, emerging from various kinds of somnolent depths, that would shape young Tennyson.

2

Somersby

So is it still possible to recover a young Tennyson? Just plain Alfred Tennyson, a tall, clever, gangling boy from Lincolnshire, with wild surprising hair and an astonishing voice full of strange music? He was far from solitary: the middle one of a large, chaotic family of eleven country children growing up on the remote, windswept coast of Eastern England, on the edge of the grey, storm-tossed North Sea which stretches away far northwards from Europe, all the way to Scandinavia and the frozen Arctic.

Tennyson was born at his father's rectory in Somersby, a remote country village in Lincolnshire, on 6 August 1809. This was not only the same birth year as Charles Darwin and Edgar Allan Poe, but also of Edward FitzGerald. Somersby was fifteen miles from the sea, in a valley dip of the Wolds, with a small bright river and tall dark elm trees. The rectory was large and rambling. The dining room had Gothic vaulting and stained-glass windows; wild woodbine climbed into the bay window of the upstairs nursery; downstairs the library was packed with 2,500 books and furnished with collapsing sofas and yellowing curtains. At the top of the house was a labyrinth of small attic bedrooms which the children shared. Outbuildings also housed a coachman, a housekeeper, a cook, a maid and several gardeners. The big lawn outside, used for family gatherings throughout the summer, was 'overshadowed on one side by wych elms, and on the other by larch and sycamore trees'.

The large family consisted of seven brothers and four sisters. Closest to Alfred were two older brothers, Frederick and Charles, and a younger sister, Emily. His father the rector, Dr George Tennyson, was a clever Cambridge-educated man, well-read in the Greek and Latin classics, and also 'a Hebrew and Syriac scholar' and a competent mathematician. He played the harp and had an unlikely passion for wood-carving. He oversaw much of his sons' education, at times tutoring them himself, particularly in classics and the new genre of popular scientific books. At the age of seven Alfred could recite Horace's Odes by heart, and use a telescope to study the moon.

But Dr George Tennyson was a disappointed man too. He had been left out of a family inheritance in favour of his younger brother Charles, being 'rude and ungovernable' as a child, and apparently unstable and extravagant as he grew up. How far this was true is not clear. But in consequence of such a perceived and rankling injustice, Dr Tennyson's rich and powerful father, Alfred's grandfather, was always mockingly referred to by the Somersby Tennysons as 'the Old Man of the Wolds'.

The Old Man lived nearby on the large Lincolnshire estate of Bayons Manor near Tealby (originally purchased from the aristocratic d'Eyncourt family). The Old Man's gifts to his younger son Charles allowed Charles to pursue a brilliant career in Parliament, supporting the Reform Bill of 1832, and eventually inheriting the whole Bayons estate and adopting the supposedly aristocratic name of Charles Tennyson d'Eyncourt. It also helped that the Old Man's daughter Elizabeth made a rich marriage to a coal millionaire Matthew Russell, and settled comfortably in London's newly fashionable West End.

By contrast, George was given no alternative career but the Church, taking on the small country parish of Somersby, with a clerical stipend and a paternal annuity of £700, barely enough to support his rapidly growing family. It was a situation of mounting humiliation and frustration for him. Moreover in future years, the Old Man of the Wolds, together with 'rich Uncle' Charles of Bayons Manor, and 'rich Aunt' Russell of Chelsea, would have a considerable financial influence on Alfred's education, beyond his control, both benign and otherwise.

Over the years of Tennyson's childhood, the rector became increasingly depressed, isolated and bullying. Ill-tempered with his family, violent with the servants, furious with his wife, Dr Tennyson ultimately descended into alcoholism and had to be despatched abroad in his sixties for a cure. Meanwhile the domestic chaos included outbursts against his sons, regular stand-up rows with the coachman, and a tragic incident – never fully explained – when the cook caught fire in the kitchen and burned to death. It is clear that Dr Tennyson felt cheated by life, and besieged by his large family. He once refused to have his rich London sister Elizabeth Russell to stay by sarcastically announcing: 'We are three and twenty in the family, and sleep five or six in a room. Truly we have great accommodation for Mrs Russell and her suite.' Yet Aunt Russell later became a great supporter of her nephew Alfred, of whom she strongly approved.

Dr Tennyson's wife Elizabeth Tennyson, once a beauty of the county, resigned herself to her abusive marriage, and became a sweet, patient and adoring mother, defending her children wherever she could. She was especially close to Alfred. He in turn worshipped her as 'one of the most angelic natures on God's earth', and praised her as 'innocent and tender-hearted' towards animals. He wrote poetry about her, and later described her as winning her way with 'extreme gentleness', and with a tender gaze that was always 'clear, without heat, undying'. His early poem 'Isabel' gave her a kind of martyred sainthood, but was shot through with irony.

A courage to endure and to obey;
A hate of gossip parlance, and of sway,
Crown'd Isabel, thro' all her placid life,
The queen of marriage, a most perfect wife.

All the children were deeply affected by this dysfunctional upbringing, producing various degrees of brilliance, eccentricity, hypochondria and mental instability. 'We Tennysons are a black-blooded race,' said Tennyson later, though with a distinct touch of pride. The whole family was known locally as strange and clever, but extremely

unconventional in their behaviour. All the children ran wild in the
surrounding countryside. They were tall and awkward and thin, and
retained strong Lincolnshire accents. They were also bookish and
famously stand-offish with outsiders. Every one of them wrote poetry
(especially sonnets) or kept journals, including the girls. Mary Tennyson, for example, a year younger than Alfred, collected '132 Sonnets
and Fugitive Pieces'. Seven of the Tennyson children (besides Alfred)
had their work published as adults in one form or another. In their
twenties, Frederick and Charles were considered serious poets in
their own right. Altogether the Somersby Tennysons were a brilliant
and remarkable family, of huge promise.

It is all the more striking that the later careers of the adult men
were so unhappy and unfulfilled. Dr Tennyson's paternal influence on
most of his sons turned out to be deeply damaging. One quickly
exiled himself and drifted for thirty years through Italy; another settled on a remote Lincolnshire estate and became an opium addict; a
third succumbed to depressive illness and alcoholism; and a fourth
spent the last fifty years of his life in Lincoln County lunatic asylum.
Only the four daughters, protected by their loving mother Elizabeth,
achieved happier and more normal lives, though they were all regarded
as 'decidedly eccentric' in the neighbourhood. Three of them married
successfully, though notably late. One emigrated for a time to Jamaica.
The next youngest, Matilda, unmarried and intensely religious, remained
as her mother's companion and carer until Elizabeth's death in 1865.
So out of the wreckage of that large, vivid Somersby household, who
could have guessed that it was Alfred who would survive best and
prosper most? He was threatened by exactly the same fatal inheritance
as his Tennyson brothers – lethargic drifting, disabling depression,
alcoholism, mental instability, or simply black spiritual despair. He
struggled with these for all his youth, and certainly for his first forty
years. But he did not succumb; he was not extinguished. He wrote
poetry instead.

Alfred was a small, shy retiring boy at first, but would grow up to
be over six foot tall, exotically handsome and when silent in a room,
faintly menacing. His striking good looks were often described as

'gypsy' or Spanish, with dark deep-set eyes and magnificent tawny hair. Physically strong, he was also short-sighted, and in his twenties would flourish a pince-nez to dramatic effect. When he did speak, his voice was low and powerful, with a Lincolnshire accent, soon to be used with hypnotic effect.

To begin with Alfred shared an attic bedroom with his two elder brothers in the gabled roof at the top of the rectory, with its own creaking stairs. He remembered its enchantments long into old age. There was a large wooden beam across the ceiling, from which the brothers would swing and tell stories: 'Knights and heroes among untraveled forests ... or on gigantic mountains fighting with dragons'. From a skylight they could glimpse the moon and stars. Tennyson would sneak down before dawn to collect fallen apples from the rec-tory orchard, and especially loved wild weather. 'Before I could read, I was in the habit on a stormy day of spreading my arms to the wind, and crying out "I hear a voice that's speaking in the wind".' That sense of infinite distances haunted him. 'The words "far, far away" always had a strange charm for me.'

From the age of ten, he revelled in long walks by the sea at Mab-lethorpe, a huge windswept beach some fifteen miles to the east. The family kept a cottage there, under the sea-bank, 'the long low line of tussocked dunes', and he would break away on his own, haunted by the 'thunderous roar' of the big breakers on the Lincolnshire shore. He wrote an early childhood prayer: 'Thou art the God of Heaven and Earth. Thou has created the immeasurable sea ... Thou givest and Thou takest life, Thou destroyest and Thou renewest.' From his boyhood, 'he had a passion for the sea, and especially the North Sea in wild weather'.

He would lie for hours alone upon the beach, or hidden up in the dunes, listening to the waves, imagining sea creatures far out in the deep, or dreaming of the ancient Greek invasion fleet that had set out long ago to besiege Troy, as recounted in Homer's *Iliad*. He learned Alexander Pope's translation by heart, before being enchanted by Byron's romantic Greece. He recalled some fifteen years later (1837):

Here often when a child I lay reclined:
I took delight in this fair strand and free;
Here stood the infant Ilion of the mind,
And here the Grecian ships all seemed to be.

And here again I come, and only find
The drain-cut level of the marshy lea,
Gray sand-banks, and pale sunsets, dreary wind,
Dim shores, dense rains, and heavy-clouded sea.

The beach at Mablethorpe continually recurs in his later poetry and letters, as a place of loneliness and bleak inspiration. The long sweep of sand and shingle was backed by a line of dunes, with a distant inland prospect of the slate grey of the Wolds. It was then a bare but beautiful region, with no buildings except some fishermen's cottages, one of which Tennyson would later rent as his writing retreat. The cottage itself was thatched and primitive, and described by a later visitor as 'an orange-coloured little house', sheltered from the sea by a forty-foot sand dune. On the other side stretched 'the cropped level of the marsh, with the faint curve of the Wolds upon the western sky-line'.

To Tennyson its dreariness and solitude were always magical, and one to which he could always return and call upon in later life. When he holed up there alone in the difficult winter of 1843, at the age of thirty-four, he drew a sketch map of the shore, showing the 'line of sand left by the tide', the reed beds and the hills beyond, and a tiny sprawled figure marked 'ME' on the beach. He asked friends to send letters and journals. 'There is nothing here but myself and two starfish . . . I send the sketch to melt your heart . . . remember me to Fitz in town.'

At another difficult time on the coast in Wales (at Aberystwyth), his thoughts went straight back to Mablethorpe. 'Anything more unlike the old Homeric "much-sounding" sea I never saw. Yet the bay is said to be tempestuous. O for a good Mablethorpe breaker!' Again and again in these years he would turn back and imagine the sound of oceans, and the geological process of erosion:

The moanings of the homeless sea,
The sound of streams that swift or slow
Draw down Aeonian hills, and sow
The dust of continents to be ...

That adjective Aeonian, struck like a gong in the centre of the third line, with all its open reverberating vowel sounds, became almost Tennyson's own coinage, used on several other occasions. Spelled here with a capital letter, it meant not only something endless or perpetual (a hundred million years in cosmology), but implied the name of an entire geological era, 'The Aeonian'. Typically, many other evocative echoes surrounded it – 'Aeolian' as in the wind harp celebrated by Coleridge, or 'Ionian' as in some distant Greek island evoked by Shelley.

Back at Somersby, the eldest son, Frederick, was despatched to Eton in 1818, where he did well and became Captain of the School. But by then the family funds had run out, and Charles and Alfred had been sent to the local Louth grammar school, which they both hated. Alfred, still only seven, was unmercifully bullied, and beaten by the cruel headmaster Dr Waite. He remembered sitting on the school steps weeping, an object of derision. He longed to escape, and annotated one of his Louth textbooks: 'A. Tennyson, Somersby, in Lincoln, in England, in the World, in the Air, in Space.'

Equally, Tennyson began to suffer from 'trances', when his whole world seemed to come to a halt. Later he started to think these might be epileptic fits, and described them in his poem *The Princess* (1847). They came 'on a sudden' and without warning, as 'weird seizures' and doubts about his own existence: 'I seemed to move among a world of ghosts.' Yet these were always offset by his endless fascination with the external nature, his feel for landscape, and especially for the expanding world-view of astronomy.

His misery was recognised by his mother, who read him poetry to console him, and finally interceded with his father. At eleven Alfred was withdrawn from Louth and was largely educated at home by his father, a good scholar when sober and not in a rage. Dr Tennyson encouraged his three eldest sons' writing, and bound up their poems into notebooks. Alfred's included translations from Greek and Latin, short scientific poems, and long Gothic verse dramas, mainly distinguished by their extraordinary length, 'The Coach of Death', 'The Devil and the Lady' and suchlike pieces running to hundreds of lines. They overflowed with his physical energy.

Tennyson later recalled: 'In my earliest teens I wrote an Epic – between 5,000 and 6,000 verses, chiefly *à la Scott* ... full of battles, dealing too with sea and mountain scenery. I used to compose sixty or seventy lines all at once and go shouting them about the fields as I leapt over the hedges. I never felt so inspired, though of course the poem was not worth preserving and into the fire it went.'

He began wanderings at night through the Somersby churchyard. He shouted his poetry out loud to the dead, and 'scared by his father's fits of despondency' melodramatically threw himself down among the graves, and longed to be dead. These wanderings actually disguised outbursts of intense childhood unhappiness, and the yearning to escape from the domestic gloom of his father's increasing drunkenness and terrifying bursts of ill-temper. Dr Tennyson's violence was 'well-known in all the adjoining villages and his horrid language is heard wherever he goes'. His Louth neighbour at Thorpe Hall, the Reverend William Chaplin, considered him 'dangerously disposed to his wife and children'. His depression and excessive drinking brought on 'repeated fits' of madness, which could only be managed by his burly coachman, while the family doctor Dr Bousfield considered certifying him as insane.

Both Frederick and then Charles were now sent off to Trinity College, Cambridge, an evident relief to them both. Alfred was promised to follow on within two years. Meanwhile Tennyson learned to ride, to explore the Wolds on his own, and secretly to smoke. 'Jackson, the saddler at Louth, once gave me one of his strong cigars when I was a boy of twelve, and I smoked it all and flung the stump into a horse-pond, and I was none the worse for it, so I was bound to be a smoker ...'

His scientific reading provided new but also troubling perspectives. One of the earliest was a children's book on physics. 'When I was a lad, a mere lad, you know, I was given a book called "Conversations on Physical Science" by a good author. The book was simple enough, but somehow, I don't know why, I felt differently after reading it. The oxygen and carbon and all the rest of it unsettled me a little, and made me less able to believe, made my faith heavier, duller.'

This book of scientific 'Conversations' was a classic of the time. Its author Jane Marcet was the talented wife of a Swiss doctor Alexander Marcet, a Fellow of the Royal Society. The Society was the premier scientific institution in Great Britain since 1660, with founding members including Isaac Newton, Robert Hooke, Thomas Hobbes and Samuel Pepys. Jane Marcet (like the astronomer Caroline Herschel) would undoubtedly have been elected herself, except for the slight hindrance that no female Fellow was admitted until 1921. Her vivacious London circle would embrace a number of other women writers, including Mary Somerville and Harriet Martineau, as well as leading scientists like John Herschel and Charles Lyell. Her special gift for enthralling children with science was remembered by the novelist Maria Edgeworth when Marcet perilously launched a thirty-foot hot-air balloon in her back garden at a London party.

Marcet's many scientific books carefully designed for 'young minds' (teenagers rather than children, and specifically including girls) were a publishing phenomenon of the Romantic age. They altered a younger generation's attitude not merely to nature as observed by science, but to the process of scientific discovery itself. Marcet's *Conversations in Chemistry* (1805) was followed by the more ranging *Conversations in Natural Philosophy* (1819) and Tennyson certainly read both. The sudden demand for both titles had run to nine editions by 1824, by which time they were well established in the Somersby library.

Marcet reinvented the dialogue form as a series of imaginary scientific lessons between a teacher 'Mrs B' (possibly based on a famous astronomer tutor, Margaret Bryan) and her two young women pupils. Emily is observant and rather serious, while Caroline is mischievous but inventive (useful qualities for a young scientist).

Caroline continually tempts Mrs B into the more imaginative aspects of science. While discussing the composition of water, Mrs B points out that oxygen has 'greater affinity' for other elements than hydrogen. Caroline instantly grasps the romantic possibilities of this: 'Hydrogen, I see, is like nitrogen, a poor dependent friend of oxygen, which is continually forsaken for greater favourites.' Mrs B starts to reply – 'The connection or friendship as you choose to call it is much more intimate between oxygen and hydrogen in the state of water' – then sees where this is going and hastily breaks off: 'But this is foreign to our purpose.' With a suppressed giggle, Caroline has discovered 'sexual chemistry' and the reader will remember forever the composition of a water molecule: two hydrogen atoms in unrequited love with an oxygen atom (H_2O). Caroline adds suggestively: 'I should *extremely* like to see water decomposed . . .'

Young Tennyson was evidently fascinated, if oddly disturbed by such exchanges. Perhaps he was also provoked by the flirtatious intelligence of the two precocious girls. At Cambridge he would later write a whole series of short poems addressed to such young women, known simply by their names. At all events, he remembered this early troubled, seductive introduction to science for years after. If they also made his childhood faith seem 'heavier, duller' at such a young age, it was also because it quickly became less naive, weighted early on with a quality of doubt and self-questioning that never left him.

He was not the only one. The finest chemist of Tennyson's generation, Michael Faraday, always recalled that it was reading Jane Marcet's *Conversations* books in 1810, when an isolated teenager in London (he was an apprenticed book-binder in Soho), that first led him to attend Humphry Davy's chemistry lectures, and thence to his own triumphant career as director of the Royal Institution. Faraday later remembered: 'I felt I had got hold of an anchor in chemical knowledge and clung fast to it.'

Similarly the young Scottish law student Charles Lyell had encountered Marcet's work before studying under the paleontologist William Buckland at Oxford. In 1827 at the age of thirty, Lyell decided to abandon the law in favour of geology. He had only published two highly technical papers in the *Quarterly Review*, but inspired by Marcet determined to launch a professional career with a short popular book. His projected title was *Conversations on Geology*, with simple explanatory illustrations like hers to match. But the subject expanded vastly as Lyell prepared it, and three years later, he launched his monumental two-volume *The Principles of Geology* instead. This book would in time have a huge impact on Tennyson. So the seed sown by Jane Marcet bore rich and varied fruit.

Already at Somersby Tennyson had begun to write scientifically observed notes on nature, scattered freely among his draft pages of verse. Apart from descriptions of plants, birds and wildlife around the village stream at Holywell, he was strangely fascinated by the different appearance of eyes in sheep, cows and pigs in the surrounding fields. He studied them with rapt, almost unnerving attention peering closely into their faces. 'The eye of a cow is of pensive cast: the pupil large and dark and engrossing almost the whole of it. The eye of the pig is very like that of the human species, handsome and of a lively black, but generally buried in fat . . .'

Similarly, he loved the scientific accuracy of fine botanical prints, then increasingly available. When shown a copy of Thomas Bewick's *History of British Birds* (1797) Tennyson 'paced up and down the Rectory lawn all the afternoon studying it with the greatest excitement'. Every so often he rushed back into the house to 'share with his brothers and sisters' his joy in the wonderful woodcuts. Many years later he inscribed some plain but touching verses into the front of the now rare copy, recording both the book and its young Somersby reader of long ago:

A gate and field half ploughed,
A solitary cow,
A child with a broken slate,
And a titmarsh in the bough.
But where, alack, is Bewick
To tell the meaning now?

He was taught to use a solar microscope on plants and insects brought in from the garden, and a telescope to study the moon from the rectory lawn at night. Adjusting the lenses showed him that he was short-sighted, which also made him wonder at the blurred infinity of the stars, and question how far they were physically real. In one of his teenage notebooks, while trying to write his long Gothic drama boldly entitled 'The Devil and the Lady', Tennyson suddenly broke off to supply the Devil with exactly these adolescent cosmic doubts and speculations:

O suns and spheres and stars and belts and systems,
Are ye or are ye not?
Are you realities or semblances
Of that which men call real?
Are ye true substance? Are ye anything
Except delusive shows and physical points
Endowed with the same repulsive potency? . . .

The passage boasts wide background reading. That odd phrase 'repulsive potency' suggests Newtonian gravitational theory, while the 'shrewd doubt' which concludes the whole passage, that the stars might not exist at all, except as 'delusive shows', perhaps celebrates the radical scepticism of Bishop Berkeley. At all events, the drama remained unfinished, though it is interesting that it was evidently shown to Dr Tennyson, since he annotated the notebook proudly that it was written when his son was fourteen.

It was during the next autumn of 1824 that Tennyson heard of Lord Byron's death at Missolonghi. All the brothers mourned, but the fifteen-year-old went out alone to carve 'Byron is dead' on the flat rocks above Holywell spring, the hidden river pool in the woods beyond Somersby churchyard. This sacred site, as it had once been, but now also the haunt of courting couples, seemed appropriate. It was 'a day when the whole world seemed to be darkened for me'.

Although he was still sharing the 'gabled attic' at Somersby, it was now partitioned off between the brothers, and Alfred increasingly acted the part of the lonely poet, though all three boys were writing. One of the rectory servants recalled in his Lincolnshire accent, ''ow he used to go upstairs to the top room and 'ang a mat over his door. I don't know what for, but they said he didn't want to 'ear no noise.' Naturally, young Tennyson was 'always dawdlin' about wi' a book'.

But the lonely poet was also high-spirited. About this time he sent a handful of poems to his sister Emily's governess, a Miss Bowesfield, extravagantly dating it from 'La Mancha' in Spain, and signing it with a flourish: 'I remain, my incomparable Dulcinea, in the truest spirit of knight-errantry, Yours ever, Don Quixote.'

He tamed a little owl, which became his familiar. Sitting at night by the open window of his own particular part of the attic ('later used as a store-room for apples'), he 'heard the cry of a young owl and answered it'. It fluttered in, came surprisingly to 'nestle in his hand', where he fed it and finally adopted it (or vice versa). It sometimes perched on his mother's head. He tenderly celebrated it in a little nonsense song, 'The Owl'. While cleverly imitating the tune of Shakespeare's song 'When icicles hang by the wall' (from *Love's Labour's Lost*), he also drew on his own experience of lying awake in the attic (the imaginary belfry) and hearing the sounds of the outside world gathering at dawn around Somersby. He would later remember this familiar creature at Cambridge.

> When cats run home and light is come,
> And dew is cold upon the ground,
> And the far-off stream is dumb,
> And the whirring sail goes round,
> And the whirring sail goes round;
> Alone and warming his five wits,
> The white owl in the belfry sits.
>
> When merry milkmaids click the latch,
> And rarely smells the new-mown hay,
> And the cock hath sung beneath the thatch
> Twice or thrice his roundelay,

Twice or thrice his roundelay;
Alone and warming his five wits,
The white owl in the belfry sits.

In a Somersby notebook, mixed up with translations of Greek philosophers and astronomical diagrams of the solar system, there is a two-stanza
fragment inspired by his looking at the moon through a telescope. The
family noted that 'The Moon' showed 'at what an early date physical science began to penetrate his verse'. But particularly striking is that
Tennyson, using the telescope's magnification, imagined himself actually on the moon's surface. He viewed the rocky and volcanic lunar
landscape towering above him, as if he were already an astronaut. When
he looked upwards from the moon's surface, he imagined the sun seeming feeble and reduced; while other planets (themselves now seeming
like moons) cast multiple and confusing moonlight shadows.

Deep glens I found, and sunless gulfs,
Set round with many a toppling spire,
And monstrous rocks with craggy snouts,
Disploding globes of roaring fire.

Large as a human eye the sun
Drew down the West his feeble lights;
And then at night, all moons, confused
The shadows from the icy heights.

'Disploding' was his own coinage, convincingly astronomical; but also
used mythologically, to suggest the activity of a potential 'monster'
('craggy snout') is already lurking in deep space, as well as the deep
sea. This was long before he could have encountered John Herschel's
A Treatise on Astronomy, not published until 1833. But he had already
heard of Herschel's work, because he once airily admonished his older
brother Frederick (then a supposedly confident Etonian), when he
was fretting about going to a neighbour's dinner party: 'Fred, think of
Herschel's great star-patches, and you will soon get over all that.'

Frederick worried about him in turn, and in July 1826 wrote to the Old Man of the Wolds: 'You have probably received from my father himself an account of Alfred's ill-health, and his intention to try the effect of sea-bathing upon him.' Frederick rightly saw that the sea itself was a release for his strange younger brother.

Frederick thought that an alternative cure to bathing in the North Sea might be the publication of a first book of poetry. Both involved a not dissimilar plunge into an icy but invigorating element. Since the friendship with his other brother Charles was still close, the three decided that same July to send a list of 101 of their combined poems to the local Louth printer, Jackson, for private publication. Dr Tennyson was somehow persuaded to pay the initial printing cost, and they brazenly asked the publisher for a royalty of £15 'should the work go into a second edition'.

The slim volume was issued as *Poems by Two Brothers* in December 1826. Alfred and Charles contributed forty-eight poems each, while Frederick (now away at Cambridge) graciously but anonymously added five more. For once the excitement was great at Somersby. Alfred and Charles hired a lightweight carriage and dashed the fourteen miles through the narrow lanes to Mablethorpe beach, where they uncorked bottles of cider and 'shared their triumph with the winds and waves'. Later Tennyson said dismissively that his were mere juvenilia, written 'between 15 and 17'. Perhaps he wanted to disavow the stagey emphasis on loneliness and misery, often presented in the poems with the greatest gusto:

> I wander in darkness and sorrow
> Unfriended and cold and alone,
> As dismally gurgles beside me
> The bleak river's desolate moan ...

Yet gradually more material from his prose notes on the natural world, and his scientific speculations, began to find their way directly into his verse. After the distant worlds of the telescope, it was the miniature universe of the solar microscope, with its surreal magnifications, which

provided new visions of nature and the possibilities of life. In one of
the most original poems Tennyson contributed to the collection,
simply entitled 'Love', he plunged directly down into the dramatic
world of the microscopic as he had done previously with the telescope. Here tiny creatures – flies, beetles, pond larvae – appeared
huge, beautiful and ferocious: a new scientific perspective on life.

> The glittering fly, the wonderous things
> That microscopic art descries;
> The lion of the waste which springs
> Bounding upon his enemies;
> The mighty sea-snake of the storm,
> The vorticella's viewless form . . .

Here in the printed collection he added a learned footnote on the tiny
'vorticella', a supposedly early form of bacterial life. He gave it scientific
authenticity by advising the reader to consult another of his favourite
scientific books in the library at Somersby, 'See Baker on *Animalculae*.'
This was Henry Baker's *The Microscope Made Easy* (1742), with large beautiful illustrations of plants and animals, which had opened up a new
world to the short-sighted Tennyson. The vorticella, a tiny organism
living in the plants or roots of ponds or streams, which under a microscope appeared to have a throbbing life of its own, was sometimes known
as the 'Bell Animalcule' because of its pulsing shape. Jane Marcet's *Conversations on Natural Philosophy* also had an entire chapter 18 dedicated to
'The Eye –Microscopes – Short Sight – Spectacles – Telescopes'. The
naughty Caroline typically remarks that the microscope makes the mites
on cheese 'look like a drove of pigs scrambling over rocks'.

This fascination with mysterious microscopic life forms halfway
between plant and animal, and their possible connection with the
origins of life itself, had already preoccupied the previous Romantic
generation, especially the Shelleys. Mary Shelley recalled the famous
discussions of summer 1816, at the Villa Diodati on Lake Geneva,
between Percy Shelley and Byron, when all three speculated for many
nights on the current biological experiments that led to her idea for

the novel *Frankenstein*. She would vividly describe this in a new introduction to a popular edition in 1831. 'Various philosophical doctrines
were discussed, and among them the nature of the principle of life . . .
They talked of the experiments of Dr Darwin . . . who preserved a
piece of *vermicelli* in a glass case, till by some extraordinary means it
began to move with voluntary motion.'

This was a reference to the biological experiments of the eighteenth-
century polymath Erasmus Darwin (not his nephew Charles). But
such general theories of changing life forms, and the possible 'transformation species', would increasingly circulate and fascinate the adult
Tennyson more and more in the 1840s. Such pre-evolutionary theories
would culminate first in a technical paper written by Charles Darwin
in 1844; and finally as 'evolution by natural selection' in his classic *On
the Origin of Species* in 1859. But before that, Tennyson would discover
the evolutionary theories of an unknown writer, Robert Chambers.

Another notable science poem from the *Two Brothers'* collection
was 'Phrenology'. It was a bouncing satire on the pseudo-science of
cranial bumps, 'this fantastic lore', in favour of genuine enquiry. It is
not clear if this was actually written by Alfred or by Charles: but
given its pastiche of Augustan couplets, probably by the latter. Yet in
either case it gives a striking impression of the scientific questions
debated at length between the brothers at Somersby. What particularly intrigued them was the new telescope observations of the
planets; but also the discoveries in geology, summarised in a nice
phrase as the 'rummage' of the rock strata.

> Rise, sons of Science and Invention, rise!
> Make some new inroad on the starry skies;
> Draw from the main some truths unknown before,
> Rummage the strata, every nook explore,
> To lead mankind from this fantastic lore;
> Solve the long-doubted problems pending still,
> And these few blanks in nature's annals fill:
> Tell us why Saturn rolls begirt with flame?
> Whence the red depth of Mars's aspect came? . . .

The following June 1827 the two young poets were even invited up to London to visit the rich but alarming Aunt Russell at 16 Berkeley Square. It was an unheard of summons, surely inspired by her reading, or at least admiring, her nephews' unexpected little booklet. They were urged by Dr Tennyson to be on their best behaviour. The visit went surprisingly well, but it was only Alfred who was invited back to stop over several nights. Elizabeth Russell was evidently charmed by his good looks and amused, rather than annoyed, by his eccentric manners. 'Alfred came a second time without his Brother and slept during the time he staid upon the drawing room sofa.' She added thoughtfully: 'I wish he had something in Life to interest him as well as his beautiful poetry. Westminster Abbey was the only thing which particularly charmed him, it suited the *pensive* habit of his Soul.' Some sort of unlikely bond of understanding was now formed between the imperious aunt, widowed and just turned fifty, and the bohemian nephew, just turned eighteen and preparing for university. Elizabeth Russell quietly decided that she would help Tennyson in the future, and from then on she followed his career closely, and he became her 'dearest nephew'.

She was less interested in the fate of the younger ones: around this time Arthur (aged thirteen), Septimus (aged twelve) and Horatio (aged eight) were all, according to their uncle Charles, destined for Royal Navy cadetships. While 'poor' Edward (aged fourteen) was already manifesting 'giddiness', and the onset of incurable mental problems which would confine him to Lincoln asylum in the next five years. Meanwhile in December Tennyson completed his Cambridge entrance exam (Latin, Greek, Algebra, Natural Theology) and an informal interview with his future tutor at Trinity College, the formidable scientific don William Whewell. He was judged 'fully competent to enter the University'.

Alfred already seemed the exception in the family. The 'pensive habit of his Soul' was not so far off the mark. Such a soul was often in evidence as Tennyson wandered through the Somersby garden that autumn, oppressed by his father's wild behaviour and by anxiety about his future, the Louth schoolboy going up to Cambridge. It emerged vividly as the voice of Tennyson's adolescent depression, in a whispering dialogue with himself, recorded with astonishing musical skill in a poem simply entitled 'Song'. This he remembered that he 'wrote on the lawn outside the rectory'. Not included in *Two Brothers*, it was possibly written a few weeks before going up to Cambridge. It was in every sense a poem of departure.

> A spirit haunts the year's last hours
> Dwelling amid these yellowing bowers:
> To himself he talks;
> For at eventide, listening earnestly,
> At his work you may hear him sob and sigh
> In the walks.
> Earthward he boweth the heavy stalks
> Of the mouldering flowers.

To this hypnotic dirge, Tennyson added a haunting chorus, whose metrical brilliance foretold other much longer and more accomplished poems, now in the making.

> Heavily hangs the broad sunflower
> Over its grave i' the earth so chilly;
> Heavily hangs the hollyhock,
> Heavily hangs the tiger-lily.

What above all he clearly needed, and desperately, was an escape from the claustrophobic atmosphere of Somersby to the intellectual freedoms and emotional promise of Cambridge. At the last moment there was one more grotesque family drama. The cook's dress caught fire in the kitchen one evening, and Dr Tennyson (possibly drunk) was badly burned trying to put it out. Dr Tennyson survived, but became more unhinged than ever – while the poor cook died a few days later. Malicious gossip flourished throughout the country neighbourhood about how this really happened. But by then Alfred Tennyson was safely away in Cambridge.

3
Cambridge

In February 1828, four months before his nineteenth birthday, Alfred Tennyson went up to Trinity College, Cambridge, to join Frederick and Charles. His first rooms were at 12 Rose Crescent sharing with Charles, immediately opposite Trinity. But he was soon established alone, at lodgings further out, past the market square and King's, at 57 Trumpington Street. Another young man from the remote east coast of England, Edward FitzGerald, born in Suffolk but partly brought up in Paris, had rooms nearby at 19 King's Parade (now sporting a blue plaque). Half-Irish by birth, charming, well-read and footloose, he would eventually become one of Tennyson's most staunch friends, but also one of his shrewdest critics.

FitzGerald had his own kind of dysfunctional family – a cold millionaire father (who was only interested in his mines in the north of England), and an extravagant and demanding mother, who shuttled between her Suffolk estates around Woodbridge and her large London house in Portland Place. Here, she attended London receptions which the young FitzGerald was dragged to, protesting. Shy but instinctively rebellious, he spent much of his life escaping these wealthy family connections, except for one beloved elder sister Eleanor, later Mrs Kerrich, with whom he would frequently stay at Geldeston Hall in Norfolk. This would become one of FitzGerald's many alternative addresses in later life.

Back from Paris in 1818, FitzGerald was posted off to boarding school at Bury St Edmunds, where he made the first of many lifelong friends, the thoughtful scholar and future biographer of Sir Francis Bacon, James Spedding. They both went on to Trinity in 1826, where FitzGerald soon teamed up with the aspiring novelist William Makepeace Thackeray, an altogether wilder figure, who had also been to Paris, and was keen to get FitzGerald back there as soon as possible. Between these two different influences, FitzGerald's idiosyncratic talents began to flower: his dreamy out-of-the-way reading; his constantly drifting travel plans; his passionate piano-playing (Handel, Rossini, Mozart, even Meyerbeer); his exquisite botanising and picture-collecting; and above all his assiduous cultivation of an intimate circle of friends, whom he orchestrated through immensely long, shrewd and tender letters. At this time he wrote (from his sister's at Geldeston Hall) to one such friend, John Allen. 'I am an idle fellow, of a very ladylike turn of sentiment; and my friendships are more like loves, I think.'

A year above Alfred Tennyson at Trinity, and moving in his own social set, their paths did not immediately cross. Yet FitzGerald quickly noticed the three tall Tennyson brothers striding across Trinity Great Court, and soon made friends with the older, wilder Frederick, who became one of his most faithful correspondents. Nonetheless, it was the youngest and most reserved brother who struck FitzGerald, with a peculiar glamour that never quite faded through a subsequent lifetime. For him young Tennyson looked 'something like the Hyperion shorn of his Beams in Keats's Poem – with a Pipe in his mouth'. Perhaps FitzGerald was thinking of Keats's lines: 'His flaming robes stream'd out beyond his heels, / And gave a roar, as if of earthly fire ...'; but if so, he characteristically added the ironic and deflating pipe in the mouth.

Tennyson's own self-image was rather different in these first Cambridge days. An early duty was to write to Aunt Russell in Berkeley Square, instead of his unstable father at Somersby. In April 1828 he thanked her for a timely gift of money, the first of many as it turned out, and his unexpectedly playful manner with her gives a sudden impression of what was becoming his infectious charm as a young man. 'My dear Aunt – I am sitting Owl-like and solitary in my rooms (nothing between me and the stars but a stratum of tiles), the hoof of the steed, the roll of the wheel, the shouts of the drunken Gown and drunken Town come up from below with a sea-like murmur.'

There is also the first rumour of the Kraken. He told his aunt that he had been looking for a long time for Edward Bulwer-Lytton's recent novel *Falkland* (1827), which contains the description of a strange, dreamlike deep-sea monster. 'He was a thousand fathoms beneath the sea . . . he saw the coral banks, which it requires a thousand years to form, rise slowly . . . and ever and ever, around and above him, came vast and misshapen things, – the wonders of the secret deeps; and the sea-serpent, the huge chimera of the north, made its resting-place by his side, glaring upon him with a livid and death-like eye, wan, yet burning as an expiring sun.'

Tennyson may already have seen and been struck by this Kraken-like passage, as he wrote: 'Those beautiful extracts of it which you showed me at Tealby haunt me incessantly.' He added that his wishes were 'like Telescopes reversed', because they only made the desired object appear more distant. Aunt Russell was delighted by her wayward nephew, and started to make him an annual grant of £100, which she continued for the next two decades. In the early years it was to be almost his only income.

Tennyson's tall and strikingly handsome figure was soon noticed by other fellow undergraduates besides FitzGerald. Initially they were the more sporting types. Charles Merivale, who rowed and played cricket for St John's College next door to Trinity (appropriate skills for the future Bishop of Ely), described surprise visits by the shy Trinity freshman Tennyson to his St John's College rooms that spring of 1828. Merivale recognised him as 'the third of the Tennysons . . . he is an immense poet, as are all of the tribe'.

Another undergraduate described Tennyson more heartily as 'Six feet high, broad chested, strong limbed, his face Shakespearian'. Tennyson also found that he impressed everyone not only with his masculine beauty, but with his tall, strong, physical presence, and the deep, powerful music of his voice. By September, a more confident Tennyson was hosting his own parties, and even rowing on the Cam.

In the winter he acted in a college production of *Twelfth Night*, a tall and disconcerting Malvolio. It was then Merivale's turn to be 'flying to Tennyson at Trinity' for lively, argumentative talk.

In October 1828, a handsome well-heeled and rather highly strung young man, Arthur Henry Hallam, arrived at Trinity from Eton, after a gap year spent touring Italy. His closest Eton friends, William Gladstone (the future prime minister) and James Milnes Gaskell, had gone to Oxford, while Hallam's ambitious father had insisted on Cambridge. Despite animated letters between these old friends, Hallam felt alone in Cambridge and was looking for new ones. Moreover he had left behind a first love in Rome, a beautiful English society girl called Anna Wintour, who (much to his father's relief) had remained in Italy. Hallam temporarily soothed himself by writing love sonnets to her (mostly in Italian) and joining an exclusive Trinity discussion group, originally called the Cambridge Conversazione Society, and now the Apostles.

Arthur Hallam was rich enough to have rooms in college so his path did not immediately cross with Tennyson, who was displaying a growing interest in natural science. At Trumpington Street he kept various animals, including a familiar monster. 'I kept a tame snake in my rooms. I liked to watch his wonderful sinuosities on the carpet.' Its uncoiling presence may have added itself to Tennyson's Kraken mythology, and makes an unexpected, sinister appearance in another early Cambridge poem, 'The Mermaid'. The mermaid presents herself as a proud, self-sufficient beauty, 'singing alone, combing her hair, under the sea'. But like all mermaids she is discontented, and continually asking, 'Who is it loves me?' She feels her beauty is so powerful and seductive that she is bound to attract a suitor to her submarine palace:

And I should look like a fountain of gold
Springing alone
With a shrill inner sound,
Over the throne
In the midst of the hall . . .

But the suitor she expects and hopes for is not human, as in most mermaid tales: instead a familiar, menacing and now clearly awakened creature.

> . . . Till that great sea-snake under the sea
> From his coiled sleeps in the central deeps
> Would slowly trail himself sevenfold
> Round the hall where I sate, and look in at the gate
> With his large calm eyes for the love of me.

That unblinking gaze summons all kind of erotic disturbance. Here Tennyson, with the light touch of an adult fairy tale, combines natural history with his developing private mythology. The poem would be published alongside 'The Kraken' in 1830.

His general scientific reading began to be influenced by his Cambridge tutor at Trinity, William Whewell. A bluff, forceful, brilliant man from the Lancaster Royal grammar school (rather than Eton), now in his early thirties, Whewell was a dynamic Fellow of the Royal Society, and already lecturing and writing extensively on cosmology and the philosophy of science. He would publish a key text, *Astronomy and General Physics*, in 1833, and speculate on the possibilities of extra-terrestrial life. His great academic rival was William Buckland, the Professor of Geology at Oxford. Tennyson followed these competitive developments closely.

Buckland was a paleontologist and Natural Theologian, whose famous excavations in the Kirkland Cave in Yorkshire during the 1820s had led to the discovery of a monstrous extinct creature known as the *Megalosaurus* (1824). This was a grotesque type of giant lizard, which had roamed the earth in unspecified antiquity, and which is now recognised as the first genus of non-avian dinosaur. Though previously identified in Lyme Regis by Mary Anning, whom Buckland knew, it was officially named the *Megalosaurus bucklandii* in 1827. Buckland's lectures and books were accompanied by the first imaginary drawings of such a monster, in their appropriate geological settings of rocky deserts or muddy shorelines, made by Buckland's wife Mary Morland Buckland.

These illustrations, sent originally to the great naturalist Georges Cuvier (1769–1832) in Paris, arriving towards the end of his life, caused a sensation. They appeared finally to prove Cuvier's dramatic and controversial theory that this giant lizard or dinosaur, alongside thousands and thousands of earlier species, had gone 'extinct'. We were living among an entire lost world of forgotten creatures. To most European men of science this theory of extinction seemed both savage and godless. (Indeed to the English, typically French.) Surely the benevolent Christian God would not destroy His own Creation? Yet, especially after the Napoleonic Wars, this radical view was developed into a general theory of 'Catastrophism', in which almost all species on earth had once – or several times – been wiped out by some single event – a global flood, a volcanic eruption, a sudden onset of glaciation, or a rogue asteroid strike.

Cuvier had first put forward this theory in a paper delivered in Paris in 1796 on his discovery of elephant fossils. He came to believe that most, if not all, the animal fossils he examined were remains of species that had become extinct. Near the end of his paper on living and fossil elephants, he wrote: 'All of these facts, consistent among themselves, and not opposed by any report, seem to me to prove the existence of a world previous to ours, destroyed by some kind of catastrophe.' Now, some thirty years later, this profoundly disturbing view, with all its metaphysical implications, was becoming difficult to ignore.

As they got into general circulation, these images and ideas evidently haunted Tennyson's imagination, and they would reappear in several of his Cambridge poems (notably 'The Two Voices'). They eventually came to occupy a central place in his great poem *In Memoriam* (No. 56), considering the savage evolution of mankind among its earliest and most terrifying inhabitants:

> . . . A Monster then, a dream,
> A discord. Dragons of the prime,
> That tare each other in their slime
> Were mellow music matched with him.

The notion that such terrifying creatures had once been all-powerful on the earth, but were now extinct (an idea simultaneously explored but challenged by Charles Lyell in his *Principles of Geology*, 1832), gradually confronted Tennyson and his whole generation with the possibilities that the all-powerful human species were likewise destined to ultimate and inevitable extinction. Destined, unless of course, humanity was protected by some divine masterplan, benevolent and mysterious, but without scientific evidence, 'behind the veil, behind the veil'. The prospect of historic extinction was new and horrifying to the young Victorians in a way difficult to recapture now, except perhaps by comparison with the modern discovery of disastrous climate change threatening the whole planet; or the destructive impact of humankind's recent activities upon so many other species and natural environments. In other words, a similar kind of deep and existential terror.

But science was still a delight and a visionary promise to Tennyson. He was soon watching one of his new Trinity friends, the fat and jolly Richard Monckton Milnes, go up in a gas balloon from Cambridge on 29 May 1829. Milnes scribbled a one-page note during the ascent to be circulated to other undergraduates, among whom was Arthur Hallam. Further excited notes were made 'on the occasion of its composition'. It became a moment of general euphoria. Tennyson described the heady experience in the draft opening four stanzas of his poem 'A Dream of Fair Women', likening it – perhaps with a touch of irony – to the ambitious ascent of a poet:

> As when a man, that sails in a balloon,
> Down looking sees the solid shining ground
> Stream from beneath him in the broad blue noon
> Tilth, hamlet, mead and mound:
>
> And takes his flags and waves them to the mob
> That shout below, all faces turned to where
> Glows ruby-like the far-up crimson globe,
> Filled with a finer air:

So, lifted high, the Poet at his will
Lets the great world flit from him, seeing all
Higher through secret splendours mounting still,
Self-poised, nor fears to fall,
Hearing apart the echoes of his fame.

It was possibly during the general excitement connected with this
balloon ascent in May that Tennyson first encountered Arthur
Hallam. He was immediately dazzled by the younger man, then
aged eighteen. The friendship was surprising, given their subtly different social backgrounds. Tennyson still felt something of the
provincial with his Lincolnshire accent, while Hallam was assumed
to be the sophisticated public schoolboy from the great metropolis.
Hallam's father Henry Hallam was a distinguished historian, a barrister and Fellow of the Royal Society, an authority on European
constitutional history, and a conventional and ambitious Victorian
paterfamilias. The family was wealthy compared to the Tennysons,
with a large London townhouse not far from Oxford Circus, at 67
Wimpole Street (a street also occupied by Elizabeth Barrett Browning's parents).

Like his father, Arthur Hallam had flourished at Eton and emerged
self-assured, clever and well aware of his blond good looks. He was
certainly charming, and girls (besides Anna Wintour) flocked to him.
Yet he was also intense, very serious-minded and intellectually earnest. That autumn he wrote to Richard Monckton Milnes (the future
biographer of Keats) on a less jolly topic than ballooning: 'I had many
grapples with Atheism, but beat the Monster back, taking my stand
on the stronghold of Reason.'

Early on in that summer of 1829 Hallam wrote a Shakespearean
sonnet (not in Italian) to Tennyson celebrating the dawn of their
friendship, and describing himself as henceforth 'knit in a brotherhood so tender' with him. Not quite a love sonnet, its tone was high,
idealistic, and unusually extravagant: except among Cambridge literary undergraduates, who were all familiar with Elizabethan parody
and pastiche:

... And well I ween not Time, with ill or good,
Shall thine affection e'er from mine remove,
Thou Yearner for all fair things, and all true.

It is clear that Arthur Hallam genuinely understood Tennyson's 'yearn-ing' for poetry, and for scientific truth. Hallam believed that poetry, by its very form, could fix human feelings beyond the fluctuations of daily life, as he later put it in an essay on 'The Influence of Italian upon Eng-lish Literature'. This idea would later shape Tennyson's concept of the unfolding stanzas of *In Memoriam*. Hallam had written: 'Rhyme has been said to contain in itself a constant appeal to Memory and Hope. This is true of all verse, of all harmonized sound.' They also wrote a short piece together, 'To Poesy', which contained the Shelleyan aspiration:

Oh might I be an arrow in thy hand,
And not of viewless flight, but trailing flame

In a different way, the friendship developed through shared jokes – both revealed themselves as fine mimics – but especially through shared confidences. Tennyson explained something of his depressions, and the family tensions at Somersby. Hallam revealed to Tennyson that despite glowing appearances he often felt ill himself, subject to sudden severe headaches, acute religious doubts and moments of suicidal gloom. These shared confidences shaped Tennyson's later poem 'The Two Voices', in which suicide and despair are openly debated.

A still small voice spake unto me,
'Thou art so full of misery,
Were it not better not to be?'

This Cambridge friendship evidently became more and more import-ant as family relations deteriorated back at Somersby. Exactly how important, or how emotionally intense for Tennyson himself, only became clear retrospectively, although much correspondence was later destroyed on both sides (by Hallam's father, and much later Tennyson's son). But initially, besides the heady mixture of high spirits and intellectual excitement, there was a powerful element of reassurance for Tennyson, amid the chaos of his relations with his father. Moreover Hallam's kindness, energy and evident good nature would soon have a benign influence – and much more – on all the older Tennyson siblings in their wildly disrupted household.

Dr Tennyson behaved in increasingly erratic fashion, and his many drunken rages got worse. In February he had his son Frederick thrown out of the rectory, with threats of physical violence on both sides. According to Mrs Tennyson, the doctor stormed around the house with a large knife and a loaded gun. 'With the knife he threatened he would kill Frederick by stabbing him in the jugular vein and in the heart.' When she remonstrated and tried to calm him, he shouted 'he would kill others and Frederick should be one'. It can be imagined what effect these scenes had on all the Tennyson children still at home. Their mother pitied her 'poor husband', but concluded 'it is not safe for his family to live with him'.

In March 1829 Mrs Tennyson herself left Somersby, taking all her daughters with her, including the eighteen-year-old Emily (to whom Tennyson was especially close), as well as Mary, Matilda and little Cecilia who was only eleven. Tennyson's younger brothers soon followed in a general family retreat to safe ground: Edward, Arthur, Septimus and Horatio (just turned ten). There was no further news of their navy cadetships. Instead the whole family had decamped to Mrs Tennyson's own father's house at Louth, while Tennyson remained safely in Cambridge with Hallam and his friends. It was from this time that both Edward, aged sixteen, and Septimus, aged fourteen, began to show signs of mental disturbance. In Edward's case his bipolar states became so severe, 'very awkward ... unmanageable', that he was sent away to board with a doctor in York, and within three years was transferred to Lincoln General Asylum, where he remained confined for the rest of his life, tragically erased from family history.

Tennyson could now look back, with a certain objectivity, at his

own sense of being forever trapped at Somersby. He could conceive it
as a dramatic scene, even as a piece of theatre. The result was expressed
with astonishing facility in Tennyson's memorable dream-like poem
'Mariana'. It is an amazingly precocious piece of work. Only the earlier autumn 'Song: A Spirit', written on the Somersby lawn, predicts
its metrical skill. It is not known exactly when it was written, but certainly within the first two years at Cambridge. At all events, it would
be published when he was still there, in 1830. Its haunting title was
taken from his Shakespeare reading, and drew on a mysterious
exchange about abandoned love in *Measure for Measure*, Act 3. The
passage ends with the phrase, 'there at the moated grange, resides this
dejected Mariana'.

This remote and encircled country 'grange', imprisoning the mysterious Mariana (about whom nothing else is known, except she was
abandoned by her lover Angelo), evidently summoned for Tennyson
a dream version of the rectory at Somersby.

All day within the dreamy house,
The doors upon their hinges creak'd;
The blue fly sung in the pane; the mouse
Behind the mouldering wainscot shriek'd.
Or from the crevice peer'd about.
Old faces glimmer'd thro' the doors,
Old footsteps trod the upper floors,
Old voices called her from without.
She only said, 'My life is dreary,
He cometh not,' she said;
She said, 'I am aweary, aweary,
I would that I were dead!'

The term 'weary' appears again and again in poems of this period.
What did it mean for the young Tennyson? Was it just a pose? If it
expressed emotional exhaustion, it also seems to hide a rebellious
spirit, a suppressed anger. This might have been resentment against
his father, or horror at the confinement of his poor mad brother

Edward. Or perhaps it reflected a more general impatience with his provincial Lincolnshire isolation, or the narrow Christian faith of his whole culture. It is a paradox that Tennyson casts this passive, sexually suppressed, weary side of himself as a frustrated young woman, who longs for life and escape into a wider world – a world indeed far beyond the 'moat'. This reappears in the longings of 'The Lady of Shalott', written in autumn 1831.

In fact several of his most important early narrators are women; and many of the shorter lyrics are addressed to young women, with simple titles such as 'Lilian', 'Adeline', 'Isabel', Madeline' or 'Margaret', which FitzGerald would later greet as 'that stupid Gallery of Beauties'. Some of his earliest reviewers would mock these as a naive or 'effeminate' style. But Tennyson soon displayed a quite different kind of poetic power in his ability to take on a whole range of heroic and mythological male voices, from Ulysses to Tiresias and St Simeon Stylites.

Once at Cambridge, among a new circle of friends, Tennyson felt encouraged to develop his gifts as a hilarious mimic. He could 'do the voices', and hold a room spellbound with his recitals, imitating Lincolnshire yokels from home or Trinity dons like Whewell from Yorkshire. He could also do a kind of silent mime, producing ludicrous gestures and faces, as for a game of charades. Not all were respectable, either. One friend remembered an evening at Trinity when 'Alfred enacted, firstly, A Teutonic Deity – secondly, The Sun Coming Out from behind a Cloud – thirdly, a Man on a Close Stool [privy] – and lastly, put a pipe stopper in his mouth by way of a beak, and appeared as a Great Bird.' Later, FitzGerald would even recall Tennyson giving comic versions of his own poetry, reciting with 'grotesque Grimness'.

In May 1829 it was agreed that Dr Tennyson would go abroad for a cure: first to Paris, then to Switzerland and finally to Italy, where he remained for over a year until July 1830, occasionally sending back wild accounts of his misadventures. His absence was an evident relief to the whole family, who – apart from Edward – regathered that summer at Somersby, where Cambridge friends could at last be invited home.

Despite all this turmoil, Tennyson had begun to work at Cambridge on ideas for the prestigious annual University Poetry Prize poem, known as the Chancellor's Gold Medal. Founded in 1813, it had once been won by William Whewell himself, and twice by Thomas Macaulay. The set subject was Timbuctoo, intended to combine romantic travelogue with radical ideas about African slavery. Many undergraduates entered, including the earnest Dean Stanley, and FitzGerald's lively friend the satirical William Thackeray. FitzGerald himself did not compete, even though he 'secretly wrote verses'. Tennyson researched the subject seriously at the Wren Library in Trinity, and discussed it at length with Hallam.

'Much I mused on legends quaint and old,' wrote Tennyson. The magic of Timbuctoo lay partly in its name, and partly in the fact that it was hitherto undiscovered. The legendary African city was located far up the River Niger, deep in the interior of Mali. For centuries it had been renowned as a mysterious centre of Islamic splendour and scholarship, and enormous wealth, but also as a perilous, forbidden city. The heroic Scottish explorer Mungo Park had tried and failed to reach it in a famous expedition of 1805, from which he never returned alive.

However in 1828 a young French explorer, René Caillié, had staggered into the French embassy in Tangier, emaciated and in rags, claiming to have just completed a solitary expedition to Timbuctoo. He had only succeeded in returning across the Sahara, he said, by attaching himself to an enormous Islamic slave trade caravan of six hundred camels, disguised as a poor Arab pilgrim who had been 'kidnapped by the French in Senegal'.

He brought with him a detailed scientific account, which he published in three volumes, *Journal d'un voyage à Temboctou*. It established him as the first European to visit the legendary city and return alive, for which he won a 10,000 franc prize sponsored by the Société de Géographie in Paris. His *Journal* was widely translated and became a bestseller throughout Europe. It also gained Caillié the Légion d'honneur, and made him an international celebrity; though fatally weakened by his travels and his deprivations, he died less than a decade later aged only thirty-eight. But the impact of his work lived on. His descriptions of the appalling bleakness of the real Timbuctoo, its primitive buildings, and the savagery of the Moors who occupied them, transformed popular romantic views of the African interior.

'I had formed a totally different idea of the grandeur and wealth of Timbuctoo,' he wrote dramatically. 'The city presented, at first view, nothing but a mass of ill looking houses, built of earth ... I saw in the streets of Timbuctoo only the camels, which had arrived from Cabra laden with the merchandise ... and a few groups of the inhabitants lounging on mats, conversing together, and Moors lying asleep in the shade before their doors. In a word everything had a dull appearance.'

In June 1829 Tennyson submitted a four-hundred-line blank verse poem, simply entitled *Timbuctoo*. It was his first public showing as a poet. His theme had become two contrasted forms of modern discovery, one geographical and the other cosmological. He may not have read Caillié's actual book at the Wren (published in English by Bentley), but clearly he had heard the expedition much discussed in Cambridge circles. In his poem he first summons traditional images of the fabled city, drawn from Milton and earlier writers, adorning it with gleaming palaces, streets of shining silver, and pagodas full of bells. Everything is dreamily reflected through river water, like an earthly mirage:

> Seest thou yon river, whose translucent wave,
> Forth issuing from the darkness, windeth through
> The argent streets o' the city, imaging
> The soft inversion of her tremulous domes,
> Her gardens frequent with the stately palm,
> Her pagods hung with music of sweet bells,
> Her obelisks of rangéd chrysolite,
> Minarets and towers?

But then a Spirit warns the poet that this mythic paradise of dreams will soon be harshly transformed by the modern colonial expeditions, and inevitably yielding to the bleak realism of 'keen Discovery':

... soon yon brilliant towers
Shall darken with the waving of her wand;
Darken and shrink and shiver into huts,
Black specks amid a waste of dreary sand,
Low-built, mud-walled, barbarian settlements.
How changed from this fair city!

Against this loss, both physical and metaphysical, Tennyson contrasts the imaginative gains of pure scientific discovery, and particularly of the new astronomy. Here there is no disenchantment, no mirage, but an observable vision of wholly new and beautiful worlds, as revealed by the increased magnifications of William Herschel's improved telescopes. Herschel's revised theories show that many apparently gaseous nebular cloud formations actually reveal themselves as solid star clusters. In a brilliant series of images, Tennyson sees many different sorts of stellar 'fair cities', gleaming far beyond our solar system. Here, as it were, is a new continent of Timbuctoos among the stars.

... The clear galaxy
Shorn of its hoary lustre, wonderful,
Distinct and vivid with sharp points of light,
Blaze within blaze, an unimagined depth
And harmony of planet-girded suns
And moon-encircled planets, wheel in wheel,
Arched the wan sapphire.

Three years later Herschel's son John Herschel published a groundbreaking astronomical catalogue in the Royal Society's series of *Philosophical Transactions* (1833), proving that he had observed over two thousand such star clusters. Indeed it could be said that Tennyson's vision of 'blaze within blaze, an unimagined depth', strikingly prophesies those astonishing deep space images revealed by Lord Rosse's great telescope the Leviathan in 1845, and eventually by the James Webb Space Telescope in 2022.

Yet at this stage Tennyson took a quite different view of the ethical temptations of astronomy, in a brief satirical poem entitled, 'A Character'. This was a portrait of a self-satisfied contemporary at Trinity, who Tennyson felt was arrogantly boasting of his absolute philosophical scepticism ('the nothingness of things') and his casual belief that nothing existed beyond the material beauty of the universe. This was certainly not a self-portrait, as Monckton Milnes identified the same intellectual who absurdly claimed 'direct contemplation of the Absolute'. In two stanzas Tennyson gently mocked this idea of a meaningless universe, and pure pantheism, in which there was no 'divinity in grass', or secret life in 'dead stones'. The latter phrase was indeed used by the astronomer Herschel. But Tennyson's sceptic is a narcissist who, while looking through a reflecting telescope, sees nothing but himself in a mirror:

> With a half-glance upon the sky
> At night he said, 'The wanderings
> Of this most intricate infinite Universe
> Teach me the nothingness of things.'
> Yet could not all Creation pierce
> Beyond the bottom of his eye.
>
> He spoke of beauty: that the dull
> Saw no divinity in grass,
> Life in dead stones, or spirit in air.
> Then looking as 'twere in a glass,
> He smooth'd his chin and sleek'd his hair,
> And said the earth was beautiful.

Tennyson's *Timbuctoo*, subtly combining both traditional and scientific imagery, won the Chancellor's Gold Medal and gained him a widening circle of young admirers within Cambridge. But some of the more conservative academics, like Wordsworth's brother Christopher (who had once won the Medal himself), thought its wild imagery was worthy of an 'Insane Asylum'. Hallam, who had also submitted a prize poem, wrote to his friend Gladstone in Oxford, cheerfully dismissing his own effort ('a queer piece of work'), but loyally praising Tennyson's 'splendid imaginative power'. Hallam now slipped happily into hyperbole about his friend: Tennyson was 'promising fair to be the greatest poet of our generation, perhaps of our century'.

4

Apostles

As news of the Chancellor's Medal circulated around the university, Tennyson found that by autumn 1829 his social isolation was over. He began to make a number of lifelong friendships among the young men around Arthur Hallam. Hallam was undoubtedly the golden boy of the group. With his ambitious father, he was understood to have a dazzling future in public life, and Tennyson always saw Hallam as 'a great man but not as a poet'. The group swiftly expanded to embrace the names of Spedding, Lushington, Brookfield and the jolly aristocratic Monckton Milnes (later Lord Houghton).

Among this circle, James Spedding, large, kindly and prematurely balding, was the stately adviser of the group, 'the Pope among us young men', thought Tennyson, 'the wisest man I know'. Born on family estates in Mirehouse, Cumberland, and a year older than Tennyson, he was sent south to attend Bury Grammar where he had befriended Edward FitzGerald. But unlike FitzGerald, his career was rooted and stable. He eventually entered the civil service, and kept a large townhouse at 60 Lincoln's Inn Fields, where Tennyson would often stay over the next decade. Spedding later spent thirty years of his life preparing a seven-volume edition of the works of Francis Bacon and writing his biography. A generous London host and an opera lover, he remained single and surprisingly rejected the offer of the Professorship of Modern History at Trinity in 1869. Spedding criticised much of Tennyson's early Cambridge poetry, and was amused by his melodramatic reading of it. For example of 'St Simeon Stylites', his monologue of the mad

penitent who retreated to the top of a pillar, Spedding remembered
appreciatively (like FitzGerald later): 'This was one of the poems that
AT would read with grotesque grimness, especially such passage as
"coughs, aches, stitches", etc laughing aloud at times.'

But the friendship had emotional depths. Spedding had a younger
brother, Edward, of whom he was especially fond. Edward had not
yet matriculated at Cambridge, though Tennyson already knew him:
'He too was a friend to me.' So he was shocked when the sixteen-
year-old Edward unexpectedly died in the summer of 1832, and the
normally equitable James Spedding went into deep mourning for his
brother. Tennyson did not quite know how to comfort his usually
genial friend, feeling that 'words weaker than your grief would make
Grief more'. But subsequently he produced a seventy-six-line elegy
for Edward Spedding, which had faint echoes of Shelley's 'Adonais'.
But it also foretold something of the cosmic images and slow four-
line mourning stanzas, to appear a year later in what would become
In Memoriam:

> Sleep sweetly, tender heart, in peace;
> Sleep, holy spirit, blessed soul,
> While the stars burn, the moons increase,
> And the great ages onward roll.

Edmund Lushington, shy, scholarly and generous, was much attached
to his family who owned a large estate in Kent and a country house
also at Mirehouse, thus emphasising the Lake District connection.
Edmund and his younger brother Henry Lushington would frequently provide Tennyson with places of retreat and Edmund would
eventually marry Tennyson's sister Cecilia. A fine Classical scholar,
Edmund was elected a Fellow of Trinity, and later Professor of Greek
at Glasgow. But his own interest in science bore fruit a decade later
in his hosting of a Science Fair in the grounds of the family home at
Park House, near Maidstone, in July 1842. The fair was a novel extension of the annual Mechanics' Institute festival, and it would provide
material for Tennyson's long poem *The Princess* (1847).

William Brookfield was the comedian and joker of the group, a wit and a mimic, famed for his manner of 'pouring forth, with a perfectly grave face, a succession of imaginary dialogues' between various invented characters, which reduced his listeners to lying on the floor, helpless with laughter. His popularity and volubility got him elected President of the Cambridge Union, the university debating society, regarded as a kindergarten for the House of Commons. In fact this was too solemn a future for the good-natured Brookfield who became instead a fashionable preacher at St James's, Piccadilly, renowned for his sparkling socialite sermons.

Tennyson was invited to join the Apostles in October 1829. Hallam of course had already been elected to this elite society in May. The Apostles (so-named because limited to twelve members, though with faintly blasphemous overtones) was a semi-secret discussion club or society, originally founded ten years previously by the muscular Christian F.D. Maurice and the charismatic John Sterling, in 1820. (Thomas Carlyle would later write a famous biography of Sterling (1851), as a parable of brilliant intellectual promise, ultimately unfulfilled, destroyed by tuberculosis and religious doubts.) It took an oath of secrecy, met on Saturday evenings, drank tea and kept minutes.

Several generations later the Apostles became the seed-bed of the Bloomsbury Group, and included such figures as the philosopher Bertrand Russell and the economist John Maynard Keynes. The club was high-minded, romantic, agnostic if not atheist, and promoted intense friendships and rivalries between its members, a few of whom were explicitly homosexual, like the biographer Lytton Strachey. In Tennyson's time it reflected a fascination with the great Romantics of the previous generation, their philosophy and beliefs. Its high intellectual tone was indicated by the list of formal questions the society debated in November and December 1829. Tennyson and Hallam's personal verdicts always agreed, as noted down from the society's secret records by a later Apostle, the philosopher of Victorian science Henry Sidgwick, who would later uphold the scientific relevance of Tennyson's poetry.

Sidgwick's notes read:

Saturday November 21ˢᵗ. On the question whether the poems of Shelley poetry have an immoral tendency? – *Hallam and Tennyson vote: no.*

Saturday November 28ᵗʰ. On the question whether there is any rule of moral action except general expediency? – *Tennyson and Hallam vote: 'aye'.*

Saturday, December 5ᵗʰ. Hallam read an essay on the question whether the existence of an Intelligent First Cause is deducible from the phenomena of universe? – *Tennyson and Hallam vote: no.*

Tennyson and Hallam's combined 'no' verdict on the first Shelley debate suggests their approval of Shelley's political radicalism, and even of his atheism; and also perhaps confirms a growing preference for his poetry over Byron's. Having carved Byron's name on the rock at Somersby, Tennyson said he grew out of Byron at Cambridge. The friends' 'aye' verdict on the second ethics or morality debate suggested their natural idealism and their rejection of any kind of utilitarian principles, or 'consequential' ethics. These would be made fashionable by Jeremy Bentham's arguments about 'the greatest happiness of the greatest number'. Their near contemporary John Stuart Mill would later debate the same subject in a series of brilliant essays in the 1830s, which came to be known collectively as *Mill on Bentham and Coleridge*. He concluded: 'Whoever could master the premises and combine the methods of both, would possess the entire English philosophy of their age.'

Their third shared verdict, where Hallam himself proposed the debate, moved into deeper theological waters. It again touched on the possibility of atheism, and raised the independent nature of scientific belief. The background was William Paley's classic *Natural Theology* (1802), still a standard text for university students of the day. It had been read and reviled by Shelley at Oxford. The book proposed that 'Intelligent design' was observable throughout the physical world, and this provided logical – indeed irrefutable – evidence for the existence of a single Designer, or divine Creator. This was the famous 'Divine watchmaker' analogy. It had already been countered by David Hume's example of the complex 'natural design' of the snowflake, in his *Dia-logues Concerning Natural Religion* of 1779. Many generations later the Design argument would again be attacked in another classic of its kind, Richard Dawkins's *The Blind Watchmaker: Why the Evidence of Evolution Reveals a Universe without Design* (1986) which also became a standard text for modern university students of his day.

Tennyson was himself asked to contribute a fourth paper on the nature of Belief, but in a more light-hearted Christmas debate on the subject of 'Ghosts'. How far and in what sense were the Apostles prepared to believe in them? Despite the 'intensity of feeling' that the idea of Ghosts aroused, were they always 'baseless' visions? And anyway, added Tennyson, 'Do not assume ... that any vision is base-less.' In the end he could not face delivering this paper himself, and instead it was read aloud by his talkative friend Charles Merivale from St John's College. A few enigmatic spooky fragments of his introduction have survived, from the point of view of a poet rather than a philosopher: 'He speaks of life and death, and the things after death. He lifts the veil, but the form behind is shrouded in deeper obscurity. He raises the cloud, but he darkens the prospect ... And forth issue from the inmost gloom the colossal Presences of the Past.'

It is difficult to catch the exact tone of these debates, tea-drinking and earnest rather than wine-drinking and rowdy. Years later Merivale was inclined to mock the solemnities of the Apostles. 'We soon grew, as such youthful coteries generally do, into immense self-conceit. We began to think that we had a mission to enlighten the world upon things intellectual and spiritual. We held established principles, espe-cially in poetry and metaphysics, and set up certain ideas for our worship. Coleridge and Wordsworth were our special divinities ... and I should have found a lofty pedestal for Kant and Goethe.' Merivale's self-deprecating attitude was perhaps explained by the robustness of his later career: he helped to inaugurate the annual Oxford and Cambridge Boat Race, and retired to the nearby deanery of Ely Cathedral where he wrote Roman history and was commemo-rated on a cathedral plaque as 'rich in learning, and caustic in wit'.

But for Tennyson and Hallam these matters of 'poetry and

metaphysics', and questions of belief, were important and alive. As a young man Tennyson evidently doubted the existence of a Christian God, though he always tried to cling to belief in some form of 'immortal soul' – not always successfully. This was also the subject of continuous academic and emotional discussions he had with Hallam. For instance in the summer of 1830, they considered at length the evolutionary implications of the work of Friedrich Tiedemann, the Professor of Anatomy at Heidelberg University, as it had been reviewed in the *Quarterly* magazine. It threw dramatic light on 'the soul and foetal development'. Perhaps there was no soul?

While the Apostles considered these subtleties, a different kind of intellectual influence continued to emanate from his tutor William Whewell. Tennyson discovered that Whewell was an amateur poet as well as professional physicist, who kept up with the latest European science, especially German, and translated Goethe. He reviewed many of the new popular science books by Buckland, Herschel and Somerville in the *Quarterly Review*. His polymathic reputation was so great that Whewell was later elected the Master of Trinity, and published a standard two-volume account of the scientific method, *A History of the Inductive Sciences* (1837). He also became known for his coinage of new scientific terms, not least the word 'scientist' itself, at a famous Cambridge meeting of the Association for the Advancement of Science in 1833. Other new terms which especially fascinated Tennyson were Catastrophism, Consilience and Carnivore.

In Tennyson's time Whewell was still in his thirties, an aggressive young Trinity don, whom the undergraduates mocked as 'Billy Whistle', because of his northern accent (the one that Tennyson mimicked). Yet Tennyson, a fellow provincial, respected him and referred to him as 'the Lion-like man'. He began taking notice of his essays and reviews and in due course buying popular science books recommended by Whewell, among them works by Charles Babbage, David Brewster, John Herschel and Mary Somerville. One by one these books were added to the family library at Somersby.

The group now formed round Tennyson and Arthur Hallam at Trinity were also obsessed by the literary work of the previous generation of Romantic poets. Byron had of course been an under-graduate at Trinity himself, with rooms in the north-west corner staircase of Great Court, already shown off to admiring visitors. But Byron risked becoming too popular and hence *démodé*. Instead there was the idealistic early influence of Shelley among the Apostles, not least since he had been thrown out of 'the other place', University College, Oxford, for atheism.

Accordingly in November 1829 Hallam and Monckton Milnes organised a special expedition to Oxford to debate the relative poetic merits of Byron versus Shelley at the Oxford Debating Society. The opposing team was led by William Gladstone. In the spirit of para-dox, each defended the 'other place's' poet. Tennyson did not attend, apparently being snowbound in Somersby. The final vote went with the Cambridge poet Lord Byron, as defended by Oxford, possibly on the grounds of milord's aristocratic pedigree. At all events it seems to have been a sort of literary version of the Boat Race.

Hallam was particularly known as 'a furious Shelleyist', partly because of Shelley's radical politics and religious scepticism. Tenny-son on the other hand thought there was 'a great wind of words in a good deal of Shelley'. Yet he admired his blank verse (as opposed to his lyrics), and above all the late autobiographical poem about Shel-ley's idealised loves, *Epipsychidion*. He also admired Shelley's interest in science, especially astronomy and his notes on William Herschel in *Queen Mab*. But the Shelley poem that would eventually work most deeply on Tennyson was the great elegy for John Keats, *Adonais*, written in Italy less than a decade previously in 1821.

Not all the Romantic poets were dead. Instead, two of them – Coleridge and Wordsworth – were still somehow alive in their antique fifties, and had gained the status of living legends. Yet their most influential poetry, with the exception of Wordsworth's *Prelude* (not yet published), had been popular fifteen or twenty years previously; and their political views were now regarded as revisionary or frankly absurd. Nonetheless they still had an immense impact on the young literary men of Cambridge, who assessed their elders with a mixture of awe and mockery.

Once again Arthur Hallam took the initiative, arranging a formal visit to the ageing Coleridge in his rooms at Highgate in the winter of 1829. Hallam was introduced by another Apostle, Robert Tennant, already a regular visitor, and accompanied by a third, his ballooning friend Richard Monckton Milnes. They were surprised by what they discovered. Coleridge was then a wild white-haired fifty-eight-year-old, not merely the renowned poet of 'Kubla Khan' and 'The Ancient Mariner' (1798), but also the philosophical author of *Aids to Reflection* (1825), then much read in university circles, and a newly revised collection of *Poetical Works* (1829).

His glamorous reputation as an opium addict had been slowly metamorphosed into that of an almost respectable and distinctly corpulent guru, the Sage of Highgate. After his many wanderings through the Lake District and the Mediterranean, Coleridge had finally come to perch in a spacious attic apartment at No. 3 The Grove, a substantial brick Georgian house in a bucolic tree-lined avenue, owned by his physician Dr James Gillman. Coleridge descended every afternoon to hold court in Gillman's large candle-lit withdrawing room. The solemnity of the occasion was spiced with touches of Apostolic mischief.

Hallam remembered Coleridge reading aloud, with 'keen delight' and at evidently stunning length, the whole of Keats's *The Eve of St Agnes* through a long winter's evening. While Monckton Milnes reported: 'Col received us as Goethe or Socrates might have done. He advised us to – Go to America if you have the opportunity – I am known there. I am a poor poet in England, but I am a great philosopher in America.' Coleridge said he had been assured about this by the young writer Ralph Waldo Emerson, and his circle of American Transcendentalists in Boston. Indeed Emerson himself would make the pilgrimage to Highgate in 1833 when he was thirty, but by then the sixty-year-old Sage was largely confined to bed, though as bright-eyed and as voluble as ever.

Tennyson would be annoyed by Coleridge's reported criticism of his own *Poems, Chiefly Lyrical* when they appeared next summer which Coleridge thought 'beautiful' but written 'without very well understanding what metre is'. It is not clear if Coleridge ever actually made this unlikely remark. Tennyson refused to go to Highgate until it was too late, and regretted his undergraduate opinion that 'Col was an ass!' But after Coleridge's death, he later spent a nostalgic evening with Coleridge's son the poet Hartley, who recalled his father with great fondness.

The other legendary Romantic figure to be subjected to the Apostolic inspection was William Wordsworth. In December the sixty-year-old poet came to stay with his younger brother Christopher Wordsworth, at the Trinity Master's Lodge. (Christopher would be replaced by William Whewell in 1841.) One evening he was entertained by James Spedding at his Trinity rooms with Brookfield and Tennant in attendance. Tennyson and Hallam were noticeably absent: they had apparently fled to Somersby the day before, precisely to avoid direct contact with the future Poet Laureate.

Tennant reported critically: 'Last Sunday Spedding gave coffee and Wordsworth in his rooms; Wordsworth was in good talking but furiously alarmist – nothing but revolutions and reigns of terror and all that.' Wordsworth particularly annoyed Tennant with his disparaging views of Coleridge's poetry. 'He also said he wished that Coleridge had not written the second part of Christabel because that required the tale to be finished. Moreover he said the conclusion of Part 1 ... was too much laboured.' This disappointed Tennant, who concluded that the great Wordsworth 'said nothing very profound or original'. At best he kept them up till one o'clock in the morning, and was evidently 'pleased with his hearers' and a flattering speech by the ever-diplomatic Spedding.

Ironically Wordsworth had expressed a high opinion of their group of Trinity undergraduates. There was, he observed, 'a great deal of intellectual activity within the walls of this College ... we also have a respectable show of blossom in poetry. Two brothers of the name of Tennyson, in particular, are not a little promising.' It is interesting that at that date Charles Tennyson, who had just published his own collection, *Sonnets and other Fugitive Pieces* (1830), was still rated equally as a poet alongside his brother Alfred. Indeed Coleridge

owned a copy of Charles's poems and filled it with his own metaphysical marginalia. He was particularly intrigued by Charles's Somersby
sonnet, 'On Shooting a Swallow in Early Youth', which must have
reminded Coleridge of his Ancient Mariner and the Albatross. Five
years later Tennyson was still too embarrassed to pay homage to
Wordsworth when he visited the Lakes.

While Hallam was influenced by Shelley's politics, Tennyson was
drawn more towards the medieval chivalric world of Keats's ballad
'La Belle Dame sans Merci' (1819) and also Coleridge's late ballad
'Alice du Clos' (1828). They shared themes of fatal love, knights riding
alone on strange quests, and brooding atmospheres of betrayal and
death. These ornate chivalric settings would soon reappear in Tennyson's own stunningly visualised and metrically self-assured ballad
'The Lady of Shalott'.

His first solo publication was originally planned as a joint collection with Arthur Hallam, evidently inspired by the Coleridge and
Wordsworth early joint publication of the *Lyrical Ballads* a generation previously in 1798. Besides 'Mariana', Tennyson's poems would
include 'The Owl', 'The Kraken', 'The Mermaid', 'A Character', and
the poem that would later transfix FitzGerald the 'Song: A spirit
haunts the year's last hours'. Hallam was first invited to Somersby by
Tennyson for Christmas 1829, and again in the following spring vac-
ation of 1830, to discuss his contributions. He thought the joint
collection would be 'a sort of seal on our friendship'.

The Reverend George was still conveniently abroad until July
1830, and in the relaxed atmosphere Hallam was immediately adopted
by the whole family. He easily formed a bond of sympathy with Mrs
Tennyson, who at once regarded him with motherly affection. He
would later send her some of his own poems, though he regarded
Alfred as 'a true and thorough Poet, if ever there was one'. He brought
London style and charm into the Lincolnshire rectory, and his light-
ness of touch particularly delighted Tennyson's two younger sisters:
Mary, the beauty of the family (with a mass of auburn hair), then
aged twenty; and Emily, the pale and passionate dreamer, a year
younger, whom Hallam described as 'a being more like Undine than

anyone I have ever known'. Yet Emily was also assertive in her own way, a courageous horse-woman, a determined walker and – so her brothers thought – 'something of a tragedy queen both in manner and temperament'.

There were rambling expeditions, walks in the fields and woods around Somersby, picnics on the lawn, and dinnertime jokes and discussions. On one long expedition they visited Mrs Tennyson's old family friends the Sellwoods at Horncastle. Henry Sellwood was a well-to-do family solicitor and had three attractive daughters. The unexpected appearance of Hallam, Tennyson's handsome friend from Cambridge, brought the young people together. On another walk round the woods of Holywell, Tennyson found Hallam drifting aside with the Sellwoods' youngest daughter Emily among the trees, and called out provokingly to her, 'Are you a Dryad or a Naiad?' This teasing, faintly amorous challenge seems to have been the first time Tennyson really took note of Emily Sellwood, who had just turned sixteen. It was perhaps at one of these picnics that Tennyson, playful and immensely strong, picked up the family donkey which was threatening to eat their food, and carried it bodily to the side of the house.

He would also recall these long summer days in idyllic retrospect:

O bliss, when all in circle drawn
About him, heart and ear were fed
To hear him, and he lay and read
The Tuscan poets on the lawn!

O in the all-golden afternoon
A guest, or happy sister, sung;
Or here she brought the harp and flung
A ballad to the brightening moon . . .

But this was long after. Meanwhile the new volume *Poems, Chiefly Lyrical*, to be published in June, was discussed with great excitement. Hallam had managed to find a London publisher, Effingham Wilson.

Yet at the last moment Hallam's father forbade the collaboration. No letters have survived to explain exactly why he took this unusual step. But apparently Henry Hallam had received mixed reports of the provincial Tennysons' wild behaviour and their drunken father. So perhaps he was thinking that his son's future career as a London lawyer at Lincoln's Inn might somehow be socially compromised? Or that the emotional poet of 'Mariana' and 'The Kraken' might be an unsuitable co-author for a lawyer's first venture into public print? Or was he simply uneasy with the evidently intense undergraduate friendship between the two? However, the most obvious explanation is that he feared his son's own poetic ambitions. Arthur Hallam wanted to publish his own extravagantly romantic poem about Italy, 'A Farewell to the South', publicly dedicated to his old flame Anna Wintour. So instead, Henry Hallam, the shrewd old diplomatic historian, privately financed a separate slim volume of Arthur Hallam's verse, where he could discreetly oversee the content.

But there may have been a wholly different reason for Hallam's father to be anxious about his son's future. Because some time this summer it became evident that something quite unexpected occurred that altered the entire Somersby connection. Arthur Hallam had fallen in love. Not with any of the Sellwood girls, but with Tennyson's younger sister Emily.

5
Adventures

Amid all these romantic excitements that summer, Tennyson and Hallam suddenly decided to dash away on a long holiday trip down through France and over the Pyrenees into Spain. It is clear that Emily longed to accompany them, but convention did not allow this, even for a Tennyson girl. (In the previous revolutionary generation Mary Godwin and Clare Claremont had accompanied Shelley on holidays to France and Switzerland, though this caused sufficient scandal even then.) Emily had to make do with Hallam's letters. The two friends seemed anxious to throw off parental influences, especially that of Hallam's father. They quietly slipped away on a ferry from Portsmouth in July 1830. Mrs Tennyson thought they were still in London staying at Wimpole Street. Henry Hallam thought they were still together in Cambridge. It seemed a secretive but light-hearted undergraduate adventure, though more serious considerations immediately became involved.

Other more political members of the Cambridge Apostles, notably their older mentors John Sterling and John Kemble, were supporting a revolutionary movement against the recently restored monarchy in Spain, under the despotic King Ferdinand, which had destroyed the hopeful Liberal constitution of the 1820s. This group of English liberal idealists were raising funds for a London Committee of Spanish exiles, much as Byron had supported the Greek London Committee less than a decade before. Tennyson was not initially keen to join the movement, and his main contribution was a poem, 'Written During the Convulsions in Spain', which he decided not to publish. But he was finally persuaded by the more enthusiastic Hallam, and they agreed to carry secret letters of support and money, provided by the exiled Spanish republican General Torrijos. These were bundled illegally into their leather travel bags, to be delivered to the Spanish resistance group known by its rousing codename 'Ojeda' (the name of a forgtten Conquistador), just over the Spanish border.

They would return safely from this holiday mission, but for others the Spanish adventure nearly ended in disaster. Of the Cambridge group, both Kemble and Richard Chenevix Trench (part of Hallam's circle, Irish poet and future Archbishop of Dublin) were arrested and questioned in Gibraltar, though scrambled back safely to Cambridge by Christmas. However, the following year General Torrijos and a group of his young English and Irish supporters attempted an armed landing at Málaga, and forty-nine of them were executed by firing squad on the beach. The whole story was retold by Thomas Carlyle in his *Life of John Sterling*. Its elements of both tragedy and farce echoed the British poets' involvement in the Greek War of the 1820s and, a century later, the Spanish Civil War of the 1930s.

On the voyage home their ship called in at Dublin, and then Liverpool, where they happened to witness the opening of the first British steam railway line, in September 1830. Here they glimpsed a different kind of drama. Tennyson, short-sighted but fascinated by the clouds of steam, pressed forward to the station platform but because of the crowds could not get close enough to see the actual workings of the engine, or the mighty wheels upon the rails. Nor did he quite see that, in the ensuing chaos, the Tory minister William Huskisson, who was performing the official opening ceremony, by a ghastly accident had his legs run over by the train and subsequently died. The obvious symbolism of this – that Progress had its perils – for the moment escaped them both. But it would appear years later in Tennyson's poem 'Locksley Hall'. Back at Somersby, he found his father returned from abroad, and so hurried back to safety in Cambridge in late October.

On their return, Hallam wrote self-deprecatingly to a Trinity friend: 'Alfred went, as you know, with me to the south of France, and a wild, bustling time we had of it. I played my part as a conspirator in a small way, and made friends with two or three gallant men.' But his most vivid memory was of their return journey over the Pyrenees and into the rugged river valley of the Cauteretz. Here they had found they were exhausted and for the first time alone, rested for several days. The Vallée de Cauterets is now famous as a skiing and moun-taineering centre, but is still remarkable for its almost unearthly series of deep wooded gorges and spectacular gushing waterfalls. Here the friends wandered, picnicked, slept, and bathed for hours in the sunlit rockpools. Now they evidently talked with an intimacy and freedom as never before. It was a time of intense excitement and happiness for them both, and in retrospect it became for Tennyson something else, a sacred Wordsworthian spot of time. Over two decades later, after he was married, he would tell Edmund Lushington that he hoped to revisit the Pyrenees 'and spend some weeks there ... in the same places where I spent some of the happiest days of my life with Arthur Hallam'.

Hallam later described Cauteretz in an exuberant travel letter, in his best Shelley manner, dramatising the otherworldly setting and fantastic effects of the wild water: 'Precipitous defiles, jagged moun-tain tops, forests of solemn pine, travelled by dewy clouds and ... waters, in all shapes, and all powers, from the clear runnel bubbling down over our mountain paths at intervals, to the blue little lake whose deep, cold waters are fed eternally from neighbouring glaciers, and the impetuous cataract, fraying its way over black, beetling rocks, which seemed as if ages had been necessary to make them yield a pas-sage to the element now so overwhelming, and so lavish in its triumphant strength.'

Hallam's special touch is evident in that word 'fraying'.

The same memory of these cold, deep, mountain waters also haunted Tennyson. Yet he wrote nothing directly about the Cauteretz part of their travel adventures that summer. However he did draft at least one sonnet and begin a long dream poem 'Oenone', with a magical opening: 'There lies a vale in Ida, lovelier / Than all the valleys of Ionian hills . . .' It had a strange passionately repeated refrain: 'Dear mother Ida, harken ere I die'. The sonnet imagines the possible arrival of new love with various striking water images:

> Clear Love would pierce and cleave, if thou werte mine,
> As I have heard that, somewhere in the main,
> The fresh-water springs come up through bitter brine.

Tennyson also imagines sitting at the top of the Cauteretz valley with Hallam, waiting for some 'new deluge from a thousand hills' to flood with 'leagues of roaring foam into the gorge' below them 'as far on as eye could see'; and suddenly being no longer afraid of death.

Many commentators have tried to interpret all these elusive passages. The great biographer Robert Bernard Martin essayed one shrewd speculation in a rare footnote: 'One possible clue, admittedly faint, to what happened to Tennyson at Cauteretz may be found in the poem he started there, Oenone. It is spoken by the nymph and tells of her grief at being deserted by Paris for the promise of Helen. Perhaps this is the unconscious reflection of Tennyson's own awareness, first recognized at Cauteretz, that Hallam's love for Emily inevitably meant the end of part of their intimacy.'

At all events, another transforming thing Tennyson brought back was sartorial: a long black Spanish cloak and large brimmed hat, a flamboyant and un-English style of dress that he adopted for the rest of his life. But of course he was exploring all kinds of new worlds with Hallam. His sister Emily, now writing her own letters to Hallam, perhaps alone understood some of this. The experience affected Tennyson deeply, though how far it affected Hallam in the same way is not at all clear. Their two written reactions seem so very different in style and emotion. At least it produced a tale of adventure they could both secretly share with Emily once back at Somersby. More surprisingly all the breathless drama and journeyings also produced the idea for a brooding new 'Mariana' poem. It was inspired not by the deep exuberant river valley celebrated by Hallam, but by the bare, flat hot landscapes after they had left the Pyrenees behind. Tennyson said drily, without further explanation: 'It came into my head between Narbonne and Perpignan.'

'Mariana in the South' is another version of Tennyson's abandoned female lover, but her damp English moated grange has become a hot remote Mediterranean farmhouse, marooned in the southern Midi, under a harsh beating sun. Its visual details are evoked with painterly care. Deep-water images are replaced by harsh sunlit ones, in vivid chiaroscuro.

> With one black shadow at its feet,
> The house through all the level shines,
> Close-latticed in the brooding heat
> And silent in the dusty vines:
> A faint blue ridge upon the right
> And empty river-bed before
> And shallows on a distant shore,
> In glaring sand and inlets bright.

Again one asks, is Emily Tennyson secretly figured in this poem? Mariana herself remains in a kind of stunned noon-day siesta throughout the poem, half-dreaming of her lover, 'I sleep forgotten, I awake forlorn'. She is sunk in one of those trance-like states that fascinated Tennyson, but which he describes indirectly, through sound and images of the southern landscape: the dry hypnotic clicking of the cicada, the sighing of a distant sea, the relentless blazing of the Midi sun. In fact, expressively, it is an emotionally cold poem, an almost scientific study in the blinding effects of unfiltered sunlight:

> The river-bed was dusty white;
> And all the furnace of the light
> Struck up against the blinding wall.

It was only thirty years later that Tennyson did directly recall the emotion of Hallam's happy valley in the poem 'In the Valley of Cauteretz', written in 1861 on a second visit.

All along the valley, where thy waters flow
I walked with one I loved two and thirty years ago.
All along the valley, while I walked today,
The two and thirty years were a mist that rolls away;
For all along the valley, down thy rocky bed,
Thy living voice to me was as the voice of the dead,
And all along the valley, by rock and cave and tree,
The voice of the dead was a living voice to me.

Here there was another kind of magic at work: a subtle underlying use of scientific theory, this time of acoustics. In a complex movement, the sound of the river water becomes the sound of Hallam's own voice, but now a 'dead voice', eventually redeeming it from the deathly state of memory and transforming it to a 'living voice' in nature, 'all along the valley, by rock and cave and tree'. The poem draws from Wordsworth's 'Lucy' poem, with its 'rocks and stones and trees'. But also from the contemporary wave theory of acoustics, described in books by Charles Babbage and Mary Somerville.

In several passages, they both argue that the pulse of a voice is carried eternally outwards on undulatory 'particles' of air, even though the voice itself can no longer be heard. For Somerville its ripple will always be out there somewhere in the physical universe as 'the vibration of aerial molecules'. She adds: 'It communicates its vibrations to the surrounding particles, which transmits them to those adjacent, and so on continually . . . like waves formed in still water by a falling stone . . .'

Babbage was quick to see the abstract, mathematical implications of this: 'The pulsations of the air, once set in motion by the human voice, cease not to exist with the sounds to which they gave rise . . . The motions they have impressed on the particles of one portion of our atmosphere, are communicated to constantly increasing numbers . . .' But he also saw the human, imaginative impact of this, and his conclusions were both romantic and ironic. 'The air itself is one vast library . . . in which stand forever recorded, vows unredeemed, promises unfulfilled perpetuating in the united movements of each particle, the testimony of man's changeful will.'

In the poem Tennyson's incantatory phrase 'all along the valley' mimes this scientific idea of endless extension, and by the magical force of repetition, gradually expands its meaning from a simple romantic location to a vista of endless time, from the local to the infinite. It becomes a metaphor of enduring human love.

The imaginative impact of such new scientific ideas was already beginning to spread through other publications, and these took Tennyson on a continuing kind of intellectual adventure. In this same year, 1830, John Herschel published his first great popular science book, *On the Study of Natural Philosophy*, which gave a sweeping new view of the role of science in society and culture. It assessed the thrilling impact of recent discoveries in astronomy, geology and physics on man's view of nature; and his place in it, and the idea of Creation itself. Its panoramic opening chapter was entitled: 'Of Man regarded as a Creature of Instinct, of Reason, and Speculation.—General Influence of Scientific Pursuits on the Mind.' On a quite different level of sophistication from Jane Marcet's schoolroom books, it is now regarded as the first serious popular study of the 'philosophy of science', and was also an early attempt to present the 'history of science' as it had developed in Europe. It soon came to Tennyson's attention, particularly when it was reviewed by his tutor Whewell in the *Quarterly* of 1834, alongside Mary Somerville's *On the Connexion of the Physical Sciences*.

Herschel's book was hugely influential on many of Tennyson's contemporaries at Cambridge. Charles Darwin, the same age as Tennyson and officially taking a Classics degree at Christ's College, wrote: 'During my last year at Cambridge, I read with care and profound interest Humboldt's *Personal Narrative*. This work, and Sir J. Herschel's *Introduction to the Study of Natural Philosophy*, stirred up in me a burning zeal to add even the most humble contribution to the noble structure of Natural Science. No one or a dozen other books influenced me nearly so much as these two.'

At Somersby that spring Tennyson had spread his books in the library, and applied himself to a rigorous personal timetable of academic study, in theory from Monday to Saturday. It included much scientific reading. He covered practical German (rather than literary French) every weekday, plus one other subject daily, usually a scientific one. The timetable was as follows: History (Monday), Chemistry (Tuesday), Botany (Wednesday), Electricity (Thursday), Animal Physiology (Friday), Mechanics (Saturday). But Sunday was of course still – Theology.

He was encouraged by favourable reviews of *Poems, Chiefly Lyrical* which finally began to appear in January 1831: one in the *Westminster Review*, another by Leigh Hunt in *The Tatler*; and a third by Arthur Hallam himself, keen to promote his brilliant friend. These were widely admired in his Cambridge circle. Though Edward FitzGerald, not yet an intimate, was first inclined to think Frederick Tennyson still the better poet. But in April he suddenly noticed and praised Tennyson's poem 'Mariana' to John Allen: 'PS. I have bought A. Tennyson's poems. How good Mariana is.'

Hallam was inspired to write a combative critical essay, coolly championing his contemporary's work. Grandly entitling it 'Some Characteristics of Modern Poetry and On the Lyrical Poems of Mr Alfred Tennyson', he managed to get it published in Edward Moxon's *The Englishman's Magazine* for August that year. In confident, elegant prose, Hallam announced that Tennyson was part of a new movement reacting against Wordsworth and 'the Lakers'. Closer to Shelley and Keats, Tennyson's work was not meditative and descriptive, but intensely beautiful and 'picturesque'. Though so far small in compass, it was powerful, fresh and exciting. 'The volume of Poems, Chiefly Lyrical does not contain above 154 pages, but ... the features of original genius are clearly and strongly marked ... we recognise the spirit of his age.'

Hallam picked out several of what he shrewdly recognised as Tennyson's 'distinctive' gifts. They obviously included 'luxuriance of imagination', but at the same time 'his control over it'. Then there was his power of 'embodying himself' in idealised peoples and places; and his 'vivid, picturesque delineations of objects', which he could hold *'fused*, to borrow a metaphor from science'. That striking 'fused' metaphor was in fact borrowed from Coleridge's *Biographia Literaria*, drawing in turn from Humphry Davy's lectures on chemistry at the Royal Society. With this long chain of association between literary and scientific concepts, Hallam carefully acknowledged Tennyson's own ability to respond to both.

Finally there was Tennyson's 'elevated' tone (not always the case), and his extraordinary skill in verse forms. In particular Hallam praised 'the variety of his lyrical measures', one that is now much harder to appreciate. The familiarity with strict metrical forms, and their huge range of possible stanzaic structures, is one that has almost been lost to twenty-first-century readers. Since the free verse of Modernism, and the open forms of T.S. Eliot, Ezra Pound and D.H. Lawrence, appreciation of 'lyrical measures' has been largely reduced to a few surviving traditional forms like the sonnet; or to comic verse. We are left with the fantastic musical skills of Auden, or the miraculous conversational stanzas of Philip Larkin, or the sublime ballads of James Fenton. Yet perhaps the New American Formalism, recently championed by the poet A.E. Stallings, who was appointed Oxford Professor of Poetry in 2022, may yet recover such lost delights.

Nonetheless, to our contemporary ears almost any other use of regular metre, let alone rhyme, still risks sounding strained and artificial – unless in a Country and Western ballad, or a pop song, or the witty parodies of Wendy Cope. What later Victorians came to appreciate as Tennyson's metrical genius, his 'matchless ear', is now paradoxically almost impossible to hear. It can simply strike us as thunderously loud. 'The Lady of Shalott' is a case in point. Can we still hear it as a delicate virtuoso weave of sound magic, not merely as a thumping nursery rhyme? Yes, if we can listen to it afresh.

Privately, Hallam himself seemed already to be aware of something like this problem. In February 1831 he wrote to another Apostle, W.B. Donne: 'An artist – as Alfred is wont to say – "ought to be lord of the five senses". But if he lack the inward sense which reveals to him what is inward in the heart, he has left out the part of Hamlet in the play.'

Hallam was very conscious of his solemn duty in first presenting 'a young poet to the public'; though he did not mention that, at twenty, the critic was younger still than the poet, and they were fellow under-graduates. But he risked concluding with a mischievous flourish. 'We have spoken in good faith, commending this volume to feeling hearts and imaginative tempers – not to the stupid readers, or the voracious readers, or the malignant readers, or the readers after dinner!' The airy dismissal of inadequate readers was much in the Apostles' style.

Meanwhile another change had taken place at Somersby. The Reverend George Tennyson, returned from his European cure, reverted to his old habits, languished, and after several increasingly disruptive months finally died, not much lamented, on 16 March 1831. What he thought of his son's poetic debut is not known. But after the funeral Tennyson said that he slept in his father's bed, 'hoping to see his ghost' and perhaps to lay it forever.

The whole family atmosphere of the house changed, the dark paternal shadow lifted. More friends and neighbours began to be invited, with large parties for meals and parlour games, and summer picnics on the lawn, and walks through the woods and along the stream running far beyond the village. Hallam was obviously the most favoured guest, his charm and high spirits continuing to delight every-one, and completely enchanting Emily. He cast a kind of magic glow around him, and towards the end of August, he sent Emily a dreamy letter conjuring fond and lyrical memories of her and Somersby.

'Beautiful this harvest moon must have been with you, and I have fancied it many a night shedding abundant tenderness of light on the garden at Somersby, whose old trees and dark, tufted corners rejoice in that lonely radiance, and seem, as the wind murmurs through them, to utter inarticulate sounds of greeting and love.' Again, the Hallam touch in those 'dark, tufted' corners.

It was during this golden autumn of 1831 that Tennyson began his ballad poem 'The Lady of Shalott'. He drew on his reading of Sir Thomas Malory and the Arthurian legends, but also, as he later said, on an Italian Renaissance tale the *Donne di Scalotta*. Most significantly Tennyson now moved King Arthur's Camelot to the Wolds of

Lincolnshire. Its landscape became strangely reminiscent of the streams and undulating fields between Somersby and the sea.

> On either side the river lie
> Long fields of barley and of rye,
> That clothe the wold and meet the sky;
> And thro' the field the road runs by
> To many-tower'd Camelot;
> And up and down the people go,
> Gazing where the lilies blow
> Round an island there below,
> The Island of Shalott . . .

Halfway between a scholarly reinvention of the twelfth-century Italian work and a pastiche of the earlier Romantic ballads by Keats and Coleridge, Tennyson started to draft it at Somersby as early as October. It eventually ran to 171 lines, divided into four dramatic parts. In the first two, the Lady is presented high in her solitary tower, furnished only with her empty bed, her loom and a magic mirror. She is forbidden, on pain of a death-curse, to look directly out of her window. She is weary (again), sick, and simply weaving away her whole life: another version of Mariana. The poem's chanting rhythms had strong echoes of Keats's 'Belle Dame sans Merci', the fatal curse, and its atmosphere of 'palely loitering'. Yet it is wholly new.

Tennyson had set himself a considerable technical challenge, with a very strict and regular rhyme scheme whose ingenuity only becomes evident upon close reading. Each of its nine-line stanzas has to start with just one repeated rhyme in the first four lines. This is followed by a second repeated rhyme within the remaining five lines. The whole stanza is then locked together with an extended couplet, always repeating the refrain words, Camelot and Shalott. But when read or recited aloud, these subtleties immediately resolve themselves into the most pure and insistent sound-music. They produce an extraordinary, pulsing and hypnotic effect, which once heard is almost impossible to forget.

But in her web she still delights
To weave the mirror's magic sights,
For often thro' the silent nights
A funeral, with plumes and lights
And music, came from Camelot:
Or when the moon was overhead
Came two young lovers lately wed;
'I am half sick of shadows,' said
The Lady of Shalott.

In the third section the Lady's world is transformed – shattered –
by a single forbidden glimpse of the beautiful Sir Lancelot, riding
beneath her window. The magic mirror explodes, and hyper-bright
reality bursts in upon her. Here the poem takes on highly original and glittering light effects, containing a premonition of the
infant art of 'photography', or the science of 'writing with light',
which was being explored throughout the 1820s by John Herschel
using silver salts as a fixative (rather than rhyme). Tennyson's
lighting burns with an unearthly incandescence, both by day and
by night.

All in the blue unclouded weather
Thick-jewell'd shone the saddle-leather,
The helmet and the helmet-feather
Burn'd like one burning flame together,
As he rode down from Camelot.
As often thro' the purple night,
Below the starry clusters bright,
Some bearded meteor, trailing light,
Moves over green Shalott.

In the fourth and final section, the Lady leaves her tower, and fulfill-
ing the curse, boards a small boat, and passively floats down the river
to her death, singing beneath the towers of Camelot.

A long drawn carol, mournful, holy,
She chanted loudly, chanted lowly,
Till her eyes were darken'd wholly,
And her smooth face sharpen'd slowly,
Turn'd to tower'd Camelot:
For ere she reach'd upon the tide
The first house by the water-side,
Singing in her song she died,
The Lady of Shalott.

The exact meaning of the curse, and the reaction of Sir Lancelot to the Lady's abrupt demise, dying for love in full song as it were, took Tennyson more than a decade to resolve. It was not until 1842 that he concluded with this heartless, but dramatically effective, dismissal:

But Lancelot mused a little space;
He said, 'She has a lovely face;
God in his mercy lend her grace,
The Lady of Shalott.'

But in his first version, published in the *Poems* of 1832, Sir Lancelot does not appear in the final section at Camelot at all, and the curse remains an inexplicable destiny, a kind of fatal attraction. The many manuscript revisions, clearly suggest it was being discussed with Hallam throughout this autumn, and it is difficult to believe that Tennyson, who loved reciting his own work at Cambridge, did not recite it to him and to Emily together, in the library at Somersby. Indeed perhaps 'The Lady of Shalott' was, in some ways, about all three of them and their tangled affections at this moment. If the poem is about the solitariness of art, it is also a poem of erotic awakening. Yet within a generation the ballad had become a popular recitation piece, almost universally known among English schoolchildren, and with its hypnotic rhymes and rhythms, often learned by heart.

The extended storyline of the ballad also begged for illustrations. These would eventually be commissioned by his publisher Moxon as part of an early collected edition in 1857. It also became the subject of three celebrated paintings by William Waterhouse. These dazzlingly coloured and detailed pictures, completed separately over thirty years, compressed Tennyson's entire story into three images, each with its own distinctive atmosphere. Shalott sitting calmly at her loom, resigned and languid (1915); Shalott leaping up on glimpsing Lancelot, tangled and desperate (1894); and finally Shalott seated in her superbly decorated boat, gorgeous and sacrificial (1888). Throughout she seems unavoidably glamorous, beautiful and doomed. Such bold pictorial treatment even suggests that Tennyson's ballad had already anticipated the possibilities of twentieth-century film adaptation, narrative by visual image alone.

Certainly it would be said that in revising Romanticism, and the poetic narrative of medieval settings, Tennyson had essentially discovered pre-Raphaelitism, a decade before the painters themselves invented it. Yet Tennyson's poem itself is not languorous: it is brisk, brilliantly bright and strikingly hard-edged – almost heartless. Its sharp visual details – and even the magic mirror itself – simultaneously suggest the objective, unflinching gaze of that other new visual form of technology, the photographic plate.

Hallam was hugely impressed by this poem and threw himself into a new role as Tennyson's hard-headed literary agent, keen to get him a serious London publisher and good financial terms. He briefed Merivale on how to make a businesslike approach. 'Call upon Mr Moxon, 64 New Bond Street, introducing yourself under shelter of my name and Alfred's, and pop the question to him: "what do you pay your contributors? What will you pay Alfred Tennyson for monthly contributions? ... and ask whether if Alfred was to get a new volume ready to be published next season, Moxon would give him anything for the copyright? and if anything, *what?*".'

To Tennyson's delight, Hallam won the Trinity College English Essay Prize in December 1831. It was their golden moment together at Cambridge, though because of his father's death and the loss of the clerical stipend, it soon became clear that Tennyson would have to go down from Cambridge without completing his degree. In February 1832 Hallam again came to stay at Somersby, and jubilantly celebrated his twenty-first birthday there.

Yet Hallam was worried about Tennyson's health and state of mind, which he had revealed in a confidential letter to Emily, written from Wimpole Street in January 1832. 'I do not suppose he has any real ailment beyond that of extreme nervous irritation; but there is none more productive of incessant misery, and unfortunately none which leaves the sufferer so helpless.' He felt Tennyson was subject to 'over-anxious thought', concentrating on his poetry, and cutting himself off from 'light mental pleasures'. Together, he implied, he and Emily might make Tennyson 'more sociable' in the future at Somersby. Perhaps their future marriage might set the seal on a new kind of family happiness.

6

Complications

Thanks to Hallam, Tennyson was not forgotten back at Cambridge. During the spring of 1832 there was much discussion among the Apostles at Trinity about Tennyson's poem in draft, 'The Palace of Art'; and also of 'The Kraken'. Brookfield, half-joking as usual, gravely told Hallam that the poem had inspired in him an opium dream in which the monster featured, rearing up in a calm sea, threatening a beautiful unknown woman. 'There was a dark-haired dreamy-looking lady in white, sailing about delicately . . . then came Alfred and became a great Kraken – the female sailed by him in safety.' In March Tennyson replied with a joshing letter to Brookfield about his 'musing and brooding and dreaming and opium eating out of this life into the next . . . Shake yourself, you Owl o' the Turret, you!'

In fact 'The Palace of Art' opened with imagery drawn directly from Coleridge's opium poem 'Kubla Khan'. But unlike Coleridge's 'pleasure-dome' floating on the waves, Tennyson's Palace was to be constructed on a remote mountainside, a great gleaming brass 'lordly pleasure-house' of learning, a place of solitude and retreat. It was to be a combination of University library (surely inspired by the famous Wren Library at Trinity), a scientific laboratory and an astronomical observatory. The Palace was topped by a great dome beneath the stars.

> I built my soul a lordly pleasure-house
> Wherein at ease for aye to dwell.

I said, 'O Soul, make merry and carouse,
Dear soul, for all is well.'

A huge crag-platform, smooth as burnish'd brass,
I chose. The ranged ramparts bright
From level meadow-bases of deep grass
Suddenly scaled the light.

Thereon I built it firm. Of ledge or shelf
The rock rose clear, or winding stair.
My soul would live alone unto herself
In her high palace there.

And 'while the world runs round and round,'
I said, 'Reign thou apart, a quiet king,
Still as, while Saturn whirls, his steadfast shade
Sleeps on his luminous ring.'

Here the stargazer is in splendid isolation. His Soul – a female presence – is content to live 'alone unto herself' in her high palace on the mountaintop. Initially all is well: he is at ease, merry, 'a quiet king'. In the Palace courts 'every Science is fair displayed'. The orbiting planets are 'steadfast' in their courses above, even if the exact composition of Saturn's ring is still a mystery. At this point astronomy still filled young Tennyson with confidence. Only later in the poem is his Soul overcome with loneliness, 'on fire within', and longs to return to 'a cottage in the vale'.

Different kinds of 'light mental pleasures' were applied in June, when Hallam inveigled Tennyson back to London, rejoicing archly to friends: 'Alfred the Great will be in Town.' He took Tennyson off to investigate the barmaid of the Star and Garter, their favourite riverside pub in Richmond. Spedding suggested the barmaid's 'garter' was especially interesting to the poet. He certainly put her into his poem 'The Gardener's Daughter'. They saw the actress Fanny Kemble (whose brother was at Trinity) in *The Hunchback of Notre Dame*, and dined with her afterwards. She described Tennyson's strange but striking good looks, 'head and face almost too ponderous and massive for beauty in so young a man'.

Hallam was now keeping Emily up to date with amusing, light-hearted notes of their London adventures. But he also wrote a long, anxious letter to Brookfield about his love for Somersby and for Emily Tennyson. 'Somersby looks glorious in the full pride of leafy summer. I would I could fully enjoy it, but ghosts of the past and wraiths of the future are perpetually troubling me. I am a very unfortunate being, yet when I look into Emily's eyes, I sometimes think there is happiness reserved for me.' Hallam's 'ghosts' of the past obviously included Anna Wintour in Italy; and his future wraiths perhaps the difficulties of a marriage to Emily. Not least among them, that his father might possibly cancel his allowance of £600.

Mrs Tennyson agreed to connive at a secret correspondence between Hallam and her daughter, and was obviously keen on the match. Hallam wanted to announce an engagement immediately, but suspected that his father might harbour social objections to the Tennysons. In an attempt to win him over, Hallam took Tennyson to dine at Wimpole Street, but it did not go well, despite some back-chat about botanical science. Alfred was 'dreadfully nervous about it: he was silent a long while, but on mention of some water-insects of his acquaintance he suddenly became eloquent ...'

Back at their beloved Somersby, Hallam wrote to another mutual Cambridge friend, their Spanish co-conspirator Chenevix Trench, and daringly made a formal – though strictly unofficial – announcement of the engagement. 'I am now at Somersby, not only as the friend of Alfred Tennyson, but as the lover of his sister. An attachment on my part of near two years' standing, and a mutual engagement of one year, are I fervently hope, only the commencement of a union which circumstances may not impair, and the grave itself not conclude.'

But the impairment and frowns from Wimpole Street continued, and by July 1832 the friends had become restive again. In a faintly rebellious mood Tennyson and Hallam took off on the spur of the moment for a summer tour down the Rhine. Once again Emily was excluded from the adventure, but at least she was forewarned by Hallam: 'I have strange news for you, news which will make your dear eyes open wide and full ... I am going with Alfred up the Rhine for three weeks! He complained so of his hard lot in being forced to travel alone, that I took compassion on him, and in spite of law and relatives etc ...' He promised that on their return he would go 'straight to Somersby'. Quarantined on a cross-Channel boat they were, 'bug bitten, flybitten, fleabitten, gnatbitten and hungerbitten'. Hallam exclaimed: 'Alfred is as sulky as possible. He howls and growls sans intermission.'

But all was forgotten when Tennyson saw the spires of Cologne Cathedral, and discovered the Titians and Raphaels in the city's art gallery. Their adventures included being nearly run down by a Rhine barge, and tipping over into a ditch in their carriage, after which 'fifty Belgians came promptly to their assistance'. Hallam's long humorous accounts were sent to Emily, and also to James Spedding. Hallam wrote: 'Alfred swears the Rhine is no more South than England, and he could make a better river himself.' But they were immensely enjoying themselves.

When they returned, Tennyson continued to work on his new collection, *Poems*, to be finally published at the very end of the year, December 1832, though most copies would carry the date 1833. Hallam advised on the contents, and edited the proofs. Tennyson wrote briskly to Moxon: 'The title page must be simply, Poems by Alfred Tennyson, (don't let the printers esquire me).' The book would include a first version of 'The Palace of Art', 'Mariana in the South', 'The Lotos-Eaters', and above all 'The Lady of Shalott'. There was also to be a long blank verse poem 'The Lover's Tale', but as this would cover some sixty pages, in the end he decided to withdraw it. Hallam voiced lively objections: 'By all that is dear to thee – by our friendship – by sun, moon and stars – by narwhals and seahorses – don't give up the Lover's Tale.' But Tennyson did not reinstate the *Tale*.

Hallam spent all Christmas and New Year 1833 at Somersby, and started reading Jane Austen's *Emma* with Emily. He again confided in Brookfield about his passionate love for Emily, whom he now

considered his bride-to-be, despite all Tennyson's worries about the
disapproval of such an engagement at Wimpole Street. 'Every shadow
of – not doubt, but uneasiness . . . that Alfred's language . . . sometimes
cast over my hope is destroyed in the full blaze of conscious delight
with which I perceive that she loves me. And I – I love her madly: I
feel as though I had never known love until now . . . I feel above consequence, freed from destiny, at home with happiness . . .' For Hallam
this love for Emily seemed both powerful and liberating, of a kind
'never known until now'. Moreover 'at home with happiness' strongly
suggests he sought this at Somersby.

It is possible he still feared that Tennyson felt some remnant of
mild jealousy – 'not doubt, but uneasiness' – about this new love for
his sister. Was this the same sense of the loss of their old intimacy
which Tennyson had perhaps experienced at Cauteretz? At all events,
Hallam added firmly that Tennyson's doubts and uneasiness really
stemmed from anxieties about the new book of poems. 'Alfred is, as I
expected, not apparently ill, nor can I persuade myself anything real
is the matter; his condition is altogether healthier. He is fully wound
up to publication.'

Through spring 1833, Tennyson stayed at Somersby to arrange
family affairs. Here he began a curious historical monologue of the
eccentric Christian hermit 'St Simeon Stylites', who banished himself
to the top of a pillar, where he did daily private penance, but remained
on public show. 'The Watcher on the column till the end, /I, Simeon,
whose brain the sunshine bakes.' It made a deliberate and striking
contrast with the permanent afternoon languor of 'The Lotos-Eaters',
'resting their weary limbs at last on beds of Asphodel'.

In April Tennyson and Hallam, now almost his brother-in-law,
embarked on a 'science week' in London together. Emily pursued
them with frequent fond messages from Somersby, while Hallam
replied with letters addressed to 'My dearest Nem', and later to 'my
own pet' or 'ma doucie amie'. When she was not well enough to join
them in London, he wrote with painful concern: 'My dearest Nem . . .
There is something horrible sad in your desiring me to have as much
fun with Alfred as usual, while you are evidently incapable of

anything approaching fun.' Later when it was his turn to be ill, Hallam wrote longingly to her: 'Pity me, in my long feverish nights pining for the garden at Somersby, the green garden, and the slope down to the gurgling brook, and the cool shades of Holywell.' The woods of Holywell had been the scene of their courting.

The London science visit included the newly founded London Zoo in Regent's Park, the Gallery of Practical Science in Piccadilly, and displays of new powerful microscopes, which particularly fascinated Tennyson, revealing the tiny microscopic creatures within their own miniature world: 'lions and tigers which lie *perdus* [hidden] in a drop of spring water'. Other microscopic objects appeared like beautiful moth's wings, mysteriously beating. Tennyson remarked thoughtfully to Hallam: 'Strange that these wonders should draw some men to God – and repel others.' Then added briskly: 'No more reason in one than the other.' Years later he would tell the story of 'a Brahmin destroying a microscope because it showed him animals killing each other in a drop of water'. He and Hallam agreed that this was a 'significant' kind of unscientific response: 'as if we could destroy facts by *refusing to see them*'.

Hallam wrote delightedly to Emily: 'Tuesday we went again to that fairy palace, the Gallery of Practical Science, saw the wonderful Magnets, and heard the Steam Gun . . .' With Hallam's encouragement, Tennyson's scientific reading and reflections continued to expand and to pour into his poetry. One of the outstanding publications of 1833 was John Herschel's *A Treatise on Astronomy*, now widely discussed. This summarised the latest theories of the solar system, the distribution of the planets, how they got their moons, and how the whole structure of the universe was produced by the clustering of enormous circular star systems. Herschel threw new light on the centrifugal development of the so-called 'nebulae', or observable galaxies, originally proposed by the French mathematician Pierre-Simon Laplace in his famous chapters on the 'nebula theory' in his *Mécanique Céleste* of 1799.

The forty-year-old Herschel had just been knighted for his revolutionary astronomical work at his father's great observatory at Slough in Berkshire. Here he had used a series of enormous ten- and twenty-foot reflector telescopes to produce the latest star maps of the northern hemisphere. He had recorded more about the newly discovered moons orbiting Saturn and Uranus; drawn up an immense catalogue of double stars, which might be used to measure the depth (and hence the age) of the Milky Way; and made new detailed observations of our own moon's mountains and craters. Herschel was about to depart for a decade's observation of the southern hemisphere at Cape Town in South Africa.

Drawing on his own telescope observations, and the earlier work of his father Sir William Herschel, the younger Herschel's book also explored the philosophical implications of such evolving star systems: the nature of their formation, their growth over immense and previously inconceivable periods of time, and finally their slow but inevitable extinction in every case. These ideas of so-called deep time and deep space were gradually transforming the whole notion of the material universe. It was stranger and vaster than previously thought, and yet more vulnerable – and paradoxically, more temporary. There were no Biblical eternities any more.

Several of Tennyson's new Cambridge poems reflected this increasing urge to grapple with the implications of what was in effect Herschel's new cosmology. These included 'The Palace of Art' now being continually revised and expanded with these new scientific ideas and imagery. After the first version appearing in the 1833 volume, there was another in 1842; and a final one not until 1853.

Early versions of 'The Palace of Art' had been circulated in a three-hundred-line manuscript among his friends at Trinity. They were much argued over, especially by the Apostles, and effectively launched a debate over the respective powers and authority of art and science. One of the Apostles, Trench, had accused Tennyson of becoming a mere dilettante: 'Tennyson, we cannot *live* in Art.' Tennyson's animated riposte was to celebrate the new astronomy.

In the early 1833 version, where Tennyson was still consciously exploring Herschel's cosmological theories, he threw in a section of vivid and powerful astronomical imagery. He imagined his Soul climbing at night into her observatory, at the top of the Palace, and adjusting a large telescope, the 'optic glass'. The astonishing celestial panorama that now confronted her was introduced with the following prose note.

If the Poem were not already too long, I should have inserted in the text the following stanzas, expressive of the joy wherewith the soul contemplates the results of astronomical experiment. In the centre of the four quadrangles [of the Palace] rose an immense tower:

> Hither, when all the deep unsounded skies
> Shuddered with silent stars she clomb,
> And as with optic glasses her keen eyes
> Pierced thro' the mystic dome,
>
> Regions of lucid matter taking forms,
> Brushes of fire, hazy gleams,
> Clusters and beds of worlds, and bee-like swarms
> Of suns, and starry streams.
>
> She saw the snowy poles of moonless Mars,
> That marvellous round of milky light
> Below Orion, and those double stars
> Whereof the one more bright
> Is circled by the other.

These glowing cosmological images, inspired by John Herschel's various telescopic papers, dramatised the new 'nebular' theory of star clusters. Tennyson envisioned an essentially dynamic universe, in which enormous gatherings of fiery primal 'lucid matter' were continually condensing and spinning out to form the great wheels of star clusters visible in the Milky Way 'the marvellous round of milky light'. Within these clusters, these 'bee-like swarms of suns', a myriad of other solar systems besides our own must surely be forming.

The striking image of 'bee-like swarms' was eventually taken up word for word by the great twentieth-century astronomer Edwin

Hubble: 'Our stellar system is a swarm of stars isolated in space. It drifts through the universe as a swarm of bees drift through the summer air.' The poem sent out other echoes into the twentieth century, including the 'Two Cultures' debate, launched by C.P. Snow, the author of the Rede Lectures of 1959. (Snow, incidentally, was another Trinity man.) Though a different kind of doubt about 'The Palace of Art' would be expressed in an essay by W.B. Yeats, who felt that Tennyson's natural lyricism was damaged by intellectual 'impurities' – Yeats instanced politics, science and religion – and that the poem was an early example of 'that brooding over scientific opinion that so often extinguished the central flame in Tennyson'. At the end of his life, Yeats changed his mind, and felt the 'central flame' had been revived.

In fact the existence of other galaxies outside our own Milky Way would not be definitively proved until Hubble's observations using the huge Mount Wilson telescope in California in 1919. While the first 'exoplanets' – planets belonging to star systems beyond our own sun – would not be definitely observed until the 1990s. (Currently over five thousand exoplanets are known.) But Tennyson was already astonished by the scale and remoteness of Herschel's observations, and he first accepted these revolutionary scientific theories in a spirit of optimism and romantic wonder. He would begin to reflect on them more deeply in *The Princess* (1847).

Much later in his fragmentary poem 'Parnassus' (1889), Tennyson would even revert to the dark, traditional view that the Victorian sciences of astronomy and geology could be sources of pure terror. They opened up such appalling new vistas of uncertainty. These two sciences menaced the very springs of the poetic imagination itself. In classical terms, they threatened to overshadow the Castalian Spring, which for the Roman writers was the traditional source of all poetic inspiration. This sacred spring or fountain is still to be found at Delphi, on the southern slopes of Mount Parnassus in Greece, home of the Muses.

> What be those two shapes high over the Sacred Fountain,
> Taller than all the Muses, and huger than all the Mountain . . .?
> These are Astronomy and Geology, terrible Muses!

It may also have struck Tennyson that the spring was guarded by another huge, lurking subterranean monster, the Python. It was eventually drawn to the surface and killed off by Apollo, the god of poetry and sunlight. This myth is subtly recalled in D.H. Lawrence's poem 'The Snake' (1924). Indeed the Python now has its own modern statue in Barcelona.

In the 1842 revision of 'The Palace of Art', Tennyson moved even more radically from astronomical to biological evolution. He added two striking stanzas drawn from a new German biological theory, put forward in the paper by Professor Friedrich Tiedemann that he had once discussed with Hallam. The subject was foetal development of the human brain in the womb, and how it appeared very similar to brain development in other animals, and the implications of this. The linking idea was progressive development through the whole of the natural world, with echoes of Schelling's *Naturphilosophie*: 'All nature widens upward.'

'From shape to shape at first within the womb
The brain is moulded,' she began,
'And through all phases of all thought I come
Unto the perfect man.

'All nature widens upward. Evermore
The simpler essence lower lies,
More complex is more perfect, owning more
Discourse, more widely wise.'

Tennyson was humiliatingly brought back to earth from his star-gazing and microscope-peering, when he read J.W. Croker's long and crushing review of his new *Poems* in the *Quarterly Review* of April 1833. This was the same John Wilson Croker who had attacked young John Keats's first volume of poems *Endymion* so viciously, just fifteen years before, and which Byron claimed had 'extinguished' Keats as a poet. Now Croker – a peppery secretary to the Admiralty Board, who spent his spare time annotating the Poetical Works of Pope – relentlessly mocked Tennyson as 'a new prodigy of genius' and a brighter star in 'the *milky* way of poetry' following Keats.

He refused to 'maim' the 'Lady of Shalott' by any 'analysis', and described 'Mariana in the South' as 'a dreamy tissue' from which nothing 'intelligible' could be extracted. Most woundingly, Croker accused Tennyson of sentimentality, naive romanticism and effeminate 'lady-like' verses. In a fond, lullaby-like poem about Tennyson's 'Darling Room' at Somersby, the one that had meant so much to him in childhood, Croker howled to find 'soft and white' rhymed with 'exqui*site*'. He likened the first intrusion of love into the narrative poem about the Miller's Daughter 'to the plunging of a water-rat into the mill-dam'. Mr Tennyson himself was 'we presume, a dreamy Lotos-eater, a delicious lotus-eater'. Tennyson was not extinguished by this critical broadside, yet would not publish again for nine years (1842).

Later there would be many defenders. Sterling, the adventurous co-founder of the Apostles, had taken the young philosopher John Stuart Mill to visit the elderly Coleridge at Highgate just before he died. They had discussed the similarities between Keats and Alfred Tennyson, and on this occasion Coleridge apparently praised the new poet. As a result Mill subsequently wrote a formative article for the *London Review* in July 1835, elegantly defending both Tennyson's *Poems, Chiefly Lyrical* of 1830, and the new collection of *Poems* from 1833, and airily dismissing the old-fashioned, cynical criticism of Croker as 'abundantly hackneyed'.

While lightly admitting the unfortunate juvenilia like 'O Darling Room' ('which we could have wished omitted'), Mill picked out the poet's characteristic powers of musicality and visual precision, 'the art of painting a picture to the inward eye'. Tennyson produced 'not mere pictures, but states of emotion, embodied in sensuous imagery'. In direct contradiction to Croker, Mill specifically praised the 'high degree of excellence' in such pieces as 'Mariana', 'The Lady of Shalott' and 'The Lotos-Eaters'. He picked out 'The Kraken' and linked it with 'The Dying Swan' describing them as 'struggling upwards to shape this *sensuous* imagery to a spiritual meaning'.

Altogether Mill felt Tennyson was potentially a poet full of 'philosophical speculations'. In his work 'whatever is comprehensive, whatever is commanding, whatever is on a great scale, is poetical'. His longest poem, 'The Palace of Art', with its rivalry between artistic and scientific knowledge, was an unusual 'attempt of this sort', rich in details, even if not 'wholly successful'. His many admirers felt that with 'further effort on his part . . . he would secure to himself the high place in our poetic literature for which so many of the qualifications are already his own'. This was in effect a formal challenge to Tennyson to press on with his new poetry undismayed, issued by one of the great philosophers of the next generation.

In July 1833 Tennyson went to visit Cambridge friends in Scotland, while Hallam remained in London, claiming that he was languishing with a feverish headache. This may have been true, but he was certainly well enough to make a flying visit to Emily at Somersby, significantly the first time he had ever gone on his own. With Mrs Tennyson's evident approval, he took another dreamlike summer walk with his unofficial fiancée. Once again there was perhaps a little touch of guilt, when he wrote fondly to Tennyson, that 'a sort of yearning for dear old Alfred comes upon me'. Then abruptly at the beginning of August, Hallam was summoned to join his father on a business trip to Europe. The plan was to stay in Vienna and Salzburg, departing almost immediately on 3 August. Neither Emily nor Alfred Tennyson could be included in what was evidently a strictly father-son expedition, with some museums and operas thrown in for light relief. It seems that Henry Hallam wanted to have a serious conversation with his son about his future.

Taken by surprise with news of this imminent departure, Tennyson made a headlong dash back from Scotland to London in two days. He was just in time to join Hallam's farewell party. Also invited were his new publisher Edward Moxon and the celebrated journalist Leigh Hunt, and after an extended supper with other Cambridge friends, Tennyson gave a poetry recital of 'glorious fragments' from 'The Gardener's Daughter'. It kept them up till four-thirty in the morning. Then promising faithfully to write from Austria, Hallam whirled away on the Dover mail with his father to the Continent.

True to his word, Hallam soon wrote from Salzburg, but it was a letter addressed primarily to Emily. He had fallen in love with the city and had 'sworn in his secret soul' that one day – God willing – he would come back with Emily herself ('not alone, Nem'). They might eventually make Salzburg their home. Meanwhile he seemed impatient with his father. As for Alfred, was he 'mad' not coming 'instantly' to Salzburg and 'staying as long as he possibly can?' Hallam signed off fondly to Emily – 'Adio, Carissima'.

Hallam sent back one other message, this time addressed especially to Tennyson. It was about the dazzling paintings by Titian he had found in the state gallery in Vienna. 'Oh Alfred such Titians! By heaven that man could paint!' He was particularly struck by one of Zeus making love to Danaë, and descending upon her languid, naked body in an explosive shower of gold. It was based on a famous provocative episode from Ovid's *Metamorphoses*, and Titian had painted no fewer than six versions of it between 1540 and 1560. Hallam was equally obsessed by it, and added a command to his friend: 'Do you write just as perfect a Danaë!' Perhaps he thought of it as a kind of Epithalamium for his future marriage to Emily.

Tennyson never forgot this request. He noted quietly: 'Zeus came down to Danaë when shut up in the tower in a shower of stars.' Shut up in a tower, like the Lady of Shalott, then; but gloriously released into life not death. Some fifteen years later he would slip the Danaë image into one of the most beautiful 'Songs' in his dramatic poem, *The Princess*. It was a love sonnet, which produced the inspired erotic lines:

... Now lies the Earth all Danaë to the stars
And all thy heart lies open unto me.

For the rest, neither Emily nor Alfred Tennyson heard anything further from Arthur Hallam all that summer; though there were rumours that the weather in Vienna was unexpectedly wet and chill.

7

Extinctions

It was probably during these suspended summer months of 1833 that Tennyson first read early reviews and essays about the astonishing work of the young Scot, a revolutionary thinker on the earth sciences, Charles Lyell. Born near Edinburgh, and having studied at Oxford under the controversial paleontologist William Buckland, Lyell had abandoned a humdrum career in law, and in the spring of 1828 embarked with his friend Roderick Murchison on an adventurous twelve-month tour of geological sites in central France and northern Italy. He brought back formidable fossil evidence, not merely of long dead creatures, but of actual 'extinct' animals whose species had died out at some unimaginable period in the distant past. They led him to entirely new ideas of geological formation, and of deep time itself. Extending not over thousands of years, but over millions, and maybe billions: difficult indeed to grasp in human terms.

Lyell and Murchison had explored the wild uplands of Aurillac in the Cantal, the remote department of central France, truly *la France profonde*, famous mainly for its thorny hillsides, deep valleys and its rumbustious cheeses. They had stayed in the local auberges, and started out at four each morning. They kept extensive geological notes on the constantly surprising freshwater rock strata of the area. Their quiet observations were a revelation. 'For if sedimentary deposits went on as slowly, formerly, as they do now, they must have required vast periods, and no period equally vast can have transpired since the time of History.'

On his return, Lyell decided to launch himself as a professional geologist, though as yet he had no university position, and began publishing a series of papers which eventually became *The Principles of Geology*, Volume 1, 1830; Volume 2, 1832. This formidable book, several times republished and updated, gradually popularised a hugely influential account of 'Deep Time', and the slow, implacable forces which have shaped the surface of our planet over millions of years – by rivers, earthquakes, volcanoes, glaciers and weather erosion. Besides his Cantal researches, the book contained memorable and troubling descriptions of the extinct Auvergne volcanoes; of the continuing sulphurous activity of Mount Etna; the sinister erosions of the Norfolk coastline; and the relentless riverbed rock carving of the Niagara Falls.

Lyell discussed the chronological implications of shells and fossils in deep geological strata. He did not accept Evolution or the transformation of species, in the sense that Darwin would define. It was 'an hypothesis ... wholly unsupported by geological evidence'. Yet he praised Alexander Humboldt's 'global' vision of an ever-expanding biosphere in *Cosmos*. Accordingly Lyell developed a polemic against biblical geology ('the narrative of Moses'), and the 'speculations' of English clerical scientists like the 'zealous' Buckland (his erstwhile tutor) who still believed in a historical Flood. He attached 'no value whatever to the hypothesis of ... Professor Sedgwick' about 'mighty waves desolating whole regions of the earth'.

Altogether Lyell proclaimed that geology, through his own work and that of Cuvier, James Hutton and John Playfair, had emerged as a fundamental and independent science, a magnificent new field of enquiry, now comparable to astronomy, and essential to any modern world-view. 'Never, perhaps, did any science, with the exception of Astronomy, unfold in an equally brief period, so many novel and unexpected truths, and overturn so many preconceived opinions.'

This new vision was infinite. 'In vain do we aspire to assign limits to the works of creation in *Space*, whether we examine the starry heavens, or that world of minute animalcules which is revealed to us by the microscope. We are prepared, therefore, to find that in *Time* also, the confines of the universe lie beyond the reach of mortal ken.' Such discoveries might provide 'clear proofs of a Creative Intelligence, and His foresight, wisdom, and power'. But the attributes and existence of any such 'an Infinite and Eternal Being' lay firmly 'outside the reach of our philosophical enquiries, or even our speculations'. Here Lyell withdrew decisively from any kind of 'Biblical' ideology or even Natural Theology. He expressed no personal faith at all, and throughout, he signally avoided using the reassuring word 'God'.

Gradually Tennyson became fully aware of the revolutionary implications of the new geology. In Tennyson's early reading, Astronomy was generally an area of intellectual hope, with optimistic views of discovering an ever-expanding universe and a glowing plurality of worlds. While Geology soon became an area of spiritual terror, with claustrophobic glimpses of a cruel, meaningless, godless physical world closing in on the heels of humanity. It was full of monsters, dust and extinctions, making nonsense of any idea of progress or individual immortality.

This deep division eventually became the two sides to Victorian science, an intellectual and spiritual schizophrenia, which Tennyson had somehow glimpsed in the slumbering Kraken, and would go on to express so powerfully and explicitly in his later poetry.

Such radically unsettling ideas and doubts struck Tennyson and most of his Cambridge contemporaries with tremendous and gathering force. But few other friends would be more receptive to the implications of this new geology, and its specifc challenge to poetry, than Edward FitzGerald.

The overall impact on their whole generation would be huge. Lyell himself, but only in his private journal, noted with profound dismay the relentless undermining of mankind's moral and metaphysical status on the planet. 'The existence of life millions of years before Man was the first shock to our exclusive reference of all things to us; the next is the approximation of Man to the inferior animals, lowering our pretensions. The next is the Antiquity of Man as helping in the belief in his long existence in a Lower state ... His instincts, the religious especially, may point to Worlds beyond – but Superstition is imperfect instinct ...'

Looking back, Tennyson's grandson remarked: 'The choice, the progress, the whole intellectual and moral life of the human race shrivelled into infinitesimal proportions. Man found himself violently dethroned from his position at the centre of the Universe and relegated to an obscure position on one of its least important elements.'

It was now, reading Lyell for the first time, that Tennyson grasped the significance of the fossil record, and even glimpsed the idea of universal extinctions. Having written 'The Kraken', he must have been tremendously struck by Lyell's similar imagining of the return of Jurassic creatures, especially the Ichthyosaurus from the depths of the ocean. In a virtuoso passage, Lyell imagined them saved from extinction by some 'geographical mutations', and an unprecedented alteration in the global climate. 'Then might those genera of animals return, of which the memorials are preserved in the ancient rocks of our continents. The huge Iguana might reappear in the woods, and the Ichthyosaur in the sea; while the Pterodactyl might flit again through the umbrageous groves of tree-ferns.' This was a prophetic vision to be later revived with tremendous effect in the film *Jurassic Park* (1993).

Tennyson remembered this passage so vividly that he recalled it twenty years later while visiting friends near Plymouth. He astonished their young daughter (later the novelist Mrs Rundle Charles) with a science fiction version of Lyell's vision. In mid-conversation Tennyson 'turned to Geology', and remarked that much of southern England and the Wealds of Kent had once been largely underwater, the 'delta of a great river'. 'Conceive,' he said to the young woman, 'what an era of the world that must have been, great Lizards, marshes, gigantic ferns!' He sometimes imagined, 'standing by a railway at night, the engine must be like some great Ichthyosaurus'.

But the question remained what exactly had caused these Jurassic extinctions? Was it through some single catastrophic event – a biblical flood, a volcanic eruption, or even a meteor strike – as Cuvier had clearly implied? And which of course could suddenly happen again. Memories of the eruption of the Tambora volcano in Indonesia in 1816, which killed over seventy thousand people, reddened skies across the whole of Europe, ruined the harvest by veiling the sun with volcanic ash, temporarily chilled the planetary climate, and inspired Byron's poem 'Darkness', were little more than a decade old.

Or was it through the slow, cumulative, 'vast periods' of geological change that Lyell now identified? It was Tennyson's tutor Whewell, the master of scientific theory and ideology, who immediately grasped and defined this distinction, and its implications, in a brilliant extended review of Lyell's work in the *Quarterly* for November 1833. He also revelled in his love of technical word-coining, which Tennyson always imagined spoken in Whewell's long-drawn-out Yorkshire vowels.

'Have the changes which led us from one geological state to another been, on a long average, uniform in their intensity, or have they consisted of epochs of *paroxysmal and catastrophic* action, interposed between periods of comparative tranquillity. These two opinions will probably for some time divide the geological world into two sects, which may perhaps be designated as the *Uniformitarians* and the *Catastrophists.*'

From now on Tennyson would talk about Lyell's work with a mixture of wonder and horror. On occasions he could also joke about it, and the threat of climate change. In a letter of 1836 to Monckton Milnes he referred to 'the great Geological winter' prophesied by Lyell in his chapter on 'Earthquakes and Volcanoes'. In preparation they should all eat heartily and put on weight, as the earth would swell at the equator and thin at the icy poles: 'in as much as a round belly is better than a white head'. But it was an uneasy joke.

Tennyson would continue to incorporate many other images directly from Lyell into his poetry, over the next thirty years. They would appear as late as his Arthurian poem of 1859, 'Merlin and Vivian'. Here Merlin's dark presentiment is compared to the relentless

incoming tides, which eroded the basalt columns deep into Fingal's
Cave at Staffa, as Lyell had vividly described during his own visit
of 1817.

> So dark a forethought rolled about his brain
> As on a dull day in an Ocean eve
> The blind wave feeling round his long sea-hall
> In silence.

It was in this summer of 1833, during Hallam's absence in Austria, that
Tennyson began working on a new kind of intensely self-questioning
poem: 'The Two Voices: Thoughts of a Suicide'.

> A still small voice spake unto me,
> 'Thou art so full of misery,
> Were it not better not to be?'

> Then to the still small voice I said:
> 'Let me not cast in endless shade
> What is so wonderfully made.'

He would continue drafting it for nearly a decade, and it would eventually run to over 460 lines. Tennyson used strictly rhyming three-line
stanzas, or triplets, to create the rapid back-and-forth of a genuine
debate or battle in his own mind, the terms of which were ever-
expanding. A final version was not published until 1842. He had evidently
discussed its major themes – personal depression, philosophical doubt,
the impact of science on religious belief – earlier and at length with
Hallam. Yet despite its private and personal nature, early versions were
widely circulated among the Apostles, who were fascinated by its arguments. It seemed to express the concerns of their entire generation.

> 'Yea!' said the voice, 'thy dream was good,
> While thou abodest in the bud,
> It was the stirring of the blood.

'If Nature put not forth her power
About the opening of a flower,
Who is it that could live an hour?

'Then comes the check, the change, the fall,
Pain rises up, old pleasures pall.
There is one remedy for all.'

The one remedy was of course death. Why were these young men all fascinated by a poem about suicide? Because, for all their privileges and advantages, Tennyson felt their futures were uncertain, their grounds for self-belief fragile. In June 1833 his friend John Kemble wrote to a fellow Apostle that Tennyson had showed him 'some superb meditations on Self-destruction called Thoughts of a Suicide wherein he argues the point with his soul and is thoroughly floored. These are amazingly fine and deep, and show a mighty stride in intellect ...' Another enthusiastic and impatient comment came from Spedding: 'The design is so grand, and the moral, if there is one, so important that I trust you will not spare any elaboration of execution. At all events let me have the rest of it ...'

Edward FitzGerald, who only saw it later in 1835, thought it was still being 'composed as he walked about the Dulwich meadows'. He concluded that it explored 'the conflict in a soul between Faith and Scepticism'.

In this conflict it is science rather than religion, which confirms the reassuring concept of resurrection throughout the natural world. This power of renewal in nature is given immediate beauty and conviction. One controlling image goes right back to Tennyson's botanising as a child along the stream at Somersby. But it also involved the traditional Greek idea of the butterfly emerging from the chrysalis as a symbol of immortality.

... To which the voice did urge reply
'Today I saw the dragonfly
Come from the wells where he did lie.

'An inner impulse rent the veil
Of his old husk: from head to tail
Came out clear plates of sapphire mail.

'He dried his wings, like gauze they grew;
Thro' crofts and pastures wet with dew
A living flash of light he flew.'

There were other bleaker, more daring stanzas from the early version, though many were later rejected. In one of these the Second Voice reviews the benign pantheism of the Romantics, as expressed in Wordsworth's calm vision in his Lucy poem 'A slumber did my Spirit seal'. Here the human dead lie peacefully in nature's maternal embrace. Wordsworth had written:

No motion has she now, no force;
She neither hears nor sees;
Rolled round in earth's diurnal course,
With rocks, and stones, and trees.

Transforming the same rolling image, Tennyson saw the human dead as part of a fiercer evolutionary history, and raised painful questions about mankind's utter insignificance in the cosmos, and probable extinction. Where did man fit among a world of monstrous creatures, alien landscapes, and unfathomable tracts of space and time?

... When Mammoth, in the primal woods,
Wore, trampling to the fountain-floods,
Broad roads through blooming solitudes –

Where wert thou with the other souls?
Rolled where the equal trade wind rolls?
Or wheeled about the glimmering poles?

It is not clear how much of 'Two Voices' was written before the autumn of 1833, but it was certainly extended and rewritten before being fully published in 1842. In the light of events later that autumn, Tennyson recalled bleakly: 'When I wrote the Two Voices I was so utterly miserable, a burden to myself and to my family, that I said, "Is life worth anything?"' For suddenly, abruptly, on 15 September 1833, Arthur Hallam died from a stroke far away in Vienna.

8

Memorials

Arthur Hallam's death aged just twenty-two was appallingly unexpected. His father had gone out to a business meeting one morning, and left his son behind in their Vienna hotel rooms, lying on a chaise longue, with a slight headache. When Henry Hallam returned, he found his son apparently asleep. But on trying to wake him – disbelief followed by horror – he was stunned to find him dead. Reports only reached Tennyson by a letter on 1 October, which happened to be handed to Emily in the Somersby garden.

Two days before, her sister Matilda had claimed she had seen a mysterious, pale figure wandering beneath the trees beyond the lawn. It was an otherworldly story (Matilda later became a mystic and member of the Church of Jerusalem) which at least suggests how much Hallam had occupied all their thoughts. The autopsy revealed a brain haemorrhage. Henry Hallam, all his paternal hopes for Arthur now wrecked, arranged for his son's body to be brought home by ship from the port of Hamburg to the docks at Bristol, the following month.

The news was devastating to Tennyson, who lost the greatest friend of his youth; and in a different but no less painful way to Emily Tennyson, who lost her unofficial fiancé. The whole household at Somersby went into mourning. Henry Hallam wrote to Tennyson from London, hoping for some kind of obituary or memoir from him. But Tennyson simply could not face it. Instead he began to draft several short elegiac poems, which he kept fiercely secret, until most of them appeared seventeen years later in the opening sections of *In Memoriam*. He remembered bleakly: 'the earliest jottings begun in 1833 . . . in a long, butcher-ledger-like book'.

Not all these first and most agonised jottings were saved; some were too raw, and remained unpublished in his papers until long after his death. These included a desperate, melodramatic fragment,'Hark, the dogs howl! The sleet winds blow', where his pain is open and his loss is physical.

> I seek the voice I loved – ah where
> Is that dear hand that I should press,
> Those honoured brows that I would kiss?
> Lo! the broad Heavens cold and bare,
> The stars that know not my distress.

At some point Tennyson secretly visited Hallam's house at 67 Wimpole Street before dawn, and stood outside it in the rain. He did not go in. Now the darkness of the Kraken had taken on an entirely new meaning. The experience was recorded, with bleak simplicity, in one of the very first poems that he placed at No. 7 in *In Memoriam*. Here again the holding, or clasping of hands takes on an agonised significance which recurs through the entire sequence.

> Dark house, by which once more I stand
> Here in the long unlovely street,
> Doors, where my heart was used to beat
> So quickly, waiting for a hand,
>
> A hand that can be clasp'd no more—
> Behold me, for I cannot sleep,
> And like a guilty thing I creep
> At earliest morning to the door.
>
> He is not here; but far away
> The noise of life begins again,

And ghastly thro' the drizzling rain
On the bald street breaks the blank day.

The image of the long city street would later also come back transformed in one of his extraordinary set of Evolution poems, such as
No. 123: 'there where the long street roars, hath been the stillness of
the central sea'. While the intimate image of the clasped hands would
reappear in no fewer than twenty-nine other poems.

Tennyson would continue to add to this fragmented graveyard of
intensely private elegies from 1833 to 1849, never conceiving of them
as a unified collection until 1847. The first poems were written during
the early time of mourning and 'magical thinking', when Hallam still
seemed alive or at least present, especially at Somersby. Their response
feels extraordinarily direct and intimate, indeed a quite modern reaction to the death of someone beloved, without any of the expected
Victorian formalities or comforts. But what is exceptional is their
intensity. They seem like lover's poems. If they belong to a formal
poetic tradition like Milton's *Lycidas* or Shelley's *Adonais*, many of
Tennyson's descriptions might be equally compared to certain
moments in C.S. Lewis's *A Grief Observed* (1961), or in Joan Didion's
mourning for her husband, *The Year of Magical Thinking* (2005); or in
Julian Barnes's mourning for his wife, *Levels of Life* (2013).

But gradually over the years, an extraordinary transformation would
take place. The overwhelming theme of personal mourning would change
and broaden out. Grief would become anger, doubt would become
intense enquiry. New philosophical and speculative materials would
flood in, which concerned Tennyson's whole generation. From the
terrible idea of Hallam's own death, an individual extinction, so hard
to accept or even comprehend, Tennyson's thoughts would be led
relentlessly on, to consider death in general, and the meaning of
'Extinction' throughout nature. Many of these reflections would be
drawn from his continuous reading in astronomy, geology and evolutionary studies.

The manuscript ledger book was carelessly treated, and was later
nearly lost in his decade of wanderings, but eventually recovered from

his Camden Town lodgings in Mornington Place in 1849. The collection continuously evolved, and only towards the very end did Tennyson try to impose on it a traditionally religious, even specifically Christian theme: Death, Mourning and Rebirth. As he would later tentatively explain: 'It begins with death and ends in a promise of a new life. A sort of Divine Comedy.'

But initially *In Memoriam* was clearly kept more like an agonised diary. In that sense its beginning was intensely autobiographical. Individual entries or sections seem to have been jotted down just as they occurred, starting daily in October 1833, and then gradually becoming weekly or monthly. They record the devastating experience of mourning, the sense of loss and initial desperation, offset by the hallucinatory or magical feeling that the lost beloved, the dead one, is somehow still alive and present; or can somehow be joined underground in death.

A modern evolutionary biologist, Stephen Jay Gould, has aptly described it as: 'a wonderful and deeply truthful account of the psychology of mourning ... the swirling and swinging moods: the anger, the despair, the emptiness, the search for answers, the exultation of temporary resolution ... the renewed despondency, the final acceptance without real answers'. Here is the grim opening of what became No. 2 in the sequence, set in the imagined churchyard (though in fact Hallam was buried in the family vault at Clevedon in Somerset).

Old Yew, which graspest at the stones
That name the under-lying dead,
Thy fibres net the dreamless head,
Thy roots are wrapt about the bones.

The seasons bring the flower again,
And bring the firstling to the flock;
And in the dusk of thee, the clock
Beats out the little lives of men ...

O, not for thee the glow, the bloom,
Who changest not in any gale,

> Nor branding summer suns avail
> To touch thy thousand years of gloom:
>
> And gazing on thee, sullen tree,
> Sick for thy stubborn hardihood,
> I seem to fail from out my blood
> And grow incorporate into thee.

Tennyson effectively buries, or plants, himself in the earth alongside Hallam's body – or actually within it. The verb 'incorporate' is stripped back to its literal meaning, and carries an agonising pun on the word corpse. Here Shelley's gentle healing idea from *Adonais*, of the dead 'becoming one with Nature', takes on a different and far harsher meaning.

The following thirty sections seem to form one distinct group, obsessively turning on the idea of Hallam's physical body, its return by ship for burial, and its lying in the earth. The possibilities of his spirit or soul – whatever that might mean – in some way surviving, seem to fluctuate, until the desperate jollity of the Christmas Eve poem of December 1833 (No. 30). Here the act of holding of hands takes on a momentary, but perhaps illusory reassurance of human endurance.

> At our old pastimes in the hall
> We gambol'd, making vain pretence
> Of gladness, with an awful sense
> Of one mute Shadow watching all.
>
> We paused: the winds were in the beech:
> We heard them sweep the winter land;
> And in a circle hand-in-hand
> Sat silent, looking each at each.

With No. 31 – the so-called 'Lazarus' poem, about the possibility of resurrection – the sequence gradually expands out to more general, and eventually philosophical, questions about the difference between religious and scientific concepts of mortality. From then on Tennyson began to consider the development of life itself on planet earth, and the creation or evolution of the entire universe.

Since Tennyson later completely restructured the running order, probably in the years between 1845 and 1849, it is impossible to know exactly the original chronology of composition. But in general intense autobiography moves towards more speculative forms of elegy and meditation, though with sudden floodings back of remembered incidents at Somersby and Cambridge. This transformation is emphasised by the variations in the three main manuscripts, which are now variously held at Trinity College, Cambridge; the Tennyson Research Centre, Lincoln; and the Houghton Library, Harvard. The elongated pages of the original 'butcher's book' also allowed Tennyson to write his first drafts at the top or bottom of each sheet, and if necessary to tear out the remaining half with alternative versions.

The mourning for Hallam affected the whole household at Somersby, not merely Tennyson and Emily. Their brother Frederick wrote on 18 December of the 'affliction into which our family, especially Emily, has been plunged. We all looked forward to his society and support through life in sorrow and in joy, with fondest hopes ... His loss therefore is a blow from which you may well suppose, we shall not easily recover.' He noted the same feelings among many of their Cambridge circle, and there was a move to put up a memorial tablet in Trinity College chapel. Elsewhere, around Somersby, there were more mixed feelings about the loss of the brilliant and stylish young interloper from London. A local clergyman, Dr John Rashdall, the new curate of the nearby hamlet of Orby, wrote sententiously in his diary: 'Hallam is dead! – such is life: the accomplished-vain philosophic Hallam dead, suddenly – at 23.' It is perhaps revealing that Rashdall – another Cambridge man – thought Hallam vain (and got his age wrong). But he was more sympathetic about the Tennysons themselves. After a later pastoral visit to Somersby Rashdall noted: 'Emily T ... does not look ill, but fearfully soul sick'; while 'AT has evidently a mind yearning for the joys of friendship and love. Hallam seems to have left his heart a widowed one.' It is a striking phrase, possibly even quoting Tennyson himself.

Emily was unable to express her feelings openly, but found some consolation in formal conventions. Later that winter another visitor remembered waiting for her to appear in the subdued drawing room at Somersby. 'She came down to us at last, dressed in deep mourning, a shadow of her former self, but with the white rose in her black hair as her Arthur loved to see her.'

Yet years later Emily was able to reveal her powerful feelings of bereavement. She was writing formally to console her Uncle Charles on the death of his beloved son in 1842, when she burst out: 'In these overwhelming griefs the effect is frequently too stunning in the first instance for the full truth to bear with all its ghastliness of power upon the mind, but when day after day passes away and nothing breaks the mysterious silence ... then by degrees the full weight of the direful blow is felt crushing one's very soul.'

Tennyson himself questioned the value of turning such deeply personal grief, what Emily called its 'ghastliness of power', into poetry at all. Did words themselves fail and become ephemeral? In the long run was the most passionate poetry doomed to become waste paper? In what became *In Memoriam* No. 77, he wrote:

What hope is here for modern rhyme
To him, who turns a musing eye
On songs, and deeds, and lives, that lie
Foreshorten'd in the tract of time?

These mortal lullabies of pain
May bind a book, may line a box,
May serve to curl a maiden's locks;
Or when a thousand moons shall wane

A man upon a stall may find,
And, passing, turn the page that tells
A grief, then changed to something else,
Sung by a long-forgotten mind.

One answer was to avoid the personal lyric completely, and to write in another mode altogether. Tennyson now did what seems an extraordinary and resolute thing. He turned to the dramatic monologue, adopting the voice of one of Homer's most dauntless heroes. With astonishing speed on 20 October 1833, not yet a month after the news of Hallam's death, Tennyson found the voice to begin his poem 'Ulysses' which strikes a wholly unexpected note, a refusal to be crushed or paralysed by grief. Tennyson later reflected: 'There is more about myself in Ulysses, which was written under the sense of loss and all that had gone by, but that still life must be fought out to the end.'

On 25 October Edward FitzGerald (now based in London) heard that Arthur Hallam had died and that Alfred Tennyson, whom he did not yet know well, was writing furiously. 'Tennyson has been in town for some time: he has been making fresh poems which are finer, they say, than any he has done.' As well as 'Ulysses', these must have included further sections of 'The Two Voices', and also 'The Morte d'Arthur'. But very few understood what Tennyson was going through. In December a family acquaintance, Francis Garden, wrote piously to Tennyson to sympathise with his grief over Hallam's death, but reproached him for 'the principles of doubt I have heard you apply to Christianity'. He urged him to have faith in divine love, which apparently he could grasp through the 'concentrated contemplation you could give it throughout your immortality'.

The funeral preparations advanced slowly and with painful formality. In January 1834 Arthur Hallam's body was solemnly buried at Clevedon Church, near Clevedon Court in Somerset. A host of grieving family and Cambridge friends attended, but Tennyson himself did not actually visit Clevedon 'till years after the burial'. Yet the earliest group of the *In Memoriam* elegies remained strikingly direct and vividly located around this shipping of the body across the sea, and its burial in the earth. Following the 'Old Yew' poem (No. 2), and the 'Dark House' poem (No. 7), now came funeral poems full of restless maritime imagery and deep-sea sounds: 'Fair ship, that from the Italian shore' (No. 9), 'I hear the noise about thy keel' (No. 10), 'Tonight the winds begin to rise' (No. 15) and 'The Danube to the Severn gave' (No. 19). There is evidence that Tennyson circulated some of these to close Cambridge friends, and they went the rounds in private letters.

In February John Heath told Tennyson that the 'Fair Ship' poem and two others had been passed to his sister Julia, and she had in turn circulated it, keeping it to read 'till the silent midnight hour'. (Heath, a junior Apostle, so treasured the poem that years later he bequeathed it, and several others copied into his commonplace book, to the Fitz-william Museum, Cambridge.) It would be the only poem to mention Hallam's actual name twice, 'my lost Arthur'; and to refer to himself as 'widowed', just as his sister Emily was widowed. It was also the phrase that the visiting clergyman Dr Rashdall had used of Tennyson's heart.

> My Arthur, whom I shall not see
> Till all my widow'd race be run;
> Dear as the mother to the son,
> More than my brothers are to me.

That same month Tennyson told Henry Hallam that he had failed to write the promised 'Memoir of A H'. But he never mentioned standing outside Wimpole Street in the rain. Later that year Henry Hallam himself published *Remains in Verse and Prose of Arthur Henry Hallam* with his own plain prose memoir. He wrote that he would not publish certain of Hallam's own writings which included any love poems about Anna Wintour. The Hallam family were evidently devastated, but each expressed this in their own way. Henry Hallam made Emily Tennyson a generous life annuity, though with the characteristic reservation (those bohemian Tennysons) that it would cease if Emily ever married. His mother Julia Hallam kept one of her beloved son Arthur's unfin-ished Cambridge notebooks as a keepsake for the rest of her life. She turned it into a commonplace book of mourning, collecting extracts of poems, hymns and sermons. The notebook had originally contained Hallam's lively plans for his secret trip abroad with Tennyson to Spain, including their itinerary and a very practical list of shared costs.

Throughout this dark winter and the spring of 1834, Tennyson shut

himself off at Somersby, and wandered alone over the Wolds, through
the woods, across the reed beds of the marshes, and out to the North
Sea. Hallam's death set off powerful resonances, which deepened as the
months passed. Viewed in retrospect, the joyful undergraduate friendship, the passionate philosophical arguments, the wild trips abroad and
the family celebrations at Somersby, the Cambridge jokes and the
London excitements, the publishing schemes, and above all the romantic plans for Emily, all became utterly transformed. Through the alchemy
of memory, they were changed into something more intense, something
more like a love affair.

This truth, this intensity of longing, was perhaps previously unacknow-
ledged by Tennyson himself, on the evidence of the surviving letters, and
certainly not by Hallam. Though it is impossible to know since Henry
Hallam subsequently burned so much of their correspondence. On the
contrary, it seemed to burst upon Tennyson alone, in the very act of writing and remembering. Nothing in either of their letters prepares for the
first relentless grief of *In Memoriam*. What was previously unspoken, or
suppressed, burst forth in a great outpouring of poetry, the most intense
of his whole career. Walking alone on the beach at Mablethorpe, Tennyson tried to capture the blind, relentless grief for his lost friend in a
four-stanza poem of classical simplicity. It conjures up the incoming beat
of North Sea waves on a winter's afternoon. 'Break, Break, Break . . .'

> Break, break, break,
> On thy cold gray stones, O Sea!
> And I would that my tongue could utter
> The thoughts that arise in me.
>
> O, well for the fisherman's boy,
> That he shouts with his sister at play!
> O, well for the sailor lad,
> That he sings in his boat on the bay!
>
> And the stately ships go on
> To their haven under the hill;

But O for the touch of a vanish'd hand,
And the sound of a voice that is still!

Break, break, break
At the foot of thy crags, O Sea!
But the tender grace of a day that is dead
Will never come back to me.

This poem was not to be included in *In Memoriam*, and expressed his grief in a different way. It has the deceptive simplicity of a song lyric, with its repeated refrain. The sighing exclamation O! appears five times, each time gathering force. The refrain 'break, break, break' itself, initially addressed to the sea, seems to change its meaning by the final stanza. What is breaking now is not the sea, but his heart.

For Tennyson the happiness of the three other figures in the bay seems simple. Yet it is also somehow cruel in its unfeeling joy. The fisherman's boy can shout to his sister in noisy delight, as Tennyson cannot shout to Emily. The sailor's lad can sing freely in his boat, while Tennyson remains inarticulate and silent, speechless with misery. The distant stately ships only serve to remind him of Hallam's body being brought back for burial from Germany. Nature itself is loud but unresponsive, offering no solace. It presents a bleak world of harsh breakers, grey stones and battered crags. The scientific vision of a world of erosion, loss and inhuman timescales is already present. This ceaseless melancholy of the tides, this eternal background soundtrack of extinction, would echo throughout his life. Though typically, Tennyson later implied that the poem came to him in a few moments, walking away from the Mablethorpe beach: '"Break, break" was made one early summer morning, in a Lincolnshire lane.'

Tennyson continued to have doubts about the value of such intensely private and confessional lyrics. He realised they might be just a form of indulgent therapy, or even a kind of drug-taking, a 'narcotic' like his brother Charles's opium-taking, or even Coleridge's at Highgate. Painfully self-aware, he confronted this in another poem in the early *In Memoriam* sequence, placed at No. 5.

I sometimes hold it half a sin
To put in words the grief I feel;
For words, like Nature, half reveal
And half conceal the Soul within.

But, for the unquiet heart and brain,
A use in measured language lies;
The sad mechanic exercise,
Like dull narcotics, numbing pain.

In words, like weeds, I'll wrap me o'er,
Like coarsest clothes against the cold:
But that large grief which these enfold
Is given in outline and no more.

But there was another way of wrapping himself against the ice of grief.
Even while writing these intensely personal lyrics of the grieving process, Tennyson now developed the alternative emotional resource of the
dramatic monologues. Having begun 'Ulysses', he added another piece
drawn from classical myth, to pair and contrast with it, which became
'Tithonus'. Both were long, vividly expressive blank verse poems, with
their own distinctive voice and rhythm. They exploited the very powers
of mimicry – of 'doing the voices' – he had once shared with Hallam.

Now he found he could insulate his grief through two quite different classical legends. The subjects are unexpected, though they
draw on his previous fascination with Homer and the heroes of the
Odyssey. But each seems to represent one aspect of Tennyson's own
attempt to face up to the catastrophe of Arthur Hallam's death.

The first was 'Ulysses', a robust, almost self-vaunting version of the
great Greek sailor and adventurer out of Book XI of Homer's *Odyssey*. 'I
cannot rest from travel: I will drink / Life to the lees . . .' But this is Ulysses in old age, stubbornly refusing retirement, announcing his heroic
determination to persist against the odds, and sail on and on with his
companions, beyond any horizon. Now all private grief and meditative
rhyme is replaced by the assertive almost swaggering blank verse.

> . . . for my purpose holds
> To sail beyond the sunset, and the baths
> Of all the western stars, until I die.
> It may be that the gulfs will wash us down:
> It may be we shall touch the Happy Isles,
> And see the great Achilles, whom we knew . . .

The plangent note then stiffens, with an almost muscular effort, to a refusal of tears, a sort of setting of the jaw, to reach its final fixed resolution. This is gathered into four monosyllabic verbs in the last ringing pentameter line (now so often quoted on its own):

> . . . Tho' much is taken, much abides; and tho'
> We are not now that strength which in old days
> Moved earth and heaven, that which we are, we are;
> One equal temper of heroic hearts,
> Made weak by time and fate, but strong in will
> To strive, to seek, to find, and not to yield.

That heroic 'striving' was presented throughout in terms of the classical image of the sea voyage. But Tennyson added one very striking image from modern science, the Newtonian analysis of the rainbow's arch. This had once appalled Keats for its reductive effect: 'Newton had destroyed all the poetry of the rainbow, by reducing it to a prism.' But Tennyson transformed the rainbow back to an image of poetic infinity, and also drew on John Herschel's spectacular description of the rings of Saturn from his latest *Treatise on Astronomy*. 'The rings of Saturn must present a magnificent spectacle from these regions of the planet which lie above their enlightened sides, as vast arches spanning the sky from horizon to horizon, and holding an invariable situation among the stars.'

Herschel had also discovered the infra-red expansions of light beyond the known and visible margins of the solar spectrum, at least to human eyes. In Tennyson, this endless moving of the Newtonian rainbow as the observer moves, and this infinite fading of visible light at the margin of Herschel's solar spectrum, combine to become a brilliant new image of the fundamental notion of what 'beyond' really means, of what 'an infinite horizon' signifies in human terms. Tennyson now wrote:

> I am part of all that I have met:
> Yet all experience is an arch where through
> Gleams that untraveled world whose margin fades
> For ever and for ever where I move.

The second dramatic monologue, by extreme contrast, was the melancholy story of the Trojan 'Tithonus', also taken from Homer, 'originally a pendant to the Ulysses'. Tennyson's source was adapted from the Homeric 'Hymn to Aphrodite'. Tithonus was the hapless son of King Laomedon, and brother of King Priam, whose divine lover the beautiful Aurora (Dawn) had obtained for him the fatal gift of immortality, without including the vital accessory of immortal youth. In consequence Tithonus is now old and in agony, and longs for the release of death. Unlike Ulysses, he passively accepts his unalterable destiny.

Far from pressing onwards beyond that infinite horizon, Tithonus sinks down exhausted at precisely the 'quiet limits of the world'. The whole of nature shares his autumnal exhaustion. Yet the result has a dazzling and poignant beauty of its own, with its different kind of infinite, metaphorical light.

> The woods decay, the woods decay and fall,
> The vapours weep their burthen to the ground,
> Man comes and tills the field and lies beneath,
> And after many a summer dies the swan.
> Me only cruel immortality
> Consumes: I wither slowly in thine arms,
> Here at the quiet limit of the world,
> A white-hair'd shadow roaming like a dream
> The ever-silent spaces of the East,
> Far-folded mists, and gleaming halls of morn.

The mood, and slow tremulous rhythms of Tithonus's voice, barely making it to the end of each verse line, could not be more different from that of the resolute, emphatic Ulysses. But the poem may also have expressed quite another kind of reaction to grief and despair: that of Tennyson's sister. Emily Tennyson's particular kind of depression and mourning seems to have involved a terrible kind of passivity (a loss of agency) for many months. She may have felt that with the loss of Hallam her fiancé, her whole social future had been withered and frozen. She was doomed to be marginalised, and stranded at the 'quiet limits' of her life. Emily once whispered to her brother at this time: 'What is life to me! If I die (which Tennysons never do).'

Indeed Hallam's whole family worried so much about Emily's mourning and state of mind that Henry and Julia Hallam invited her to spend the whole of the winter of 1834–5 with them at Wimpole Street, where they treated her with 'truly parental kindness as the daughter-in-law she had so nearly become'. Here Emily formed a passionate friendship with Arthur's younger sister Ellen Hallam – 'O Darling that thou art!' she would write – which seemed to make up for something of her emotional loss. This slow recovery was also marked by her taking up singing. Meanwhile Tennyson continued to mourn in his own secret way with his poems.

As many poems written at this time seem to demonstrate two utterly opposed responses to Hallam's death, so Tennyson's own mood and health appeared to fluctuate violently during these months. Sometimes he was almost suicidally depressive, as when he continued work on 'The Two Voices'. Sometimes he seemed almost hectically active, as when he returned to the bitter 'St Simeon Stylites', also completed in draft by 27 November. Here grief keeps him tingling and insomniac:

> Or in the night, after a little sleep,
> I wake; the chill stars sparkle; I am wet
> With drenching dews, or stiff with crackling frost.

But his most important resource was his 'butcher's notebook' of fragmented pieces, which would eventually run to 131 poems. Enormously
varied in length, they all settled into the same four-line stanza form, so
becoming a kind of repetitive litany, the 'sad mechanic exercise', a phrase
later picked up by Edward FitzGerald. But also a banister or railing, or
literally a 'hand hold' to prevent him falling into shapeless, irretrievable
and chaotic misery. The poems still had no dates and no titles, but were
simply given numbers as he wrote them. Privately he referred to them as
his 'Elegies for Arthur Hallam'. In No. 13, he again expressed the 'widower' feelings of physical loss, this time in an image of striking intimacy
and tenderness, as if a husband waking alone in an empty marital bed at
night, might instinctively reach for the lost wife no longer at his side:

Tears of the widower, when he sees
A late-lost form that sleep reveals,
And moves his doubtful arms, and feels
Her place is empty, fall like these.

It is an infinitely touching image, but also a mysterious one. What
was its source? It is difficult not to speculate. Did Tennyson intend a
deflected image of himself, remembering sharing a traveller's bed
with Hallam, perhaps somewhere in a French auberge during their
great Continental adventure of 1831? Or alternatively, an image of his
sister Emily imagining lying in bed with Hallam in some hotel in
Salzburg in 1833, agonising over the loss of her future husband? Or
nothing so autobiographical and confessional as these, simply a repetition of the Rev. Rashdall's perceptive but conventional expression
that Tennyson clearly suffered from a 'widowed heart'? Whatever the
case, it is one of the most vivid moments of magical thinking in this
early sequence, and drenched in unappeasable tears.

As time went on there is evidence that some of his closest friends –
certainly including Spedding and Lushington – grew uneasy at so
much intensity, so much time writing about his grief, so much dedication to mourning for Hallam. Why was he continually casting his

poetry in the form of elegies? Was it ultimately a kind of self-indulgence? Was there nothing else to write about? Tennyson was aware of such criticisms, and they may have been one of the reasons he kept the writing of *In Memoriam* increasingly secret after 1834. Such doubts were raised as early as *In Memoriam* No. 21:

> Another answers, 'Let him be,
> He loves to make parade of pain
> That with his piping he may gain
> The praise that comes to constancy.'

Or was he abandoning the social responsibilities of a poet?

> 'Is this an hour
> For private sorrow's barren song,
> When more and more the people throng
> The chairs and thrones of civil power?'

Above all, was Tennyson neglecting his old fascination with the outward world of nature and scientific discovery, once shared with Hallam? As Edward FitzGerald would also urge, was this:

> 'A time to sicken and to swoon,
> When Science reaches forth her arms
> To feel from world to world, and charms
> Her secret from the latest moon?'

The 'latest moon' referred explicitly to William Herschel's successive discoveries of the several moons orbiting Uranus, which were later confirmed and poetically named (for example Ariel, Titania and Oberon) by his son John Herschel through the 1830s and 1840s. Even more moons were discovered orbiting Saturn (currently more than 240 are known). There was a whole new universe out there to be studied. So was it time for Tennyson to turn his own intellectual telescope back to the heavens?

Gradually these self-questionings would lead Tennyson out of the claustrophobia of private grief, the tragedy of a single life extinguished. It would lead inexorably, over the next few years, to the far greater challenge posed by general extinction in nature. This might involve, as he slowly discovered, the extinction of whole species, and perhaps of mankind itself. Death became, unavoidably, both a new kind of philosophical and a scientific problem, urgent and inescapable, and impacting the whole of Victorian culture.

Out of his dark grief and mourning, Tennyson began to look around again to confront the new light revealed by his continuous reading in contemporary science. The intellectual ferment that had begun at Somersby, and continued at Cambridge with Hallam, now came in some sense to rescue him emotionally. He was kept alive by Herschel's astronomy, Lyell's geology, and the general idea of biological evolution. These truths were often seen to be bitter, reductive and deeply menacing. They raised profound and even disabling questions about human mortality, and mankind's ephemeral presence in the 'secular abysm'. But they would eventually produce a group of some of Tennyson's greatest short poems of evolutionary awareness in the 1840s, placed at the very widowed heart of *In Memoriam*.

9

Romances

Later in the summer of 1834, by a kind of wild but not unexpected reaction to his misery, Tennyson's solitary walks around Somersby turned into a series of neighbourly visits. Melancholy rambles became brisk horse rides round the district. First came chance teadrinking, then invited lunches, then party-going, and finally several big summer dances over the next two years. These led on to to flirtations, of various degrees of intensity, with three very different young women living within a ten-mile radius of Somersby. These were the exuberant Sophie Rawnsley, the beautiful Rosa Baring, and lastly the rather shy and serious Emily Sellwood.

Tennyson was particularly welcomed and cheered by the Rawnsleys, a large family living nearby at the country town of Horncastle, whom he had known since boyhood. A specially cheering attraction was their pretty teenage daughter Sophie, now sixteen, unsophisticated but playful and mischievous. It also turned out that Sophie was a very good dancer, and was keen to teach some new steps to the handsome, but distinctly strange and saturnine neighbour, with his secret grief.

More seriously Tennyson also found himself pursuing another sort of romance with a quite different neighbour, three miles' ride away in the opposite direction, at the village of Spilsby. Here was the large family estate of Harrington Hall, owned by a branch of the rich banking family of Barings. Tennyson's introduction into this wealthy squirearchy had originally been made through Arthur Hallam, and his father's contacts with the Barings in London. Here Tennyson first glimpsed Rosa Baring. She was the eldest daughter of Sir Francis Baring, a chairman of the East India Company, not yet twenty-one and already renowned as one of the beauties of the county, as well she knew.

Harrington Hall was famous for its splendid dining room, its walled garden, and little chapel with a crusader's tomb. Through the course of the year Tennyson attended several house parties and dances there. Echoes of these enchanted but often emotionally fraught gatherings frequently appear in later poems, such as 'Audley Court', 'Locksley Hall' and notably long after in *Maud* (1855). But almost everything Tennyson wrote about this joyful time was doomed to survive only, or largely, in melancholy retrospect.

> Alas for her that met me,
> That heard me softly call,
> Came glimmering thro' the laurels
> At the quiet evenfall,
> In the gardens by the turrets
> Of the old memorial hall!

Rosa Baring herself had far more light-hearted memories of flirting with young Tennyson, and enjoying easy social exchanges between the rustic vicarage at Somersby, her own splendid Harrington Hall, and Sophie Rawnsley's friendly townhouse at Horncastle. They were all within a couple of hours' horse or carriage ride of each other, and mutual visits seem to have been regular, if on the spur of the moment. The Rawnsleys remembered Rosa well: 'She would tell of how she and one of her girl friends, in admiration of the young poet, would ride over to Somersby, just to have the pleasure of pleasing him or teasing him as the case might be.'

It was clearly Rosa's choice whether to please, or to tease. Her social standing, as well as her wealth and beauty, put her in command of the relationship, whether at summer dances or winter suppers, or playful picnics in the springtime woods. (These picnics seem reminiscent of the picnic in Jane Austen's *Emma*, much favoured by Emily Tennyson and Hallam.)

Rosa had one rather sharper memory of Tennyson's awkward style of courting, which captures something of her own imperious charm, though dating from a piece of gossip fifteen years later. 'Alfred, as we all called him, was so quaint and chivalrous, such a real knight among men, at least I always fancied so; and though we joked one another about his quaint, taciturn ways, which were mingled strangely with boisterous fits of fun, we were as proud as peacocks to have been worthy of notice by him, and treasured any message he might send or any word of admiration he let fall.' The emphasis on Tennyson's 'quaintness', as well as Rosa's own 'peacock' feelings, is suggestive of their awkwardness together.

For Tennyson, these meetings with Rosa were transformed twenty years later into a lovers' tryst, in the famous plangent lyric beginning 'Come into the garden, Maud'. It appears to be a dawn love song or aubade, later much favoured by sentimental Victorian performers (particularly the musical setting composed by Michael Balfe in 1857, and famously recorded by Count John McCormack). Yet as originally placed in his poem *Maud*, it is also subtly haunted by images of decline and death. The garden roses are blown, Venus the planet of love is fading, and the idea of death and 'fainting' occurs three times.

> Come into the garden, Maud,
> For the black bat, night, has flown,
> Come into the garden, Maud,
> I am here at the gate alone;
> And the woodbine spices are wafted abroad,
> And the musk of the rose is blown.
>
> For a breeze of morning moves,
> And the planet of Love is on high,
> Beginning to faint in the light that she loves

In a bed of daffodil sky,
To faint in the light of the sun she loves,
To faint in his light, and to die.

At the time Tennyson certainly wrote several sonnets both to Rosa Baring and to Sophie Rawnsley, which each girl valued enough to keep. One to Rosa, dated 1836, opens lavishly: 'Sole rose of beauty, loveliness complete . . .' Full of conventional compliments, it nonetheless contains one strikingly awkward and physical image of passion beating in the blood:

But all my blood in time to thine shall beat,
Henceforth I lay my pride within the dust
And my whole heart is vassal at thy feet

This same image of blood pulsing underfoot would recur, but with violent and literal force, in *Maud* twenty years later.

My dust would hear her and beat,
Had I lain for a century dead,
Would start and tremble under her feet,
And blossom in purple and red.

The sonnet was not only kept by Rosa, but also crisply annotated by her: 'Lines written by Alfred Tennyson after a quarrel at a Horncastle or a Spilsby Ball, 1836'. But it was perhaps indicative of her own casual feelings that Rosa could not actually remember the exact location of their quarrel.

The sonnet to Sophie, also dated 1836, strikes a franker more intimate note, and though equally clumsy, is revealing in a more tender way. The imagery is of snow, not blood.

To thee, with whom my best affections dwell,
That I was harsh to thee, let no one know . . .
Though I was harsh, my nature is not so:

A momentary cloud upon me fell:
My coldness was mistimed like summer snow ...

Sophie Rawnsley's own recollections also have a different kind of appreciation. 'He was a splendid dancer, for he loved music, and such time, but, you know he would say, "we liked to talk better than to dance together", at Horncastle, or Spilsby, or Halton.' Then Sophie added something revealing about Tennyson's talk, with a glimpse of its seriousness and range. 'Most girls were frightened of him. I was never afraid of the man, but of his mind.' Why should Sophie have been *frightened* of Tennyson's 'mind'? Surely because he touched on his secret depressions, his religious doubts, his scientific speculations, and his intense mourning for Arthur Hallam. His was not, as she put it, the chatter of a 'dapper young gentleman of the ordinary type, at a ball or a supper party'.

Tennyson himself later told Sophie that he remembered darker, withdrawn moments at these dances. 'Sometimes in the midst of the dance a great and sudden sadness would come over me, and I would leave the dance and wander away beneath the stars or sit on gloomily and abstractedly below stairs. I used to wonder then, what strange demon it was that drove me forth and took all pleasure from my blood.'

Such challenging thoughts and speculations continued to gather force in Tennyson's mind. A year earlier, in February 1835, he had ordered a copy of his old tutor William Whewell's recently published Third Bridgewater Treatise, *Astronomy and General Physics*, a survey of contemporary science and its apparent confirmation of Natural Theology. Tennyson was looking for philosophical reassurance during these months of secret turmoil, and partly found it in his old tutor's astronomical reflections, especially the chapter in Book 3, 'On the Vastness of the Universe'. Whewell accepted that the 'plurality of worlds' was indeed possible: that unknown modes of being, and specifically that 'organization, intelligence, life' might exist in many planets beyond our own solar system. There might well be advanced civilisations 'that leave our powers of conception far behind'. But how

Alfred Tennyson, portrait by Samuel
Laurence, c.1841. Painting especially
commissioned by Edward FitzGerald, to
celebrate the publication of Tennyson's
landmark *Poems in Two Volumes*, 1842.

Tennyson's Birthplace Somersby.

The Old Rectory, Somersby: Tennyson's birthplace in 1809 and home until 1837. His childhood room was in the attic at the back.

Mablethorpe beach, Lincolnshire, on the North Sea. The deserted sands and dunes were the site of Tennyson's many solitary walks.

Great Court, Trinity College, Cambridge. Tennyson was an undergraduate here from 1828 to 1832, and made many lifelong friends among the Cambridge 'Apostles'.

Alfred Tennyson as an undergraduate, sketched by his friend James Spedding, 1831.

Edward FitzGerald as an undergraduate, also sketched by James Spedding, c.1832. The great friendship between Tennyson and FitzGerald began later in 1835, in the Lake District.

Arthur Hallam in his twenties. A memorial bust in marble by Sir Francis Chantrey, photographed 1893. Based on a retrospective drawing made in 1836.

William Whewell, Tennyson's
influential scientific tutor at Trinity
College, author of *A History of the
Inductive Sciences* (1837), and later
Master of the College. Tennyson would
fondly mimic his Lancashire accent.

The Lady of
Shalott by Holman
Hunt, illustration
commissioned
by Moxon for
the Illustrated
Tennyson, 1857.

The Lady of Shalott
by John William
Waterhouse, 1888.
A painting inspired
by the end of
Tennyson's poem,
showing the Lady
under the fatal curse,
as she floats downriver
to her death under the
walls of Camelot.

The Lady of Shalott by John William Waterhouse, 1894. This earlier scene from Tennyson's poem, shows the Lady after her forbidden sight of Sir Lancelot, desperate to escape her fate, but tangled in the 'magic web' of her own weaving.

Edward FitzGerald in late middle age, 1873. A rare studio photograph, place and photographer unknown. 'I had no truer friend,' said Tennyson. FitzGerald's kindly, thoughtful nature is almost hidden by the melancholy expression he has put on. He hated photography.

far such 'plurality' was compatible with the strict Christian dogma –
which taught that only planet earth had a single Divine Saviour – was
far from clear to Whewell.

Tennyson came to incorporate Whewell's philosophical dilemma
into the passionate poetic debate with which he now expanded his
poem 'The Two Voices'. It had been originally bleakly entitled
'Thoughts of a Suicide', and retained its first form as an agonised dialogue of the soul, but now vastly extended over 463 lines. These
dramatised an inner battle between traditional belief and modern
scepticism. Each voice struggles for mastery over the course of a
single sleepless night, until reaching an exhausted dawn. This scenario
pictures an inner world very far from Tennyson's social flirting and
dancing, and suggests why Sophie Rawnsley might well have been
'afraid' of his mind.

It gradually becomes clear that the First Voice's survival – will the
speaker commit suicide or not? – depends on the conclusion of the
debate. The sceptical Second Voice ('the still small voice') often has
the strongest emotional arguments against conventionally limited
religious views of the universe. And though small, it can speak powerfully and fearlessly:

> Thereto the silent voice replied:
> 'Self-blinded are you by your pride:
> Look up through night: the world is wide.
>
> 'The truth within thy mind rehearse,
> That in a boundless universe
> Is boundless better, boundless worse.
>
> 'Think you this mould of hope and fears
> Could find no statelier than his peers
> In yonder hundred million spheres?'

Tennyson also took courage from John Herschel's more liberal views
in *A Treatise on Astronomy* (1833), which regarded the distant worlds

of the 'nebulae' and 'island universes' (or galaxies) as places of limitless beauty and imagination, and 'an inexhaustible field of speculation and conjecture'. Herschel continued to champion the expanded cosmology of his father, Sir William Herschel, in *Outlines of Astronomy* (1849). These glimpses of redeeming beauty in the universe were also celebrated in 'The Two Voices':

> 'Some vague emotion of delight
> In gazing up an Alpine height,
> Some yearning toward the lamps of night . . .
> Moreover, something is or seems,
> That touches me with mystic gleams,
> Like glimpses of forgotten dreams . . .'

Indeed Tennyson's mood seemed to lighten by the spring of 1835, and so it seems did his sister Emily Tennyson's, back from Wimpole Street. In April, they held a big dance at Somersby, with seventeen-year-old Sophie Rawnsley as the surprise star. 'Alfred is delighted with her,' wrote Emily with relief. 'I sometimes fancy she is the prototype of his [poem] "Airy Fairy Lilian".' Then in May, James Spedding invited Tennyson to a recuperative holiday in the Lake District at his large beautiful Jacobean house at Mirehouse on Bassenthwaite. Spedding also shrewdly asked his old schoolfriend Edward FitzGerald to join them. One of the most important, but unlikely new friendships of Tennyson's later life was thus established, at the very time – as the wise Spedding perhaps foresaw – that Tennyson was trying to recover some sense of balance, and even joy, in his life.

Tennyson had not got to know FitzGerald any better during his last year at Trinity. FitzGerald had matriculated earlier, remained closer to the older footloose Frederick Tennyson, and had disdained the intellectual intensity of the Apostles set. Immediately after Cambridge he had disappeared to Paris for several weeks with William Thackeray, who led him into a wild Left Bank life, but described him as 'a very good fellow but of very retired habits'. Correspondence with Frederick and Thackeray would continue for the next thirty years, much of it telling his anecdotes – and his anxieties – about Tennyson's wonderful poetry and his equally odd behaviour.

Now, walking and boating in the Lakes, and talking long into the night, they struck up their unexpected but historic friendship. Next to the tall, noble, golden Tennyson, 'Fitz' – as he soon became – made an almost comic physical contrast: comparatively short, fat and eccentric, with chaotic clothes, gleaming spectacles and wild hair that quickly departed in his thirties; they seemed hardly compatible. But gradually, under Spedding's genial supervision, suspicions melted and common grounds emerged. FitzGerald played the piano well, which soothed Tennyson. He collected rare books and pictures, and visited ancient battlefields, accounts of which intrigued the poet. He also had a store of funny stories about his eccentric family, to which Tennyson soon found himself contributing.

Having grown up in East Anglia, FitzGerald understood Tennyson's fascination with the North Sea well. Much later he would own a series of small yachts at Aldeburgh – one called 'the Scandal' – and even go into business with a local fisherman there, another sort of scandal. His mother, Mary Frances FitzGerald, that domineering woman of 'striking beauty and imperious manner', held on to him once back from Paris and increased his allowance to £800 per annum, which was more than enough to support a bachelor life provided he stayed in England. She also gave him the use of country lodgings at Geldeston Hall in Norfolk, and later a thatched cottage in the grounds of Boulge Hall, near Woodbridge in Suffolk.

At Bassenthwaite the weather was wet and cold; Tennyson was at first very 'gruff and unmanageable', and stubbornly refused to visit Wordsworth at Rydal Mount. He struck FitzGerald as remarkably 'sulky'. But despite this unpromising start, Tennyson was successfully teased out of his gloom by an exuberant FitzGerald, who announced that he had admired his early poetry from Cambridge days. He proceeded to charm Tennyson with a vivid account of how he had first read 'The Lady of Shalott'.

It happened when he was on a flying visit to Trinity in summer 1833. The poem, with its hypnotic rhythms and shimmering images, had come upon FitzGerald like a revelation, and he never forgot it. He had read it, surrounded by his trunks, while waiting to catch the night mail to London. He had uneasy 'visions' of it in the coach 'all the way to London, through the June night – most of all in the *dawn*, cold even at that time of year, and inclining one to sleep and dream uncomfortably'.

At this Tennyson's melancholy melted, as it would always do in FitzGerald's company. Early one morning the three of them hired a boat and went rowing far out on Windermere, and amid the glassy silence Tennyson burst into poetry, as FitzGerald recalled. 'Resting on our oars, and looking into the lake quite unruffled and clear, he quoted from the lines he had lately read us from the Ms of Morte d'Arthur, about the lonely Lady of the Lake and Excalibur.' The exact lines, sent echoing over the waters, were these:

'King Arthur's sword, Excalibur
Wrought by the lonely maiden of the Lake,
Nine years she wrought it, sitting in the deeps
Upon the hidden bases of the hills' ...

For a moment FitzGerald was left speechless: admiring, sceptical, enchanted and embarrassed. But after a solemn pause, Tennyson innocently exclaimed: 'Not bad that, Fitz, is it.' So the friendship was sealed in a burst of laughter that echoed long after the poetry far over the lake.

A successful literary encounter arranged by FitzGerald took place at a local tavern where they met Hartley Coleridge, now a middle-aged alcoholic but an exquisite sonnet writer. 'After the fourth bottle of gin', Hartley talked fondly of his father Samuel Taylor Coleridge, and claimed that he was delighted with the younger poet's work. Tennyson remembered it long after: 'I met poor Hartley one night at the Salutation Inn, Ambleside ... He promised to correspond with me and took my direction [address] ... which I found next morning under the table ...' Tennyson also heard that Mill was about to review his *Poems* in the new *London Review*. He returned to London much cheered.

Sometime on this expedition Spedding executed a surprisingly romantic sketch of Tennyson. He is seen in half-profile, lounging in the open air, apparently lost in thought, peering down through lowered eyes at the stem of a long clay pipe. His thick, tousled hair falls richly over his brow and round his face casually down to the collar. The mouth is unexpectedly dark and full. But it is the lowered eyes, with their lightly raised eyebrows and long heavily fringed eyelashes, rendered with great tenderness, which give the unmistakable impact of his youthful beauty. In other sketches by Spedding, Tennyson appears equally tousled but decidedly more robust and even burly, in a heavy greatcoat and gravely reading a book. All the sketches record an essential truth about Tennyson's remarkable physical presence.

Back at Somersby in June, Tennyson wrote the first of his many long letters to 'Fitz'. He recalled their rowing on Lake Windermere, their many arguments with Spedding, and his own general grumpiness. He swiftly added: '... but grumpy to *you* – never, as I hoped to be saved ... The sooner I see thee again the better ... It would have been no small refreshment to me in this land of sheep and squires to have looked upon your *sonsie* face once more.' His momentary use of 'thee' was reserved only for closest friends.

FitzGerald remembered Tennyson reading or reciting out loud from the 'Morte d'Arthur' many times in 1835. The manuscript then had no introduction or epilogue, but the poem broke dramatically into the opening lines:

So all day long the noise of battle rolled
Among the mountains by the winter sea;
Until King Arthur's table, man by man,
Had fallen in Lyonnesse about their Lord,
King Arthur ...

FitzGerald never forgot the impressive effect and the astonishing sound of Tennyson's voice. 'Mouthing out his hollow *O-es* and *A-es*, deep-chested music ... with a broad north-country vowel. His voice, very deep and deep-chested, but rather murmuring than mouthing, like the sound of a far sea or a pinewood . . .' FitzGerald's admiration for Tennyson was instinctively displayed in this characteristically dashed off description. The striking idea that Tennyson's voice was like 'a far sea or a pinewood', a whole sweeping soundscape or land-scape in itself, also suggests his affection. Yet FitzGerald was always careful to modulate his praise. 'There was no declamatory showing off in AT's recitation of his verse; sometimes broken with a laugh, or a burlesque twist of voice, when something struck him as quaint or grim.'

Despite his reservations, from now on FitzGerald was convinced of his new friend's genius, and determined to advance Tennyson's career as a poet, in any practical way possible. In a letter of July he tactfully offered him money, which he claimed was spare from the large new maternal allowance. The offer was thrown off casually, in the inimitable FitzGerald manner. 'Dear Tennyson, though I am no Rothschild, my wants are fewer than my monies; and I have usually some ten or twelve pounds sitting very loosely in my breeches pocket . . . It is very difficult to persuade people in this world that one can part with a Banknote without a pang . . .' There is some evidence that FitzGerald repeated this generous offer 'on occasions', through-out the next difficult decade; though Thomas Carlyle's much later suggestion that FitzGerald regularly 'used to give him £300 out of his annual income' was surely exaggerated. Nevertheless, FitzGerald had a penchant for taking on protégés, and eventually gave much larger sums to other needy friends like the young farmer, William Browne and his family, he had met in Wales and virtually adopted; and later the independent Persian scholar Edward Cowell whom he helped to send to Oxford.

FitzGerald also took over something of Arthur Hallam's role as literary agent, encouraging the publishing connection with Edward Moxon, and especially promoting (and protecting) the next collec-tion of poems that did not emerge until 1842. It was through him that Tennyson made his first significant contact with the literary world in Paris. In May 1835 FitzGerald's colleague Thackeray had spotted a short but very favourable article on Tennyson in *Le Voleur*, by a young French journalist Hippolyte Lucas. Lucas had initially been regarded as an upstart provincial poet from Rennes in Brittany, but he had just made a splash in the Paris literary world with a first novel, *Le Cœur et le Monde*. His opinion mattered at that moment (he was later praised by Victor Hugo, but damned by Gustave Flaubert). Thackeray forwarded the article from Paris to FitzGerald in London, who swiftly passed the cutting on to Tennyson, urging him to bestir himself – for once – and actually respond. So an unlikely correspondence sprang up between the two poets across the Channel.

Tennyson exclaimed that he was delighted to find 'a poetical spirit that can sympathise with me on the other side of the broad sea'. He continued a little self-consciously to Lucas, referring to their shared *métiers* with a grand capital letter: 'Poets, as you say, are – or at least they ought to be – bound together in an electric chain – for a Poet does not only speak to his own countrymen but wishes his words and feelings penetrate wherever there exists a brother to echo them.' It is striking that Tennyson imagined this literary connection with Europe as a sort of undersea Poet's telegraph line beneath the Channel. Lucas himself went on to write fashionable comedies, some based on English plays, and became the official director of the famous Bibliothèque de l'Arsenal in Paris.

That July Tennyson finally read John Stuart Mill's favourable essay on his poetry in the *London Review*, countering the earlier sneering attacks by 'the egregious Croker'. He was delighted to find 'The Lady of Shalott' was compared by Mill to Coleridge's 'Christabel', with the appreciative observation that 'the precise nature of the enchantment is left in mystery'. He noted Mill's shrewd reflection that the best of the poems were 'states of emotion, embodied in sensuous imagery', a remark close to Hallam's earlier appreciation. And Mill had concluded that despite his lyric strength, Tennyson was essentially attracted to 'philosophical speculation'.

The long-awaited death of Tennyson's distant and domineering grandfather, George Clayton Tennyson, 'the Old Man of the Wolds', occurred in July at his large estate, Bayons Manor, at Tealby. There were hopes of legacies, though none of the Somersby Tennysons attended the funeral. With the bulk of the Old Man's inheritance going to his younger son, Tennyson's uncle Charles d'Eyncourt (not to be confused with Tennyson's elder brother Charles), he was henceforth regarded as a social climber by Tennyson. The hostility was freely returned by d'Eyncourt, who wrote of the Somersby Tennysons: 'I know no reason why any of that family should be dissatisfied . . . Alfred I find has been so violent (though he gets the Manor of Grasby . . .).'

In fact Tennyson's mother initially received only a small annuity of £200 from her father-in-law's estate, though she (and her sons) hoped this would allow her to continue living at Somersby for at least a few years more. As for his elder brothers, both received small legacies. Frederick used his to go abroad to Corfu and then Italy, where he drifted, though keeping in touch through frequent letters to FitzGerald, and rarer ones to Tennyson himself. Charles, who had adopted the surname of Tennyson Turner (according to another inheritance), combined his capital to settle into the life of a Lincolnshire squire, acquiring a fine old Georgian house at Caistor, but also an opium habit.

Tennyson himself was left a small, rundown farm at Grasby, which he would never visit; and the promise of a capital investment worth £3,000. The rents brought him a slight income, but he was still painfully reliant on Aunt Russell's annuity of £100, and any money quietly supplied by FitzGerald. He remained at Somersby, which he had come to love and associate with Arthur Hallam's memory, and the early years of walking, picnicking and writing; though also with his father's madness and gloom. In practice Somersby also gave him the domestic responsibility of his mother, and the remaining eight younger siblings, including his four sisters (with Emily still in mourning), and his unstable brother Septimus, with his gloomy weeping fits.

Tennyson wrote acidly to his amusing Cambridge friend Brookfield about the varied inheritances, knowing he would appreciate the joke. But he also revealed something of his own bitterness. 'You ask after Charles: he lives at Caistor (a wretched market town which looks more like the limits of the civilized world ... about 9 miles from Tealby): he is however the first man in the place, an owner of books and pictures, a water closet with a recumbent Venus in it, and a house with a stone front. The country about him is dreadful – barren rolling chalk-Wolds without a tree: he talks of travelling for three years, and afterwards taking a wife.'

Throughout the mid-1830s at Somersby Tennyson remained immersed in social duties and unexpected romances. In May 1836 his brother Charles, spurred into action sooner than predicted, courted and swiftly married the youngest of the Sellwood daughters at Horn-castle, Louisa. Tennyson was best man, and Louisa's older sister, Emily Sellwood was bridesmaid. For the first time since Arthur Hal-lam's visits, he noticed Emily properly, and when he wrote a poem to celebrate the marriage, addressed it to 'The Bridesmaid' rather than the bride, and began to pay mildly flirtatious attentions to Emily.

> For while the tender service made thee weep,
> I loved thee for the tear thou couldst not hide,
> And prest thy hand, and knew the press returned,
> And thought, 'My life is sick of single sleep ...'

But the romantic auspices were uncertain. Charles's marriage quickly ran into difficulties, his opium habit revealed itself at Caistor, and after a few months, Louisa was making regular returns home to her parents at Horncastle which continued throughout the next decade.

All this complicated Tennyson's new attentions to the dark and delicate Emily Sellwood, now aged twenty. He learned that she owned her own copy of his *Poems* of 1833, and had read 'The Lady of Shalott'. He later recalled several magical moments with her, in par-ticular the time when he came back from London unexpectedly to find her sitting alone on 'the iron garden chair' on the Somersby lawn, surrounded by wild flowers. 'You were in a silk pelisse, and I think I read some book with you.' He claimed he could not remember the book, only Emily in her silk dress.

In June Tennyson dined at the Sellwoods' with Sir John Franklin (Emily's uncle), the famous Arctic explorer. He was about to depart to take on the governorship of Tasmania. Perhaps showing off to Emily, Tennyson embarrassed the entire company by flirting with Franklin's darkly pretty niece, Catherine Franklin. He sat close beside her as she played the piano, peering at the jewels round her neck through his pince-nez, and announcing they were 'fit for a Princess'. Told that Catherine had in fact been brought up in Madras, he then asked Emily provokingly: 'Is she a Hindoo?' Catherine herself defused the situation by remarking mildly that she had been considered 'somewhat Eastern-looking as a girl'. But from the beginning Tennyson could be an uncertain suitor.

Again, at the Spilsby Ball of July 1836, Tennyson had the row or 'tiff' with both Rosa Baring and Sophie Rawnsley which set off his summer of sonnet-writing to the two young women. But in retrospect this also seemed to be the result of his attentions to Emily Sellwood. Much of this confusion was reflected in various short uncollected love lyrics of the time, and years later in *Maud*. It is even possible that all three young women became shadowy identities emerging in some of the Somersby poems for *In Memoriam*. Tennyson was sufficiently unsettled to think of leaving England that autumn, and joining his older brother Frederick in Italy, though Frederick was himself writing despairing love letters to his cousin Julia Tennyson d'Eyncourt. An atmosphere of romantic tragicomedy seemed to surround Somersby, and add strange lustre to the darker memory of Hallam.

Tennyson's friends in London were aware of his frustration and unhappiness, but also puzzled by its complex causes, and sometimes frankly impatient. Fitz always remained sympathetic, and Spedding philosophical. But Monckton Milnes – who was briskly trying to get Tennyson to contribute poems to his new magazine *The Tribute* – openly complained to another contributor George Darley. Darley replied: 'But what can T have to drown him in the slough of despond, as you tell me, if he retain the free use of his faculties . . . But what can be the worm at Tennyson's heart? Is he in the flames, furies and tortures, agonies and excruciations of the "tender passion"?' It was an apt question.

Certainly Tennyson was still haunted by the memories of Arthur Hallam's many visits to Somersby during the three long summers between 1831 and 1833. He remembered how those happier, simpler parties went on in the rectory garden long after dark. At some moment he took out Hallam's letters to reread, as recorded in *In Memoriam* No. 95:

By night we linger'd on the lawn,
For underfoot the herb was dry;
And genial warmth; and o'er the sky
The silvery haze of summer drawn . . .

But when those others, one by one,
Withdrew themselves from me and night,
And in the house light after light
Went out, and I was all alone,

A hunger seized my heart; I read
Of that glad year which once had been,
In those fall'n leaves which kept their green,
The noble letters of the dead . . .

But his scientific reading also continued. During the early months of 1837 he was deeply immersed in the third volume of Lyell's *Geology*. He also read Babbage's newly published *Ninth Bridgewater Treatise: A Fragment*. This was intended as a highly unorthodox addition to the eight official Bridgewater Treatises, originally sponsored by the thoroughly orthodox Earl of Bridgewater in 1833. The pious purpose of the series was to demonstrate 'the Goodness of God, as manifested in the Creation'; in other words an up-to-date scientific exposition of Paley's *Natural Theology* of thirty years before. Eight Fellows of the Royal Society had contributed, the first being by Thomas Chalmers on the 'Moral Adaptation of External Nature', the third being Whewell on astronomy, and the sixth William Buckland on geology.

The series had officially concluded with No. 8, 'On the Function of Digestions considered with reference to Natural Theology' by William Prout MD. Not surprisingly it was generally regarded as very worthy, but very dull.

Babbage's No. 9 *A Fragment* was on the other hand written in a quite different spirit. It mischievously deployed mathematics and the logic of Babbage's new calculating machines to challenge and satirise many of the traditional 'arguments by design' of Natural Theology. Published in a popular edition by John Murray, it opened with a mocking quotation from Whewell himself, who disputed the authority of 'mechanical philosophers and mathematicians of recent times' to have any relevant views of 'the administration of the universe'. Responding to this academic gibe, the mathematician Babbage went on the rampage.

Reading Babbage's intellectual onslaught, Tennyson became aware of the sharpness of scientific disagreement among his contemporaries on the most fundamental issues. Quoting both Herschel and Lyell, Babbage insisted that the Book of Genesis was not to be taken literally, and that both astronomy and geology had demonstrated quite different patterns of physical evolution. He summarised drily, 'whilst the testimony of Moses remains unimpeached, we may also be permitted to confide in the testimony of our senses'. In fact Lyell was so shocked by Babbage's irreverent tone that when he read the proofs of the *Ninth Treatise*, he wrote urgently to the Anglican geologist Adam Sedgwick in Cambridge hoping he might help to suppress 'objectionable passages'.

Early drafts of Tennyson's new poem 'The Golden Year' included explicit fragments of this scientific reading. Babbage had written of the formations of star clusters in the Milky Way, as observed telescopically: 'Others there are, in which the apparently unformed and irregular mass of nebulous light is just *curdling*, as it were, into separate systems ... each, perhaps, the splendid luminary of a creation more glorious than our own.' He had also speculated on the idea of the physical universe as a vast calculating machine, with the stars organised by a kind of divine computer program. In Tennyson's draft these passages produced several vivid images of cosmological movement, within an impersonal 'clockwork' mechanism. (They included the Audenesque line, 'the complicated clockwork of the suns'.) All this led to his anxious metaphysical questioning of the idea of astronomical 'motion' itself as any kind of inevitable 'progress' or 'gain', and the whole uncertainty of man's restless destiny within an 'adverse' nature:

> The noiseless ether curdling into worlds,
> And complicated clockwork of the suns.
> Motion: why motion? were it not as well
> To fix a point, to rest? again, it seems
> Most adverse to the nature of a man
> To rest if there be any more to gain.

This whole period of writing was emotionally as well as intellectually unsettled. He continually looked back on his unhappy love affairs (whatever they were) with Rosa Baring and Sophie Rawnsley, partly reflected in the winding narrative of 'The Gardener's Daughter'. He still had flashbacks of Hallam at Somersby. Yet he had made a kind of secret engagement with Emily Sellwood, and began an extensive secret correspondence with her, increasingly intimate, though nearly all of it would later be lost.

Nonetheless there were unexpected glimpses of a suddenly extrovert Tennyson, thoroughly enjoying bachelor life in the city, in the taverns and theatres around Holborn and Fleet Street. So in mid-April 1837 Spedding wrote to Monckton Milnes in epicurean mood: 'Yesterday I dined with Alfred Tennyson at the Cock Tavern, Temple Bar. We had two chops, one pickle, two cheeses, one pint of stout, one pint of port, and three cigars.' A week later, on 12 April, Tennyson was whisked off to a popular theatre, the Olympic on the Strand, to attend what sounded like a rather risqué show. It was a burlesque version of Alexander Pope's satirical poem *The Rape of the Lock*, a comedy of boudoir flirtations and seductions (though not an actual rape) set in old pre-Napoleonic high society. It evidently involved an inordinate amount of dressing and undressing, which Tennyson mocked in a note to FitzGerald. 'What a face you would have made if you had been with me at the Olympic last night, and seen the lumpish sylphs hanging plumb-heavy over their toilette ... Pope in every way travestied.'

In a similar spirit, Monckton Milnes attempted to get Tennyson to contribute risqué verse to his other new venture, the *Annual Magazine*. The result, again unexpected, was a burlesque ballad of the 'exotic' type, about a beautiful black maiden from Haiti. Entitled 'Anacaona', it was metrically ingenious – almost a calypso – and in a provocative style that evidently appealed to Milnes. It set off for Haiti with a fine flourish:

> A dark Indian Taino maiden,
> Warbling in the bloom'd liana,
> Stepping lightly flower-laden,
> By the crimson-eyed anana
> Wantoning in orange groves ...

But it soon degenerated into 'naked and dark-limbed' bathing and dancing, and Tennyson had second thoughts about publication. He may also have discovered the true historic circumstances of Anacaona's life: that she was imprisoned and murdered by Spanish conquistadores in the sixteenth century. At any rate he withdrew the poem, saying simply that its bad 'botany' would be corrected by any Caribbean sailor. 'It would be confuted by some Midshipman who had been in Hayti latitudes and knew better about Tropical Vegetables ...'

Evidently aware of potential racism, he added that the whole tone of the poem was unsuitable: 'I cannot have [Anacaona] strolling about the land in this way – it is neither good for her reputation nor mine.' To Milnes's surprise he replaced it instead with an entirely different and heart-breaking love lyric, which began:

> 'Oh! That 'twere possible,
> After long grief and pain,
> To find the arms of my true-love
> Round me once again ...'

He would much later incorporate this beautiful, timeless song into his long poem *Maud*, somehow summoning up all his romances. The switch between the mood and content of the two poems vividly suggests Tennyson's wildly fluctuating spirits at this time.

Rebuked in no uncertain terms, Milnes himself turned his attentions to something more serious, and began the editorial work that in 1848 produced his landmark edition of *The Life and Letters of John Keats*. Nevertheless, he did continue to collect various kinds of erotic and exotic literature in private, and eventually bequeathed a famous pornographic archive to the British Library. Evidently this was not Tennyson's style at all.

There was one other major upheaval. The Tennyson family was forced to leave their beloved Somersby home in spring 1837, as the new rector was at last installed. Unwillingly they searched for a more practical house, closer to London. Tennyson felt bereft at leaving Lincolnshire and more especially the rectory at Somersby, which held such powerful memories, of the enchanted circle that held him, for three magic summers in a triangular bond of love, with Hallam and his sister Emily. He recorded this several times in *In Memoriam*, such as No. 102:

> We leave the well-beloved place
> Where first we gazed upon the sky;
> The roofs, that heard our earliest cry,
> Will shelter one of stranger race.

> We go, but ere we go from home,
> As down the garden-walks I move,
> Two spirits of a diverse love
> Contend for loving masterdom.

10

Asylum

Tennyson initially disliked the new house at High Beech, a village on the edge of Epping Forest, in Essex. It all seemed very genteel and southern compared to the wild Wolds. To his friends he was inclined to dismiss Beech Hill House as 'a cockney residence', though he had in fact chosen it himself, as suitable for the whole family, and available for rent. It was a substantial three-storeyed Georgian house, one of the largest in the neighbourhood, with its own walled grounds, vegetable gardens and an ornamental lake. It had a creeper-covered veranda, with rolling lawns below, and a distant prospect of the forest itself. But Tennyson liked to mock the suburban garden with potting sheds and greenhouses, and its domestic view over Waltham Abbey. There were 'no sounds of Nature and no society; equally a want of birds and men'. He claimed the only advantage in it was that 'one gets up to London oftener'.

Such a substantial property, though only rented, had become affordable through the capital bequested by the Old Man of the Wolds. Despite Tennyson's complaints about his grandfather's meanness, the probate and sale of these investments had eventually provided the Tennyson children with £3,000 each; and their mother twice that. These were not insignificant sums if wisely invested, but could be easily absorbed by any kind of extravagance. The daughters' portions were held in trust, but what would the sons do with theirs? Frederick made his go far by living in Italy; while Charles had moved from Tealby to the larger estate at Caistor. Meanwhile their uncle, the wealthy Charles d'Eyncourt, continued to issue financial advice to the whole family, while grandly restoring the splendours of Bayons Manor. Initially Tennyson was undecided what he, or his younger brothers (for whom he felt a new responsibility), should do with their inheritance. But it remained clear that for him, the sole profession was poetry.

One first effect of the move to Beech Hill House in June 1837 was that domestic conventions reasserted themselves after the bohemian disorder of Somersby. Mrs Tennyson was again able to afford servants, including a cook, a liveried handyman and a housemaid. The family was still large, now consisting of the three remaining younger brothers – Arthur, Septimus and Horatio – and the four unmarried sisters, Emily, Mary, Matilda and Cecilia. They were summoned to regular meals by the novel addition of a brass gong, though Tennyson was usually late, or else had disappeared into Epping Forest. They bought a 'nice black pony' for Tennyson's sisters to ride. 'We have given 9 pounds for it and Alfred thinks that it is well worth it,' wrote Cecilia, who also kept her own dog Ariel (named after Shakespeare's spirit, or perhaps after Shelley's fatal boat in Italy). The pony, a reminder of Somersby days, cheered the whole family, being 'full of spirit' and 'never seems wearied'.

It seems that Elizabeth Tennyson and her daughters were glad to escape the isolation of Lincolnshire. Their neighbours were 'sufficiently hospitable', and there was a certain amount of dining out with 'venison and champagne'. Tennyson occasionally enlivened the house with 'some wild fun', telling family jokes and brilliantly mimicking the voices of old friends and new neighbours. It was the sort of mimicking he had done in his Cambridge days. Cecilia told a friend about one evening after their arrival: 'We sate up till one o'clock, Alfred amusing us all the time by taking different characters. He made us laugh so much you should have heard him – would have amused you so.' Possibly one of the voices was that of his older brother Charles when under the influence of opium, since Cecilia added: 'Has Charles been in any of his ridiculous Moods lately?'

Tennyson was soon embarking on long rambles in Epping Forest. In the winter he found he could skate on the lake, 'with his long blue cloak blowing out behind him'. But the new domesticity was a trial. At least he could afford to hire a set of rooms at Mornington Place, in Camden Town, so he could make good his escapes to London, though he told Emily Sellwood that his mother was afraid if he went to town 'even for a night'. How could they get on without him 'for months'? Meanwhile he described the capital city to Sophie Rawnsley, perhaps temptingly, as 'this great black Babylon'.

Yet secretly Tennyson was restless, and missed Somersby intensely. He wanted to be 'cheered with the sight of a Lincolnshire face'. Later that December he sent another of his rare surviving letters to Emily, full of complaints. 'I have been in this place [High Beech in Epping Forest] all the year, with nothing but that muddy pond in prospect, and those two little sharp-barking dogs. Perhaps I am coming to the Lincolnshire coast, but I scarcely know. The journey is so expensive and I am so poor.' But it was a curious excuse, as he was actually better off now with the legacy.

Instead of Lincolnshire, Tennyson began an endless shuffle between High Beech and various friends in London, notably Spedding with his large chambers in Lincoln's Inn, and 'Old Fitz' nearby in Mornington Crescent. The distance was just over twelve miles, an easy coach trip, though he sometimes preferred to walk, even at night. Then, in heady anticipation he would see above the first dark, silent city streets, 'the light of London flaring like a dreary dawn', and his spirits would leap. He kept his friends up late into the night – or early in the morning – smoking, drinking, talking and reciting his latest poems. Sometimes he slept over on their sofas, or began composing in the mornings in his 'butcher's books' at their dining room tables.

In truth, despite the family house at High Beech, and the rooms in Camden, Tennyson was still homeless in many senses, frequently ill and depressed. His London friends noted he continually 'complained of nervousness', and smoked obsessively: 'the strongest most stinking tobacco out of a small blackened clay pipe on average nine hours every day'. Secretly he feared he was doomed to fall victim to 'the black blood of the Tennysons' like Arthur, or Septimus; or even to complete insanity, like poor Edward. Yet there was now a possible cure at hand.

High Beech contained one unusual feature that became increasingly important to Tennyson, and must have influenced his original – and apparently strange – choice of location. It was the site of an extremely advanced kind of healing centre, which offered services quite different from the traditional lunatic asylum. Ever since poor Edward's banishment to the grimness of Lincoln General Asylum, Tennyson had been deeply concerned about the mental health of two of his younger brothers: Arthur who was drinking to excess, and Septimus who was chronically depressed. High Beech Institution offered a possible solution.

The centre was run on enlightened lines by a charismatic and eccentric Scottish doctor, Matthew Allen, then in his fifties. Allen's ideas were based on those pioneered at The Retreat, a revolutionary Quaker establishment for healing, founded at York in 1792, by the philanthropist William Tuke. The old eighteenth-century idea of the madhouse, or Bethlehem, as a grim and punishing prison for 'lunatics' – originally those suffering from 'moon madness' or epilepsy – was revolutionised by Tuke's policy of humane treatment and practical kindness. It was inspired by the Quaker ideals of moral respect, simplicity and natural equality. The Retreat still flourishes as a hospital to this day.

Born in 1783, Allen had studied in Edinburgh, and claimed to have qualified as a medical student in Aberdeen. He was appointed apothecary to the York Lunatic Asylum, an important post, at the age of twenty-six in 1819. His advanced system, like The Retreat, was based on occupational therapy, group discussions, minimal restraint and maximum freedom, with much gardening and artwork, good food and walking in Epping Forest.

Allen had just published a highly original book on the enlightened treatment of mental patients, *Essay on the Classification of the Insane* (1837). In it he also set out his professional achievements, his experience in this novel field, and his personal commitment to the work. 'As early as 1807 I visited lunatic asylums *con amoré*; and that in

1816, 1817, 1818 and 1819 I was engaged in lecturing on Mind and its Diseases.' He stressed that among his patients 'not more than about 3 percent suffer personal restraint'. Patients included people from all walks of life: labourers, farmers, naval men, house servants, governesses, lawyers, artists and even poets. As well as a full staff of attendants, Dr Allen gave each patient individual attention, much of it based on regular periods of private conversation.

Sometimes Allen's methods were unorthodox and counter-intuitive. With one particularly violent patient (Case No. 372), Allen recorded how he spent an hour each day, at the same time, simply sitting down talking quietly with him. The subject was not the patient's life and childhood memories; but *Allen's own life*. Allen would 'detail to him a history of my own life, always contriving in the style of the Arabian Nights Entertainments, to break off suddenly at some point of interest . . .' This apparently succeeded in establishing new calm, confidence and hope for the future (and possibly, one might think, therapeutic boredom). Allen claimed that it extracted the patient from his enclosed monomania, and gradually drew his attention back to the normal outside world. To this treatment Allen added simple physical work in the garden, 'work with a spade', and the specific task of digging a new path between the asylum's different buildings entirely on his own. Allen then did something inspired. The path was formally given the patient's *own name* in a little ceremony. This gave him a new sense of identity and achievement. It was a thought-provoking method: a combination of psychology, occupational therapy and the charms of a mountebank.

All this presented Tennyson, continually worried about his own mental state, with an entirely new approach to helping both the alcoholic Arthur (now aged twenty-one) and his depressed brother Septimus, aged twenty-two. Problems with Septimus had begun early in his teens at Somersby. He had not been thought clever enough for university, but apprenticed at sixteen to a local land agent in Louth; and then at eighteen sent as a doctor's assistant to Lincoln. Finding he was unable to live on his own, Septimus had soon abandoned this attempt to establish a profession and returned in misery to Somersby. Overcome by depression and tearful nervous states, and beginning to rely on alcohol like his father, he then rejoined the family at High Beech.

Tennyson had always felt specially protective towards Septimus (the neglected middle 'seventh' in that large family). As a small child, Septimus was very beautiful and delicate, so Tennyson used to take him to bathe in the Somersby stream below the rectory. On the way back he and Frederick would 'chair' the little boy between them, and race round the meadow with wild shouts of delight. But by his mid-teens Septimus was increasingly 'subject to fits of the most gloomy despondency accompanied by tears' which would last for whole days, very like his brother Edward, by then permanently confined.

Tennyson had himself taken the unusual step of writing urgently about Septimus to his despised but rich and successful uncle Charles d'Eyncourt in January 1834, begging him to find Septimus a job and position 'removed as far as possible from home'. He thought his brother might recover his energy and independence in 'some bust-ling active line of *life remote from the scene of his early connexions*'. Otherwise it seemed as if his mind would prove 'as deranged as Edward's'. For a time it looked as if placing Septimus with Dr Allen might avoid this gloomy fate, or what Tennyson called 'so miserable a termination'.

It was evidently at High Beech that Tennyson became truly con-scious of the psychological problems that all the brothers had inherited from their childhood at Somersby. Something in its rural isolation, in the long shadow cast by the tyranny and instability of their father, and perhaps also in the 'angelic' neediness of their mother, who clearly clung to her sons, was subtly destructive. He confided to Charles: 'I have studied the minds of my own family, I know how delicately they are organized . . .' The idea of madness, both its causes and its treatment, began to preoccupy him at High Beech and even-tually have a powerful influence on his later poetry, notably on *Maud, or The Madness*, not finally published until 1855. He gradually became fascinated with Allen's new therapeutic system, his adapted Quaker idea of 'moral treatment', and the lives of the patients themselves.

He was immediately struck by Dr Allen's refusal to use such

conventional terms as 'Asylum' or 'Keeper'. Instead he referred to 'Places of seclusion for exhausted Minds'. Allen's establishment at High Beech, whatever it called itself, was physically divided into three separate buildings set back from the village – Fair Mead for men, Springfield for women and Leopard's Hill for a few severe cases. Scattered across several acres, the houses were connected by a 'double-thorn' hedge, which protected them from the village and allowed patients to wander in privacy through the large enclosed grounds, and go through a gate into Epping Forest itself. Each detached house received about a dozen paying patients. In the first two, they had their own rooms, with early types of central heating ('Sylvester Patent Air Stoves'), medical vapour baths, and most notably, some indoor water closets (still a metropolitan rarity), as well as a separate outdoor lavatory block. Many of the residents were in fact short-term voluntary cases, who paid large fees of five guineas a week to live in considerable comfort and recover from various forms of mania and depression.

Matthew Allen had gained many influential London supporters, including Coleridge's friend William Sotheby, Wordsworth's friend Basil Montagu, and later Thomas Carlyle. His wife Jane Carlyle had visited in October 1831:

'One day I went with Mrs. Montagu to Epping to visit Dr Allen, a Scotsman, who has a lunatic establishment in the midst of the forest – a place where any sane person might be delighted to get admission. The house, or rather houses are all overhung with roses and grapes and surrounded by gardens, ponds and shrubberies without the smallest appearance of constraint. The poor creatures are all so happy and their doctor such a good humane man ... I am going to pack and stay some day. Dr Allen is an old friend of Carlyle's and his wife is an excellent woman.'

The High Beech asylum so impressed Tennyson that he quickly decided to send the gentle, but increasingly confused and anxious Septimus to board there as a paying guest, in the healing comforts of Fair Mead. Arthur on the other hand, ever more violently drunk just like his father, was a harder case to manage. Eventually he had to become a 'voluntary' patient far away in Edinburgh at the notorious

Crichton Institution for alcoholics. Even the infinitely forgiving Mrs Tennyson acquiesced in this extreme measure 'to conquer the ruinous and destructive habit of drinking'; though she managed to obtain him a 'private sitting room' at an extra £50 a year, 'which is absolutely necessary for his health'. Only the youngest brother, Horatio Tennyson, made good his escape from this family doom, by going even further away, in fact to the other side of the globe, to farm in the wild hillsides of Tasmania.

Septimus seemed to improve under Allen's enlightened supervision, though the atmosphere was not always calm. Once while Tennyson was lunching with him and Allen at Fair Mead, the doctor temporarily left the dining room, and another male 'guest' rose to his feet, picked up a knife from the table, and confronting Tennyson demanded, 'Why do you wear that *glass thing* in your eye?' Tennyson, obviously used to his brothers' outbursts, was not perturbed. Gently taking off his monocle and waving it apologetically, he replied with a smile: 'Vanity, my dear Sir, sheer vanity.' Calm was immediately restored. Perhaps odder than the guest's behaviour is the idea of young Tennyson wearing a monocle.

The asylum attracted several other literary clients including a confused, alcoholic and deluded poet from Northamptonshire. This was John Clare, the author of *The Shepherd's Calendar* (1827) and most recently *The Rural Muse* (1835), then in his forties. Clare began his residency at High Beech in July 1837, just at the time of the Tennysons' arrival, and remained there for four years, initially happy ones. Encouraged by Allen to continue his poetry, he wrote among many other pieces 'A Walk in the Forest', which evokes the pastoral atmosphere of the place. Clare also mentions another literary figure, not Tennyson, but the son of the Scottish poet Thomas Campbell, author of *Gertrude of Wyoming*. Clare longed for such companions, though remaining essentially solitary.

I love the forest and its airy bounds,
Where friendly Campbell takes his daily rounds,
I love the breakneck hills that headlong go,

And leave me high, and half the world below
I love to see the Beech Hill mounting high,
The brook without a bridge, and nearly dry . . .

Typically, Allen attempted to organise a fund for Clare's support, and
wrote about him to *The Times*: 'It is most singular that ever since he
came . . . the moment he gets pen or pencil in hand he begins to write
most poetical effusions. Yet he has never been able to obtain in conversation, nor even in writing prose, the appearance of sanity for two
minutes or two lines together, and yet there is no indication of insanity in any of his poetry.' The fund was never fully established, but
subscriptions provided £60 for at least one year of Clare's upkeep, also
an indication of Allen's relatively high fees. Septimus must have paid
a similar sum, though he could well afford to after his inheritance.
Allen did not overlook this either.

Whether Tennyson was ever formally introduced to John Clare by
Allen, or met him casually wandering through Epping Forest (as they
both loved to do), has never been established. But it seems highly
likely. Clare was already a well-recognised figure in literary circles,
and would have been something of a celebrity at High Beech. When
the London journalist Cyrus Redding visited High Beech in 1841, he
simply asked to see 'John Clare, the Peasant Poet', and was taken directly to meet Clare who was peacefully standing in a field of beans,
hoeing and smoking a small pipe. Redding thought he looked happy
and had put on weight. Yet shortly after, Clare ran away to Northampton, and left manuscripts describing himself as the poet 'Boxer
Byron', Matthew Allen as 'Doctor Bottle-Imp', and High Beech as
'Allen's Hell'. He never returned, but settled instead in Northampton
Asylum, and remained there for the rest of his life.

Over the next two years Tennyson grew increasingly fascinated by
Dr Allen and his enlightened work among the supposedly insane.
Tennyson's friends in London, FitzGerald and Spedding, followed
this development with some scepticism when they eventually learned
of it. James Spedding remarked in August 1840 that Tennyson 'has
been on a visit to a madhouse for the last fortnight (not as a patient),

and has been delighted with the mad people, whom he reports the most agreeable and the most reasonable persons he has met with. The keeper is Dr Allen ... with whom he has been greatly taken!'

Over several years High Beech asylum supplied Tennyson with extraordinary psychological materials, versions of which appeared much later in *Maud, or The Madness*. The gardens at High Beech, the paths and the shrubberies, the wandering figures of the insane, muttering to themselves, recur like a remembered tune fifteen years later.

> None like her, none.
> Just now the dry-tongued laurels' pattering talk
> Seemed her light foot along the garden walk,
> And shook my heart to think she comes once more:
> But even then I heard her close the door,
> The gates of heaven are closed, and she is gone.

It is possible that Tennyson even drew on specific cases described in Dr Allen's *Classification of the Insane*. In one marked Case No. 22, Allen mentioned a patient 'whose mind was instantly wrecked by the female of his heart unexpectedly marrying another, the very day previous to that on which she had promised to be made his own for ever'.

In *Maud*, the unnamed lover who speaks throughout this strange 'monodrama' is driven temporarily insane through unrequited passion. The scene is set in a Lincolnshire landscape, with a recognisable version of Harrington Hall and its famous gardens (perhaps overlaid with those of High Beech), in which the lover hopes for a tryst with Maud – who never appears. One of the controlling psychological tropes throughout the poem is a geological one: that of the lover being buried alive. In the garden he is buried beneath Maud's feet, and in that surreal image previously noted, his blood comes bursting up through the turf:

> My heart would hear her and beat
> Were it earth in an earthy bed;
> My dust would hear her and beat,

> Had I lain for a century dead,
> Would start and tremble under her feet,
> And blossom in purple and red.

This desperate, urgent voice would be quite unlike anything else in Tennyson's earlier poetry. It has none of the languor of 'Mariana', or the intellectual poise of 'The Two Voices'. It was only here at High Beech that Tennyson first heard the voices of the alienated, the psychically buried or socially lost. His narrator would use or mimic them, as his extraordinary poem of grief and rejection unfolded.

> O me, why have they not buried me deep enough?
> Is it kind to have made me a grave so rough,
> Me, that was never a quiet sleeper?
> Maybe still I am but half-dead;
> Then I cannot be wholly dumb;
> I will cry to the steps above my head
> And somebody, surely some kind heart will come
> To bury me, bury me
> Deeper, ever so little deeper

His interest in the imagery of madness, as well as its scientific explanation, continued in his reading over the next few years. Much later he would say that the idea of obsession or insanity as a kind of psychological burial, 'deeper, ever so little deeper' beneath the surface of normal life, was also confirmed by his discovery of an American story by Edgar Allan Poe, 'The Premature Burial'. Poe's tale, first published in 1850, begins as a serious non-fiction essay on historic cases of mistaken burials of people still living, then slides seamlessly into the realms of Gothic fiction, to produce a brilliant, imaginary evocation of the terrors of lying alive but completely helpless under the earth.

It may be asserted, without hesitation, that no event is so terribly well adapted to inspire the supremeness of bodily and of mental distress, as is burial before death. The unendurable oppression of the lungs – the stifling fumes from the damp earth – the clinging to the death garments – the rigid

embrace of the narrow house – the blackness of the absolute Night – the
silence like a sea that overwhelms . . . these things, with the thoughts of the
air and grass above, with memory of dear friends who would fly to save us
if but informed of our fate, and with consciousness that of this fate they can
never be informed – that our hopeless portion is that of the really dead –
these considerations, I say, carry into the heart, which still palpitates, a
degree of appalling and intolerable horror from which the most daring
imagination must recoil.

Here was a kind of suspended extinction, the 'really dead' who are
not dead, but can never be reached or saved or resurrected, even by
their dear friends. The hypnotic power of Poe's prose, and the stark-
ness of his dark images, including the paradoxical idea of an
'overwhelming sea' under the earth, spoke directly to Tennyson. It
gave a further turn to the idea of deep time, the images of *In Memoriam*, and even perhaps recalled the deep-sea eruption of the Kraken.

Tennyson later said that Poe's tale gave him the inspiration to write
the whole of one key section in a kind of trance, 'in twenty minutes'.
This section placed at the end of *Maud* Part 2, is one of an extraordinary eleven-part monologue, each stanza obsessed with the idea of
burial, both literal and psychological. Here the verse is short and
tightly rhymed, having the effect of someone talking to themselves, or
even mumbling under their breath. The first monologue begins:

Dead, long dead,
Long dead!
And my heart is a handful of dust,
And the wheels go over my head,
And my bones are shaken with pain,
For into a shallow grave they are thrust,
Only a yard beneath the street,
And the hooves of the horses beat, beat . . .

Years later, the poet William Allingham entranced Tennyson by reciting to him another of Poe's dark visions, the Gothic poem 'The Raven',
during the course of a long autumnal walk. This sinister and sprightly

dirge, telling of a ghastly night visitation by a bird of ill-omen, captured Tennyson with its obsessive memory of another lost beloved, Poe's beautiful 'Lenore'. The piece turns upon a hypnotic and repeated chorus line, both portentous and jeering, 'Quoth the Raven, Nevermore'.

Though he had never previously read the poem, this evidently touched a familiar chord in Tennyson's imagination, and resonated with the obsessive themes of *Maud*. The raven itself appears briefly in Part 1 of his poem, 'For a raven ever croaks, at my side . . .' It led him to rate Poe very high among his contemporaries. 'I know several striking poems by American poets, but I think Edgar Poe is (taking his poetry and prose together) the most original American genius.'

Tennyson's own poem *Maud* was eventually to be told by his single, mad, lovelorn narrator, but in twenty-eight wildly different lyrical sections, using an astonishing range of metrical forms. They would emerge years later in fragments, not always clearly dated. But the first section he wrote was actually the one he first published in Milnes's magazine *The Tribute*, in place of 'Anacaona' in 1837. Tennyson later said it gave him the core idea for the whole of *Maud*, and perhaps the initial desperate voice of his narrator. He explained that a friend, Sir John Simeon, had 'begged me to weave a story round this poem, and so *Maud* came into being'. Whether this was true, he later took care to hide it away deep in Part 2, Section IV of *Maud*. Here it appears as the beautiful, plaintive and completely traditional love lyric, 'O that 'twere possible', the second verse of which continues:

When I was wont to meet her
In the silent woody places
By the home that gave me birth,
We stood tranced in long embraces
Mixed with kisses sweeter, sweeter
Than anything on earth.

The 'silent woody places' might recall Somersby. But Tennyson always insisted that the 'true love', the seventeen-year-old Maud, was not to be identified with any living woman, either Rosa Baring or Sophie

Rawnsley or even Emily Sellwood. Yet they were all surely part of his inspiration. Moreover his broken narrative form, 'an *entirely new form*, as far as I know,' said Tennyson, was something he may have first learned from listening to some of the voices at High Beech asylum.

Among those voices may even have been that of John Clare him-self. Clare's own later asylum poems, written at Northampton, develop a similar, obsessive sense of absolute loss and abandonment:

> I am – yet what I am none cares or knows;
> My friends forsake me like a memory lost:
> I am the self-consumer of my woes—
> They rise and vanish in oblivious host,
> Like shadows in love's frenzied stifled throes
> And yet I am, and live – like vapours tossed

Tennyson was later told that 'one of the best-known doctors for the insane' wrote that *Maud* was 'the most faithful representation of mad-ness since Shakespeare'. He himself remarked delphically: 'I took a man constitutionally diseased and dipt him into the circumstances of the time and took him out on fire.'

As Septimus settled down under Allen's care, a previous plan to send him on a therapeutic visit to join Horatio farming in Tasmania was deemed unnecessary. Horatio himself did not have much luck with colonial life. The experiment failed after three years, and when he returned to England he seemed, as FitzGerald put it, 'rather unused to the Planet'. Meanwhile Septimus slowly improved, but had to remain at High Beech for the next four years, until preparations were made for him to join Frederick in Florence in the spring of 1844. There he would spend his time sightseeing and writing languid son-nets, 'I am in the land of beauty and the vine ...'

Tennyson's confidence in Allen grew steadily, and an unusual friendship sprang up with the engaging doctor, who would encourage it assiduously. Both FitzGerald and Spedding continued to note Dr Allen's growing influence over their friend, initially with amuse-ment but later with something like alarm. There were odd emotional undercurrents. Once when Tennyson had disappeared for too long in his London lodgings, Allen wrote: 'I shall be glad to see you again, for I have been very sad since we parted, and I can hardly tell why, unless it was because you were so out of spirits.' Indeed this friendship was not quite what it seemed initially, and was to draw the whole Tennyson family into the unsuspected entrepreneurial side of Allen's career.

Over the next two years it became clear that the doctor was seeking financial investment, not in the asylum itself, but in a related crafts scheme. This seems to have been based on the making of simple wooden furniture by some of his patients. (Did Tennyson recall his father's obsession with wood-carving?) Allen's idea was to transform this essentially therapeutic small-scale craft into a full-scale domestic furniture business. He had discovered – or actually invented – a type of mechanised wood-cutting and wood-turning machinery, which could potentially mass-produce a range of decorated tables and chairs. He gave this project the impressive name of 'Pyroglyphs' (apparently meaning something like 'fire mark' furniture), and eventually convinced Tennyson to invest almost his entire remaining capital of £3,000 in it.

Sometimes Tennyson did get back to Lincolnshire and the North Sea at Mablethorpe, though he explained to Emily Sellwood that he was 'not so able as in old years to commune *alone* with Nature'. Yet the seashore continued to inspire him. One undated Mablethorpe beach fragment, belonging perhaps to 1839, reappeared nearly fifteen years later in *Maud*, transferred to the coast of northern France. It was a short, tenderly observed poem of natural science. By lovingly contemplating a single seashell 'lying close' to his foot on the shore, the narrator seems for a moment to escape the chaos and terror of his own madness. It was evidently inspired by one of Tennyson's own lonely walks on the Lincolnshire shoreline, paddling along the edge of the strand, and peering down with his short-sighted gaze at the glistening wet sandy margins beneath his naked feet.

See what a lovely shell
Small and pure as a pearl

Lying close to my foot,
Frail, but a work divine,
Made so fairly well
With delicate spire and whorl,
How exquisitely minute,
A miracle of design!

What is it? a learned man
Could give it a clumsy name.
Let him name it who can.
The beauty would be the same.

The tiny cell is forlorn,
Void of the little living will
That made it stir on the shore.
Did he stand at the diamond door
Of his house in a rainbow frill?
Did he push, when he was uncurl'd,
A golden foot or a fairy horn
Thro' his water-world?

Slight to be crush'd with a tap
Of my finger-nail on the sand,
Small, but a work divine,
Frail, but of force to withstand
Year upon year the shock,
Of cataract seas that snap
The three-decker's oaken spine
Athwart the ledges of rock,
Here on the Breton strand.

The poem drew on Lyell's reassuring observation of individual life persisting through all the apparent chaos and violence of nature, and despite the threat of extinction. Lyell had written in the first volume of *Geology*: 'It sometimes appears extraordinary when we observe the violence of the breakers on our coast that many tender and fragile shells should inhabit the sea in the immediate vicinity of the turmoil.' Tennyson's poem seems to acknowledge this specific passage, while also rejecting the 'clumsy' or limiting language of 'learned' scientists in general. The experience is more existential for him, and the final mention of the wrecked 'three-decker' ship even suggests some comparison with Coleridge's Mariner escaping from his guilt by contemplating the otherworldly beauty of the sea creatures around him.

The poem was eventually published separately as 'The Shell', but Tennyson initially assigned the piece to Part 2 of *Maud*, where he sends his lovestruck madman to seek solace on a trip to the wild sea coast of northern France. Tennyson annotated it: 'In Brittany. The shell undestroyed amid the storm perhaps symbolises to him his own first and highest nature preserved amidst the storm of passion.' Perhaps something like this had happened to him at High Beech.

II

London

Throughout these uneasy years at High Beech, Tennyson was partly sustained by a new restless style of life, alternating between London and Epping Forest, town and country, friends and family. He was glimpsed by a still small circle of faithful companions, who never knew exactly when he would appear: morose but thoughtful, possibly drinking too much and certainly smoking too much, sometimes gruff but always hypnotic. In April 1838 he perched briefly with FitzGerald in London. 'We have had Alfred Tennyson here; very droll, and very wayward: and much sitting up of nights till two and three in the morning with pipes in our mouths: at which good hour we would get Alfred to give us some of his magic music [poems], which he does between growling and smoking; and so to bed.'

Sometime this spring he heard that Rosa Baring had become engaged, and would marry a rich local landowner in October. It was a subtle blow, and moved all his memories of Harrington Hall into the realm of fantasy and regret. He tried to inoculate himself with a savage little sonnet dismissing her beauty, which ends:

A hand displayed with many a little art
An eye that glances on her neighbour's dress
A foot too often shown for my regard
An angel's form – a waiting-woman's heart
A perfect-featured face, expressionless,
Insipid, as the Queen upon a card.

He would later put his bitterness in another surprisingly stinging verse: 'thou art mated with a clown, / And the grossness of his nature will have weight to drag thee down.' But the sense of loss and frustration would stay with him much longer, until fifteen years later he would finally succeed in disinterring some of his most desperate feelings about sexual love through the mad narrator he had discovered at High Beech and eventually released in *Maud*.

For the time being Tennyson sought movement and distraction, and relief in a more expansive kind of poetry. In the autumn he embarked on a walking tour in the West Country; but alone and not even with FitzGerald. He settled briefly at Torquay, where he wrote a long conversational poem set on another imaginary estate, 'Audley Court'. Though this country house is presented as overlooking the sea, its location is mysterious and is perhaps suggested by Lushington's Park House, or even represents some continued exorcism of Harrington Hall. Tennyson described Torquay enthusiastically, 'in the old days the loveliest village in England', and gave special significance to the light and air of its welcoming seacoast. 'I, coming down the hill over Torquay, saw a star of phosphorescence made by the buoy appearing and disappearing in the dark sea.' That bobbing between brightness and dark seemed to catch exactly his mood.

Though Tennyson was very much alone, 'Audley Court' is a poem of warm companionship, composed in a new style of sauntering blank verse, easy and relaxed. It opens cheerfully with the name of two taverns – the Bull and the Fleece – and an invitation to a well-prepared picnic lunch in the fields. Strongly reminiscent of Coleridge's Quantocks Hill Conversation poems of thirty years before, it fills a pastoral landscape with sunny fellowship, shared memories and lively contemporary gossip. It presents two old friends spending a long summer day, simply walking, talking and sprawling in the sun, in an idyllic orchard above the sea. But the encounter contains half-hidden emotions between the two. One of them, Francis Hale ('the farmer's son, who lived across the bay'), unpacks a flask of cider 'from his father's vats' and lays out a homemade feast that is rich with earthy memories and ripe with geological imagery:

There on a slope of orchard, Francis laid
A damask napkin wrought with horse and hound,
Brought out a dusky loaf that smelt of home,
And, half cut-down, a pasty costly made,
Where quail and pigeon, lark and leveret lay,
Like fossils of the rock, and golden yolk
Imbedded and injellied ...

In a light-hearted, confidential, almost joshing way they each proceed to reminisce about unhappy love affairs they have both suffered with faithless women. Francis 'claps his hand' reassuringly in his friend's, and puts his experience into a bracing song. This is also touched with geo-logical images, but this time the erosion of the heart by water and wind:

'Who'd serve the state? for if I carved my name
Upon the cliffs that guard my native land,
I might as well have traced it in the sands;
The sea wastes all: but let me live my life.

'Oh who would love? I wooed a woman once,
But she was sharper than an eastern wind,
And all my heart turned from her, as a thorn
Turns from the sea; but let me live my life.'

The narrator answers with his own tender but ironic love lyric, about a woman sleeping in her sister's arms, and forgetting all about her actual lover, 'Sleep, Ellen Aubrey, love and dream of me.' Cheered by these shared confidences, they go on casually to talk 'old matters over, who was dead/ Who married, who was like to be, and how/ The races went, and who would rent the Hall.' They discuss the corn laws, the price of grain, and the perils of soldiering, all solid masculine topics. Throughout the autobiographical note is strong. It might well be the replay of a day spent long ago with Hallam; or more recently with FitzGerald. The underlying sense of secure, male affection is confi-dent and tender. At dusk they saunter home beneath a crescent moon,

and see the glinting harbour buoy exactly as Tennyson had seen it.
The final mood is serene, almost blissful.

> So sang we each to either, Francis Hale,
> The farmer's son, who lived across the bay,
> My friend; and I, that having wherewithal,
> And in the fallow leisure of my life
> A rolling stone of here and everywhere,
> Did what I would; but ere the night we rose
> And saunter'd home beneath a moon, that, just
> In crescent, dimly rain'd about the leaf
> Twilights of airy silver, till we reach'd
> The limit of the hills; and as we sank
> From rock to rock upon the glooming quay,
> The town was hush'd beneath us: lower down
> The bay was oily calm; the harbour-buoy,
> Sole star of phosphorescence in the calm,
> With one green sparkle ever and anon
> Dipt by itself, and we were glad at heart.

Refreshed and back at High Beech, Tennyson embarked on a burst
of confessional letters to Emily Sellwood. They suggest he was taking
stock of his own heart and his old friendships, and seeking reassurance from the natural world around him. But there was no declaration
of love. 'I have dim mystic sympathies with tree and hill reaching far
back into childhood,' he told her. 'A known landskip is to me an old
friend, and continually talks to me of my own youth and half-
forgotten things, and does more for me than many an old friend that
I know. An old park is my delight and I could tumble about it
forever.'

Tennyson spent December 1838 wandering through the woods of
Epping Forest, and found time to catch up on his scientific reading.
He was given eight books as Christmas presents, including the 1837
edition of Mary Somerville's *On the Connexion of the Physical Sciences*,
inscribed 'Xmas Day, 1838'. It was possibly a gift from Emily Sellwood

who had visited High Beech shortly before, in November. If so, it was a timely one.

Having studied mathematics and astronomy in Edinburgh, Somerville had come to London where her exceptional gifts became known among a small circle surrounding the Royal Society. In 1830, at the invitation of the Society for the Diffusion of Useful Knowledge, Somerville translated Laplace's highly technical *Mécanique Céleste*. It made her name as a new kind of science writer and interpreter. In a parliamentary debate on scientific education, she was referred to as 'one of the only six persons in England who understands Laplace'. This had been while Tennyson was still at Trinity.

Her follow-up work, the *Connexion*, was published in 1834. It was a five-hundred-page book which opened up the whole field of hard sciences – astronomy, physics, chemistry, geography, meteorology and electromagnetism – to a general readership. It was an even greater success, and an immensely popular one, enthusiastically reviewed and hailed by such general magazines as the *Athenaeum* and the *Edinburgh Review*. The copy that Tennyson now read was the third and 'greatly augmented edition' of 1837.

What would have particularly struck him at High Beech was the authority and simplicity of Somerville's scientific explanations. Particularly impressive was her account of the 'undulatory motions', observable throughout the natural world. These were the wave-like forces that underlay and united so many of the apparently mysterious phenomena in nature. Her chapter 16 'Sound', opened with a plain but memorable comparison: 'The propagation of sound may be illustrated by a field of corn when it is agitated by the wind ...' This immediately recalls the 'long fields of barley and of rye' that fill the opening of 'The Lady of Shalott' with its rustling panorama.

Somerville then goes on to explain a vast range of acoustic phenomena from birdsong to waterfalls and thunder. 'The various musical instruments, the human voice and that of animals, the singing of birds, the hum of insects, the roar of the cataract, the whistling of the wind and the other nameless peculiarities of sound, at once show an infinite variety in the modes of aerial vibration, and the astonishing acuteness and delicacy of the ear thus capable of appreciating the minutest differences in the laws of molecular oscillation.'

Tennyson would frequently draw on this 'infinite variety' of vibrations in his later lyrics, notably his poem 'In the Valley of Cauteretz'. While in one of the most musical songs to be introduced in another of his future poems, *The Princess*, three kinds of wave motion follow each other – light, water, sound – with extraordinary physical impact and energy.

> The splendour falls on castle walls
> And snowy summits old in story:
> The long light shakes across the lakes,
> And the wild cataract leaps in glory.
> Blow bugle blow, set the wild echoes flying,
> Blow bugle: answer echoes dying, dying, dying.

Here simple verbs carry all this energy. Light 'shakes', water 'leaps' and the bugle 'set flying' sounds which bounce back, in repeated waves, repeatedly 'dying', but perhaps never quite extinguished.

Mary Somerville summarised the 'undulatory' or ripple effect throughout nature with another of her striking analogies, suggesting the 'connexion' between the various kinds of wave propagated in the different mediums of water, atmospheric air and solar light. 'Anyone who has observed the reflection of the waves from a wall on the side of a river after the passage of a steam-boat, will have a perfect idea of the reflection of sound and of light.'

This groundbreaking style, clear and logical, opened out into many other passages of sublime perspective, such as the description of universal gravity as a force equally present 'in the descent of a rain drop as in the falls of Niagara; in the weight of the air, as in the periods of the moon'. Somerville ranged over phenomena from stellar parallax to terrestrial magnetism, from comets to giant seaweed. Her work was itself like a university course in general science, and indeed was later recommended as such at Cambridge.

Somerville's special standing as a female author was recognised by a number of academics, not least by Tennyson's old tutor William Whewell, who had written about her with great intellectual respect in the *Quarterly Review* for September 1834. Though even Whewell could not quite forbear a mild gender witticism, describing the *Connexion* as 'a *masterly* survey – if Mrs Somerville will excuse the word'.

Tennyson's first reading of Mary Somerville's work coincided with his growing interest in an entirely new project at High Beech, in that spring of 1839. The subject was women's education, and it took shape in the challenging idea of a long poem about what was then a thoroughly controversial concept, the possibility of a 'Woman's University'. This was a long way from the world of 'The Lady of Shalott', and a subject usually approached with mockery, or at least light satire. Samuel Johnson had proposed it in *Rasselas* as one of the ideal but delusory paths to happiness. How did Tennyson suddenly arrive at it? It may have been partly inspired by his reading the *Connexion* itself, and recognising the exceptional intellectual abilities of a female author like Mary Somerville. Her career path suggested whole new possibilities in the world of national education.

But the subject of a Women's University must also have been the result of Tennyson talking it over with his sisters, and with Emily Sellwood. The four Tennyson girls – now no longer girls, but young women aged between twenty and thirty – were already widely read, wrote their own poetry, and regarded themselves as bluestockings. They had formed a light-hearted book group at High Beech that winter, which they christened 'The Husks'. The name, referring to the rejected outer shell of ripe ears of corn, might suggest the sisters' self-mockery of female powers in a house full of brothers. But perhaps it also suggested to them something more positive and challenging: their 'husky voices' as passionate speakers insisting on a woman's right to be heard. There was a kind of poetic analogy too: a salute to Tennyson's 'Two Voices', by reference to his brilliant stanza describing the emergence of the iridescent dragonfly from its restricting chrysalis:

> An inner impulse rent the veil
> Of his old husk: from head to tail
> Came out clear plates of sapphire mail.

So the sisters intended to reveal their hidden intellectual armour too. At all events, the Husks became quite serious, with approved book lists (Keats, Shelley, Coleridge – but definitely not Byron) and after-dinner meetings in the parlour. They invented their own private language for the group: for instance, to 'shuckle' meant to hold an intense highbrow discussion; while to 'sloth' meant to go to ground alone with a book (ideally in some remote corner of the garden).

Various female cousins and friends joined the Husks on visits to High Beech, and also formed a correspondence group about their favourite ('deadly' viz. thrilling) books. One of them, Louisa Lanesborough, 'later well-known as an authoress', and already keeping a lively diary, eagerly wrote up her brief exciting glimpse of literary life at Beech Hill. 'Alfred wandering weirdly up and down the house in the small hours, murmuring poetry to himself; the sisters fond, proud, cultivated, appreciative; and [in her own 'Husky' phrase] the tender mother's spirit brooding over all.' It is clear that Emily Sellwood herself briefly joined the Husks on several, possibly clandestine, visits to High Beech that winter and the following spring.

So in the newly animated atmosphere of Beech Hill House, Tennyson began to wonder if an imaginary University of Women, the sort of place that could contain Mary Somerville and his sisters, might be celebrated in quite another kind of long poem, somehow adapted to the controversial nature of the subject. Its form – discursive, narrative, or dramatic – was not yet clear. He certainly talked over the plan of the poem with Emily Sellwood in 1839, and probably discussed it with her in a series of later letters, though most of them have been lost. The earliest manuscript draft dates from this time at Beech Hill, and was simply entitled 'The New University'. The eventual title would be 'The Princess', but it always retained the elements of this group enterprise, a shared game among clever women.

At all events, Tennyson now embarked on a more intimate, self-revealing kind of correspondence with Emily from High Beech. These letters are quite different in style from anything else he wrote at this time. As before, only a few fragments have survived from these years 1839–40. They shift in a mercurial way between the pious and passionate, to the hilarious and teasing. Throughout Tennyson uses the intimate, slightly stylised form of 'thee' when addressing her (as he had once used to Arthur Hallam, and then to FitzGerald). Yet it is never clear when he is wholly serious. He wrote to her in January 1839: 'I murmured (like a hen in the sunshine) lines and half lines of some poem to thee, I know not what: but I could not think of thee, thou white dove, brooding in thy lonely chamber without movements of truest affection towards thee ...'

The light, humorous, chivalric tone rarely falters. But sometimes these letters take on a kind of pedagogical passion, and the subject of women's education begins to emerge. 'All life is a school, a preparation, a purpose; nor can we pass current in a Higher College, if we do not undergo the tedium of Education in this one.' To this he sometimes urged a metaphysical, otherworldly dimension upon Emily. 'Annihilate within yourself these two dreams of Space and Time. To me often the far-off world seems nearer than the present, for in the Present is always something unreal and indistinct, but the other seems a good solid planet, rolling round its green hills and paradises to the harmony of more steadfast laws.' He later apologised for his 'preaching'.

Until the end of this year 1839, it was understood that they were unofficially engaged, though the Sellwood family still did not feel the Tennyson family were established enough to provide a second son-in-law. Such concerns were increased by the shaky marriage of Charles and Louisa, now evidently further undermined by Charles's opium habit. Did his younger brother – an unproductive poet loitering in the woods of Essex (or disappearing into the dark lodgings of Camden Town) – promise anything more secure? Tennyson himself seemed uncertain of his own position, and there was still no record of any formal marriage proposal. On the contrary there is some suggestion that Emily's father Henry Sellwood increasingly disapproved of their actual meeting, and even of their correspondence, on the grounds of Tennyson's religious speculations and other eccentricities. In the summer of 1840 Henry Sellwood would definitively forbid it.

This suspended situation curiously echoes (in reverse) that between Arthur Hallam and Emily Tennyson, six years before. Tennyson had finally gathered himself to travel back to Lincolnshire in May 1840, and made an awkward visit to the Sellwood townhouse at Horncastle. The meeting with her father did not go well, and the engagement (if that is what it was) was definitely broken off. Tennyson withdrew to his old retreat by the sea at Mablethorpe, writing a sort of farewell letter, yet still confused and evasive: 'I fly thee for my good, perhaps for thine, at any rate for thine if mine is thine ... Sayest thou "are we to meet *no more?*" In answer I know not the word nor will I know it ... The immortality of man disdains and rejects it.'

Emily's side of this correspondence was later burned at Tennyson's own direction, and few letters are known between the two for the next eight years. What fragments do remain have been recovered by Ann Thwaite in her superbly researched biography. A last surviving fragment dated November 1840 contains the phrase, 'I scarce expect thee to agree with me in many things I have said – believe only that all is kindness to thee, to [thy sister] and ... to thy father.' It is signed off, 'Thine dear, for ever and ever' – but in Greek, which Emily did not read. Tennyson's own sister Emily continued to send the occasional confidential notes to the Sellwood household, to maintain the connection, signing off quite simply, 'All send love, thy very affectionate Emily'.

It was around this time, and surely in reaction to this rejection, that Tennyson's London contacts began to expand beyond the original Cambridge circle. Through Spedding he first met Thomas Carlyle, the Scottish polemical essayist and historian, who had recently moved from Edinburgh to a house in Cheyne Row, and was dining out on the success of his flamboyant history, *The French Revolution* (1837). Carlyle's first impressions of Tennyson, fourteen years his junior, were equally flamboyant. 'A true human soul ... One of the finest looking men in the world. A great shock of rough dusty-dark hair: bright laughing hazel eyes; massive aquiline face, most massive yet most delicate, . . . almost Indian looking; clothes cynically loose, free-and-easy; smokes infinite tobacco.'

Like FitzGerald, Carlyle was struck by Tennyson's extraordinary voice, whether talking, laughing or reciting poetry. 'His voice is musical metallic – fit for loud laughter and piercing wail . . . speech and speculation free and plenteous . . .' But he was not without shrewd doubts about the poet's health and future career, expressed in his own diagnostic manner. 'We shall see what he will grow into. He is often unwell; very chaotic – his way is thro Chaos and the Bottomless and Pathless; not handy for making out many miles upon.'

Besides Carlyle, Tennyson dined with the veteran journalist Leigh Hunt, now in his mid-fifties, who had never forgotten first spotting the young Romantics Keats and Shelley twenty years before, and agog to deepen his acquaintance with this new poetic phenomenon. He also attended the literary salon hosted by the elderly Samuel Rogers, who had once advised the publisher John Murray to take on the young Lord Byron. Rogers too was impressed. Rumours about this new young poet Tennyson – still a mysterious figure, tall, handsome, provincial, wreathed in blue clouds of tobacco smoke, faintly Spanish but with a deep English country accent – perhaps from Devonshire, or even Cumberland and the Lakes? – began to spread around the literary world of London. Perhaps here was the long-awaited inheritor of the Romantic generation?

One who heard the excited gossip was the thirty-two-year-old Elizabeth Barrett, still marooned on her chaise longue in the family house at 50 Wimpole Street, not far from Henry Hallam's address at No. 67, where Tennyson had once stood outside in the rain. She was quick enough to pick up the rumours. She eagerly quizzed her friend Miss Mitford about him and his family: 'Is Alfred Tennyson among your personal acquaintances? I have heard of him the other day as having an unduly large head, handsome features, and a fathoming eye – and that they had all settled in a cottage in Devonshire where he smoked and composed all day . . . and that he was separating from his family *because they distracted him*.'

Seeing his friend increasingly absorbed in this London life, FitzGerald felt things were sliding professionally. 'I want AT to publish another volume: as all his friends do: especially Moxon, who has been calling on him for the last two years for a new edition of his old volume: but he is too lazy and wayward to put his hand to the business. He has got fine things in a large Butcher's Account book that now lies in my room ...'

Tennyson's apparent 'laziness' exercised FitzGerald. He pointed out that their new friend Thomas Carlyle was impressively active in developing his career and his readership. Carlyle had just published a collection of his *Miscellaneous Essays* (1839) which was being widely read – 'some abusing, some praising: I among the latter'. His emphatic, 'Germanic' style of argument about public issues, combined with his Scottish accent, caused a stir. He was consciously transforming himself from a dramatic historian to an apocalyptic social commentator, and raising what soon became known as the 'Condition of England Question'.

In response to the poverty and discontent in the new industrial cities, and the violent popular demand for a series of political reforms, Carlyle was about to publish his first socially engaged work, *Chartism*. It was a brief, Cassandra-like pamphlet with the Delphic epigraph: 'It never smokes but there is fire'. Carlyle felt he could address the whole nation. 'A feeling very generally exists that the Condition and Disposition of the working Classes is a rather ominous matter at present; that something ought to be said, something ought to be done ...' Revelling in this new prophetic role, Carlyle could be seen riding briskly about Chelsea on a hired horse improving his health and his local celebrity. Shouldn't Tennyson (Fitz implies) now be doing something similar with his poetry?

In fact Tennyson was busily engaged in a more humdrum task: packing up his books at Beech Hill House and preparing to move further south, to Kent. They had been at Epping for nearly three years, and Septimus would be remaining in Dr Allen's care. The exact reason for the move is unclear, except they would be nearer the large house of his hospitable friend Lushington at Boxley, deep in the countryside near Maidstone. His mother and sisters seemed particularly keen on this. It also suited Tennyson himself, though as usual he was impatient with the disruption, and claimed he might have to sell off many of his precious books. 'I really do not know where to stow them and the house at Tunbridge is too small, a mere mousetrap.'

His escapes to London became increasingly important, and he sometimes talked wildly of even going abroad like Frederick. He apologised to one friend for dashing into a supper party 'in traveller's costume – possibly too I may have to leave you early on a steam to the Continent'. Similarly in August 1840 Spedding noted Tennyson's rushing about at Lincoln's Inn, generating urgent signals of arrival and departure, but which Spedding observed with his usual philosophic calm. 'Alfred Tennyson has reappeared, and is going today or tomorrow to Florence, or to Killarney, or to Madeira, or some place where some ship is going – he does not know where.'

Here it was FitzGerald's steady support, his gentle friendship and belief in Tennyson's genius which, despite his impatience, never really faltered. His encouragement took many forms. In 1840 he commissioned a portrait of Tennyson by the society painter Samuel Laurence. At least this made his friend sit still for a moment. The poet's striking, leonine appearance rendered in rich golden tones, clean-shaven but with heroic hair, was dramatic and deliberately romantic. It is the only finished portrait known of Young Tennyson, and FitzGerald encouraged the circulation of copies. Later a chalk engraving of it would be used to illustrate an early and influential essay on his work by the journalist Richard Henry Horne. The original would finish up in the National Portrait Gallery (though later 'improved' to Pre-Raphaelite standards by Edward Burne-Jones).

Also attempting to secure his finances for the future, in November 1840 Tennyson finally yielded to Matthew Allen's entreaties long pursued at High Beech, and sold off a share in his Grasby estate to invest a first £900 in the therapeutic wood-carving business. Further money would follow. Motivated partly by philanthropy, Tennyson may also have intended to recover his standing with Henry Sellwood as a sensible businesslike person.

Allen evidently pursued Tennyson in London, and this autumn accompanied him to Thomas Carlyle's new house in Chelsea. The three of them sat smoking in Carlyle's summer house one September evening, as Emily Tennyson excitedly (and secretly) reported to Emily Sellwood. Carlyle continued to descant on Tennyson's appearance and repeated his first impression of magnificent chaos. 'A fine, large-featured, dim-eyed, bronze-coloured shaggy-headed man is Alfred: dusty, smoky, free-and-easy: who swims, outwardly and inwardly, with great composure in an inarticulate element as of tranquil Chaos and tobacco ...' In recognition of their new friendship, Carlyle cut a special niche in his garden wall where Tennyson could keep his pipe, so he could drop in for a smoke anytime uninvited.

Yet 1840 became, as Fitz feared, a further year of Tennyson's drifting. He was thirty-one and without definite plans, either for settling in a proper London house, or for marrying, or for publishing new poetry. No further work was done on the great 'University' poem. Carlyle also observed this strange lack of direction and drive, in his otherwise praiseworthy assessment of Tennyson sent to Emerson in America. Tennyson 'preferred clubbing with his mother and some sisters, to live unpromoted and write Poems. In this way he lives still, now here now there; the family always in reach of London, never in it; he himself making rare and brief visits, lodging in some old comrade's rooms.'

Yet all this time Tennyson had been secretly trying out new forms of poetry and new subject matter. His conversations with Carlyle, who was so concerned with Chartism and social questions, evidently had their effect. While FitzGerald's urgings about unused materials in the 'Butcher's book', quite apart from the slowly accumulating elegies, had clearly spurred Tennyson on to further experiments.

Now Tennyson left aside the easy discursive style of 'Audley Court', and postponed work on the intellectual complications of the University of Women. Instead he turned again to the dramatic monologue. But this time, unlike the classical voice of the legendary 'Ulysses', he created an entirely new modern persona to address contemporary themes. He is still an adventurer, about to embark on a voyage into the unknown. Yet this speaker is a common English soldier, who has joined the ranks to fight abroad, apparently for 'the jingling of the guinea', but actually to forget a disastrous love affair. Preparing for embarkation with his regiment, he passes the ruined hall where his love had once lived (Harrington Hall again?), and turns aside to reminisce about 'the tumult' of his whole past life.

The poem opens with the soldier's memories of a happy childhood by the sea, full of his youthful hopes and dreams. They clearly draw on a version of Tennyson's own memories at Somersby and Mablethorpe, the beauty of the natural world, the big Lincolnshire sky at night, and the excitement of star-gazing. Tennyson's new, long, sweeping verse line starts off the story with a dazzling display of astronomical imagery:

Many a night from yonder ivied casement, ere I went to rest,
Did I look on great Orion sloping slowly to the West.

Many a night I saw the Pleiads, rising thro' the mellow shade,
Glitter like a swarm of fire-flies tangled in a silver braid.

Here about the beach I wander'd, nourishing a youth sublime
With the fairy tales of science, and the long result of Time;

When the centuries behind me like a fruitful land reposed;
When I clung to all the present for the promise that it closed:

When I dipt into the future far as human eye could see;
Saw the Vision of the world and all the wonder that would be.

But these hopeful memories soon shift to the history of a failed romance with a beautiful but unattainable woman, 'my Amy', 'my cousin' the 'shallow-hearted' daughter of Locksley Hall who, in a curious underwater phrase, was 'falser than all fancy fathoms'. The soldier had once been a naive young man full of passionate desires. But he has been betrayed, and embittered by his loss of both love and worldly prospects, now dreams of abandoning England and escaping abroad.

It is difficult not to see this as a reflection of Tennyson's affair with Rosa Baring at Harrington Hall, and the 'loss' of his inheritance to his rich uncle Charles Tennyson d'Eyncourt. But the disenchanted mood may also recall the departure of his brother Frederick to Italy, and even his younger brother Horatio to Tasmania. At its heart lies a fantasy of escape that Tennyson himself entertained in these years, as he often told FitzGerald in his most depressed moments. Partly begun at High Beech, it may also reflect some of the bitter and desperate stories and voices Tennyson heard at the asylum.

But his youthful 'Vision of the world' also strikes a powerful positive note at the centre of the poem. The soldier had once imagined a progressive future for society, especially aided by science. Tennyson here projected a brilliant piece of airborne science fiction, quite as daring as anything in the submarine world of the Kraken. He conjured a whole new world of flight and airships, almost in the spirit of Jules Verne. Yet Verne himself only started writing his famous series of science fiction novels – *Les Voyages Extraordinaires* – in the 1860s. In the 1840s no powered aircraft had been invented, though Tennyson had seen gas balloons at Cambridge and Park House, and probably had heard of the very earliest experiments with steam-powered airships that had begun in France.

Nevertheless he managed to construct the outline of a complete science fiction novel in a mere twenty lines. It is an extraordinary passage, almost unique in the poetry of this period. Starting with an innocent, shining vision of aerial trade and international competition, it darkens into the storm clouds of aerial warfare and airborne combat, but finally ends with the hope for a global parliament and world peace. It is addressed to his comrades, who now sound more like a gathering of hopeful Chartists than a group of disillusioned soldiers: 'Men, my brothers, men the workers, ever reaping something new …' To them he offers a brilliant panorama of scientific possibilities:

For I dipt into the future, far as human eye could see,
Saw the Vision of the world, and all the wonder that would be;

Saw the heavens fill with commerce, argosies of magic sails,
Pilots of the purple twilight dropping down with costly bales;

Heard the heavens fill with shouting, and there rain'd a ghastly dew
From the nations' airy navies grappling in the central blue;

Far along the world-wide whisper of the south-wind rushing warm,
With the standards of the peoples plunging thro' the thunder-storm;

Till the war-drum throbb'd no longer, and the battle-flags were furl'd
In the Parliament of man, the Federation of the world.

By the final lines science has suddenly achieved global peace for all mankind. The whole world would be ruled as a democracy, with a universal Parliament. This is a peculiarly Victorian vision of a peaceful world government that stretches forward not merely to Verne, but a hundred years ahead to H.G. Wells and his *New World Order* published (ironically) as late as 1940.

For all the startling force of this scientific idealism, a pure product of 'eager-hearted' youth, Tennyson's mature soldier is soon overcome by the harsher realities of life. As he laments: 'Science moves, but slowly, creeping on from point to point.' Not only has love failed him, material success too. He is disillusioned about the inequality of wealth in English society: 'Cursed be the social wants that sin against the strength of youth!' He longs to go abroad to some more free, generous and exotic world, perhaps somewhere among the Pacific islands. Here the soldier is tempted by old Romantic dreams, and Tennyson evokes these with lush conviction.

Larger constellations burning, mellow moons and happy skies,
Breadths of tropic shade and palms in cluster, knots of Paradise ...

Droops the heavy-blossomed bower, hangs the heavy-fruited tree
Summer isles of Eden lying in dark-purple spheres of sea.

This enticing place was not a West Indian colony, but more dreamily
some utterly innocent world where 'the trader' had never come, and
the 'European flag' had never floated. So his rejected love for 'Amy',
his disillusion with material progress, and his own lack of worldly
success, would carry the soldier off to some warm, liberating tropical
paradise. Moreover it would carry him away from civilisation, into
the welcoming arms of some dark native beauty. Here the modern
reader is brought up with a start.

There methinks would be enjoyment more than in this march of mind,
In the steamship, in the railway, in the thoughts that shake mankind.

There the passions cramped no longer shall have scope and
    breathing space;
I shall take some savage woman, she shall rear my dusky race.

This erotic and liberating idea of the 'savage woman' and her dusky
race might be presented as the soldier's passing fantasy. Yet it is
repeated a few lines later, with a kind of shamed dismissal – 'Mated
with a squalid savage – what to me were sun or clime?' Such 'exotic'
dreams of the soldier would now be called neo-colonial or clearly
racist; but they were openly expressed by many of Tennyson's generation, and often in much more violent form, as by his friend Thomas
Carlyle. Carlyle's notorious 'Occasional Discourse on the Negro
Question' (1849) was instantly refuted by John Stuart Mill's icy reply:
both essays published anonymously.
    Tennyson too was evidently uneasy with his soldier's dreams.
When he returned to them in a third passage, it was with increasing
doubt – 'Could I wed a savage woman, steept perhaps in monstrous
crime?' – and he later rejected this entirely. But Tennyson's evident
fascination with them betrays both attraction and discomfort, and
the whole 'dusky' sexual theme had already appeared in his suppressed

'Hayti' calypso poem. It would recur in his correspondence with
FitzGerald, who would urge him – but half-jokingly – to quit civilisation and go abroad 'among savages'. Much later it would surface in
his poem *Maud*, as a madman's disenchantment with corrupt English
civilisation. There are strange premonitions here of a whole world of
tropical escapes, which ran through the nineteenth century as far as
Joseph Conrad's Malaysia (*An Outcast of the Islands*, 1896); or Paul
Gauguin's Tahiti (*Spirit of the Dead Watching*, 1892); or in a much
lighter, more sentimental vein, to Rudyard Kipling's Burma – now
Myanmar – in 'The Road to Mandalay' (1892).

Yet Tennyson's attitude to his soldier's fantasies remains ambiguous –
'Fool, again the dream, again the fancy! But I know my words are wild.'
This uncertainty hovers uneasily over the whole of 'Locksley Hall',
which despite all its energy, gives the poem its dated and discordant
tone. It seems to offer the stark choice between Future Progress or
Tropical Dreams. On the one hand there is the industrial world of scientific invention and Chartism; on the other, the colonial world of
imperial fantasy and South Seas languor. This would continue to be a
genuine dilemma for his whole generation. But in the end Tennyson's
soldier chooses the future, finding the railway image that would become
famous among later Victorian readers:

Not in vain the future beacons. Forward, forward let us range.
Let the great world spin for ever down the ringing grooves of change.

Through the shadow of the globe we sweep into the younger day:
Better fifty years of Europe than a cycle of Cathay.

In fact the railway was fast becoming the common symbol for Victorian progress – both its triumph and its menace – most obviously in
the novels of Dickens like *Dombey and Son* (1846) and *Hard Times*
(1854). Tennyson's great admirer, Elizabeth Barrett, had adopted it, in
her protest poem of 1842, 'The Cry of the Children'. Here it is the
'rushing of the iron wheels' which prevent Parliament – or God
himself – from hearing the appeals of the little boys and girls working

in the factories and mines. It is precisely the thunderous sound and mechanical rhythm of the railways that drowns out all else. 'The wheels in their resounding' are so noisy that even prayers cannot be heard.

In this setting, Tennyson's new use of the long, emphatic verse line, with its eight stresses (rather than the five of the traditional English pentameter), has a special interest. It was more energetic than ordinary speech rhythms, deliberately so, and drove his soldier's story along at a swift, distinctive pace. This was sometimes described as a cavalry gallop; but it might equally be heard as a railway rhythm, the steady insistent beat working beneath the soundtrack of the poem.

FitzGerald would particularly admire it, using another image of drawn-out power, 'the long roll of the Lincolnshire Wave', which he heard 'reverberating' throughout the poem. To others it seemed brilliantly adapted to recitation, especially by patriotic young men now that Queen Victoria had come to the throne. But to the modern ear, it is remarkable how discordant and insistent it now sounds. Curiously Tennyson seems to have been uneasy with it too, explaining later that Hallam's father had recommended it to him, saying that 'the English people liked verse in trochaics, so I wrote the poem in this metre'.

When not at his London lodgings, or at the mousetrap in Kent, Tennyson still spent long weeks alone at the Mablethorpe cottage. Here he secretly continued writing his elegiac poems for *In Memoriam* to the sound of the restless sea. 'I walk about the coast here,' he told Fitz, 'and have it to myself, sand and sea.' He once tried sharing a glass of gin with a local fisherman, but the poor fellow 'got drunk, talked wildly about God and the Devil, fell down like a corpse'. Not much company.

Tennyson steadfastly refused to publish any of this new work. He even refused a serious offer to have his earlier poems published in America, despite repeated urgings from FitzGerald. He wrote dismissively to one prospective American publisher: 'Dear Sir, I have been residing for the last month at a remote little village on the coast of Lincolnshire – which has no regular intercourse with any post-town . . .' So he apologised for being slow to respond, but his previous two volumes contained 'many things so exceedingly crude' that it was impossible to consider a reprint. To FitzGerald he was even more emphatic. 'Dear old Fitz . . . Curse not anything till you see what it brings forth – not even Pyroglyphs . . . You bore me about my book: so does a letter just received from America, threatening, though in the civilest terms, that if I will not publish in England they will do it for me in the Land of freemen – Damn! . . . I want to know what has become of your stewardship.'

He sought out FitzGerald whenever he returned to London, and told him he was besieged by feelings of scepticism about his own work, and about the world in general. Sometimes they met at the Golden Cross Tavern in the Strand, and embarked on long rambling expeditions across the Thames into south London. Walking through the fields to Dulwich Art Gallery, on one occasion, Tennyson expressed specifically religious doubts to Fitz. He liked to think that St Paul's Cathedral was 'a Symbol that man is Immortal', but that this did not reassure him about the existence of a Deity.

Picking a daisy, he peered at it with all the attention of a naturalist (but also because he was short-sighted), and examined the beautiful symmetrical design of its crimson-tipped petals: 'Does this not suggest a thinking Artificer, one who wishes to ornament?' he murmured to Fitz. This was the very question he had debated more than a decade ago with Hallam at the Cambridge Apostles; and had continued to debate in 'The Two Voices'. But he now added bleakly: 'I would rather know I was to be damned eternally than not to know that I was to live eternally.'

FitzGerald reported that he would frequently come home to find Tennyson 'installed' on his Bloomsbury sofa. He would stay for days in FitzGerald's rooms in a very restless, uneasy state, 'being really ill in a nervous way: what with an hereditary tenderness of nerve, and having spoiled what strength he had by incessant smoking etc.' The etc clearly included drinking, and what FitzGerald was actually describing seems to have been a kind of nervous breakdown. Fitz gently asked him to call 'a truce' on all his 'complaints and complaining'.

FitzGerald took him out of London hoping to distract him, on expeditions up to Warwick Castle, and out to Stratford-upon-Avon, where Tennyson – 'seized with a sort of enthusiasm' – scribbled his name on the wall in Shakespeare's house, 'though a little ashamed of it' afterwards. 'The feeling was genuine at the time, and I did homage with the rest.' This pious literary pilgrimage was suddenly interrupted by the machinery of the modern world. On the return journey, via Leicester, Tennyson experienced for the first time what it was actually like to travel (as a third-class passenger) on a section of one of the newly opened railways. His first thrilled reaction was: 'liker to flying than anything else'. But more complex thoughts on the notion of speed would soon follow.

During all these expeditions together, FitzGerald may have glimpsed the real cause of Tennyson's unhappiness, what he called that 'hereditary tenderness of nerve'. He described his friend's behaviour as 'very perverse, according to the nature of his illness'. But exactly what *was* the nature of that illness? Some kind of deep depressive mourning, perhaps – going right back to his childhood at Somersby. To the childhood dominated by his drunken unstable father? For Cambridge and the dead Arthur Hallam? For Harrington Hall and the lost Rosa Baring? For the voices of madness overheard at High Beech? For worldly ambitions deferred, or the love affair with Emily Sellwood so strangely suspended ? Or for other far deeper things extinguished? For the Boundless Deep itself? FitzGerald, perhaps tactfully, was inclined to dismiss all such questions, and turned them aside with studied insouciance: 'Poets . . . one must allow are . . . a somewhat tetchy race.' But they remain to haunt Tennyson's whole biography, a dark under-sea swell beneath all this narrative.

Tennyson's clothes and appearance turned steadily more and more tramp-like, dirty and tobacco-stinking. He had sudden disagreements and 'growling' moods even with FitzGerald. Yet for the faithful Fitz he was still 'noble natured, with no meanness or vanity or affectation of any kind whatsoever'.

During this year Tennyson made several disappearances abroad, on apparently solitary summer expeditions to Amsterdam, to Calais, or to Paris. He may also have revisited Cambridge in these restless months, and the painful memories released of lost, happy days are simply but painfully recorded in the middle sections of *In Memoriam*, such as No. 87. Cambridge was just the same, but not the same, without Hallam.

> I past beside the reverend walls
> In which of old I wore the gown;
> I roved at random thro' the town,
> And saw the tumult of the halls;
>
> And heard once more in college fanes
> The storm their high-built organs make,
> And thunder-music, rolling, shake
> The prophet blazon'd on the panes;
>
> And caught once more the distant shout,
> The measured pulse of racing oars
> Among the willows; paced the shores
> And many a bridge, and all about
>
> The same gray flats again, and felt
> The same, but not the same; and last
> Up that long walk of limes I past
> To see the rooms in which he dwelt.

The long walk of lime trees is still there, a luminous and silent canopy of green light, eighty yards long, with water meadows on both sides. From high black iron gates, it runs up from the Backs and across Trinity Bridge over the placid Cam, and leads directly under the stone arch into the deep calm of Trinity New Court, where once Hallam had his rooms at staircase G.3.

By the autumn it was FitzGerald's turn to go in search of an increasingly disordered and shiftless Tennyson. He tracked him down to various chaotic London lodgings, not always in the rented house in

Camden. He would find Tennyson 'in certain conjunctions of the stars at No. 8 Charlotte Street with a little bit of dirty pipe in his mouth; and a particularly dirty vellum book of MSS on the sofa'. FitzGerald told Frederick (safe in Italy) that he thought Tennyson still lived 'in great intimacy, as you may have heard, with one Dr Allen, a mad Doctor at High Beech: who is very mad himself'.

FitzGerald had become convinced that the only hope for Tennyson's cure – or his salvation – lay in publication, and tried to encourage any such thoughts, but it proved difficult. London was grim that winter, with thick fogs 'the colour of sage cheese'. They passed 'some hours together every day and night: with pipes and Brandy and Water'. He added loyally: 'I hope he will publish ere long. He is a great fellow. But he is ruining himself by mismanagement and neglect of all kinds. He must smoke twelve hours out of the twenty-four.'

Tennyson now worried openly about money, and the investment with Matthew Allen's Pyroglyphs business, which as FitzGerald had feared, was not prospering. It only wound Tennyson in increasing complications. It emerged that Allen had attempted to obtain funds from Frederick in Italy; and most unprofessionally, from Septimus, who was still officially Allen's patient at High Beech. He even tried to have Mrs Tennyson sign a separate financial agreement. FitzGerald hoped that Spedding, with his legal experience, could somehow intervene.

Tennyson continued occasionally to flee north to his beloved Lincolnshire, and hole up in the cottage by the sea at Mablethorpe. From there he again put off any discussion with his publisher Moxon, by telling him the post was only delivered on Saturdays by the local muffin man. Apart from that, 'there is nothing here but myself and two starfish'. Yet it was perhaps on one of these flights to the North Sea that he wrote his poem 'The Shell'.

When he came south again he further delayed by turning aside to stay with his old family friends the Rawnsleys at Halton, and catch up with Sophie Rawnsley. He also encountered the 'Hindu' beauty Catherine Franklin again, who was now secretly engaged to be married to Sophie's brother Drummond. Catherine was struck by

Tennyson's distracted manner and wild appearance. She never forgot the moment, and later recalled: 'He looked very much like the old man of the sea, as if seaweed might cling to him, unkempt and unbrushed and altogether forlorn as to the outer man. When told he had seen me before, he looked hard at me and said: "Now who are you, and what are you? Where do you come from?"' She replied with unexpected 'curtness' that she was a 'spinster of Nottingham'. The bluntness of this answer produced a knowing laugh, and Catherine suddenly realised that Tennyson, 'with his short and keen sight', had guessed the romantic secret of her engagement. But he may also have been thinking about Emily Sellwood.

Back south again, FitzGerald firmly advised him to stay with friends out of London that winter. The year was saved by an invitation to spend Christmas in the country with the ever-hospitable Lushingtons at Park House, Kent. They managed to reunite the whole Tennyson clan on their extensive estate, not only improving Tennyson's morale but having a dramatic effect on two of his sisters. Cecilia Tennyson promptly fell in love with Edmund Lushington, while Emily Tennyson – to everyone's astonishment – revealed her secret engagement to Richard Jesse, a young naval lieutenant she had met at the Rawnsleys'.

Several friends were shocked by the speed of these developments, especially with regard to Emily and her apparent faithlessness to Arthur Hallam's memory. Hallam's sister Julia was one. She had believed that Emily 'would never dream of marrying – that she was a kind of *Nun* now, and that nothing was more *impossible* than her marrying – she had felt Arthur's death so much'. She was outraged. 'I had such a romantic admiration for her ... All my feeling about her is *bouleversé*.' Julia added pointedly: 'and Alfred Tennyson falls headlong into the abyss with her'.

Yet Tennyson was understandably pleased, and perhaps relieved, with both these romances. After all, his friendship with the Lushingtons dated back to Cambridge days, and an engagement looked imminent. And while Richard Jesse could never be Arthur Hallam, he was remarkably kind and handsome, and also brave. He had received a gallantry medal for saving the entire crew of a French vessel ship-wrecked in a North Sea storm off Margate, an action which won Emily's heart and clearly impressed Tennyson. Perhaps Jesse would prove a gallant saviour for the shipwrecked Emily too.

When they were married in London in January 1841, it was Tenny-son who gave the bride away. Emily looked sweet and 'elflike', with her long black hair now tied in youthful ringlets down her back. Even Hallam's father approved of the match, and continued his £300 annual allowance for the rest of Emily's life. Thus many of Tennyson's family responsibilities were unexpectedly lifted, and his emotional outlook subtly shifted. Both these marriages (Cecilia's took place in October 1842) would eventually introduce a new theme of redemption into Tennyson's poetry, and give a new structure to the hitherto fragmen-tary and mourning collection of pieces that had not yet become *In Memoriam*.

12

Publication

In the spring of 1842 Tennyson, at last and very reluctantly, began to consider publishing a new collection of poems, after a nine-year gap. It seems to have taken that long to get over the Croker review. FitzGerald, hugely relieved, urged him on, while his publisher Moxon noted the growing interest in America. There were influential intellectuals on the East Coast, notably Ralph Waldo Emerson and his Boston circle, who were already quoting Tennyson's earlier poetry and placing a twenty-page review of his work in the *Christian Examiner*, a leading liberal magazine which became the house journal of the New England Transcendentalists. Emerson exchanged admiring but quizzical letters with Carlyle on the subject, and referred to Tennyson's early verse – perhaps with a certain transcendental whimsy – as 'a perfect music-box for all manner of delicate tones and rhythms'.

Outside events may also have finally stirred Tennyson. The distant foreign war in which his restless soldier of 'Locksley Hall' had dreamed of finding glory, had come to pass. Britain's colonial campaign of 1841 in Afghanistan came to a humiliating end with the siege and fall of Kabul in January 1842. An army of four thousand troops and twelve thousand civilians was destroyed while retreating in winter from Kabul through the Khyber Pass to Kandahar. Apparently only one man survived, a surgeon's assistant, William Brydon, who later wrote a *Memoir* of this death march. An official army report recorded: 'Yesterday, about 1 P.M., Brydon, an Assistant Surgeon, of the Shah's Service, reached this place, (on a horse scarcely able to move another yard) wounded and bruised from head to foot with stones, and he, alone, has arrived to tell the fearful tale.'

Tennyson's 'Morte d'Arthur', though partly drafted seven years before, when he recited early passages to FitzGerald and Spedding in the boat on Lake Windermere, would be included in the new collection. It had gained a wholly new resonance. It must have seemed to Tennyson, and his readers, to echo and even prophesy this imperial disaster:

> So all day long the noise of battle roll'd
> Among the mountains by the winter sea;
> Until King Arthur's table, man by man,
> Had fallen in Lyonnesse about their Lord,
> King Arthur: then, because his wound was deep,
> The bold Sir Bedivere uplifted him,
> Sir Bedivere, the last of all his knights,
> And bore him to a chapel nigh the field,
> A broken chancel with a broken cross,
> That stood on a dark strait of barren land.
> On one side lay the ocean, and on one
> Lay a great water, and the moon was full.

This violent but dreamlike end of Arthur's Round Table – and especially when much later developed by Tennyson in *Idylls of the King* – came to seem like an early premonition of the essential fragility of the Victorian Empire. Its mood was not unrelated to Shelley's earlier sonnet 'Ozymandias' (1818) which ends on a similar note of bleak, but barren beauty: 'boundless and bare/ The lone and level sands stretch far away.' While the nostalgia for the lost island of Lyonnesse also became the delusive dream of far distant imperial lands, and was still echoed nearly a century later by Thomas Hardy, 'When I set out for Lyonnesse'.

But there was also potential disaster closer to home, within England, amid growing social unrest, and the start of a wave of strikes, as Carlyle had predicted. In May 1842 the Chartists presented their second national petition of three and a half million signatures to Parliament, demanding universal male suffrage, free education and other rights. It was rejected out of hand. During the subsequent Potteries riots, the poet Thomas Cooper was arrested and imprisoned for two years in Leicester. Carlyle's pamphlet *Chartism*, first issued in a small private edition in December 1839, was now republished by a general publisher James Fraser, and started to receive much wider attention. It pressed home the Condition of England Question: 'What means this bitter discontent of the Working Classes? Whence comes it, whither goes it? Above all, at what price, on what terms, will it probably consent to depart from us and die into rest? These are questions.' Questions that seemed to awaken Tennyson, and demand answers from his own work.

When he returned to London, he was still prevaricating about the exact choice of poems, and kept saying they would all look 'detestable' in print anyway. FitzGerald said he finally had to take Tennyson 'by violence' to see his publisher Moxon to firm up publication details. Should this include only new work? Moxon, commercially shrewd and anxious to retain the small readership Tennyson already had, firmly suggested combining both old and new. After much debate, it was decided the publication would be split into two parts. The first would be the best of the old work from the 1830 and 1833 collections, thoroughly rewritten. These included subtly altered versions of 'The Lady of Shalott' and 'The Palace of Art'. They were followed by twenty-nine new more experimental pieces, including both 'Audley Court' and 'Locksley Hall', and a greatly expanded version of 'The Two Voices'. Together they would become *Poems in Two Volumes*, an apparently simple title, but clearly inspired by Wordsworth's break-through volume of the same name in 1807.

FitzGerald 'dragged' a protesting Tennyson to Spedding's at 60 Lincoln's Inn Fields to make a final selection and start correcting proofs. Individual poems were stripped out of the long pages of his 'butcher's book'. FitzGerald remembered this final editorial process fondly, completed in clouds of tobacco smoke. Long after he teased Tennyson about it: 'It was in 1842 when you were printing the two good old Volumes – in Spedding's rooms – and the Butcher's Book, after its margins serving for Pipe-lights, went leaf by leaf into the fire;

and I told you I would keep two or three leaves of it as a Remembrance. So I took a bit of my old favourite Audley Court, and a bit of another, I forget which . . .' FitzGerald faithfully treasured these bits of manuscript, and years later proudly bequeathed them to the library of their old college, Trinity.

Moxon proposed to reprint a highly selected version of the 1833 poems in the first volume, including what he now thought of as the Cambridge poems, but also the mysterious 'Oenone', the grotesque 'St Simeon Stylites', and the languid 'The Lotos-Eaters'. There were many revealing additions. For instance, in the revised version of his ballad 'The Miller's Daughter', there was an intriguing new stanza, explaining and demonstrating the way the rhythms of particular lines of poetry would seize upon Tennyson physically. They 'beat time to nothing' and echoed repeatedly, almost painfully in Tennyson's head. His poetry, he explained, arrived like a bodily sensation. It possessed him like a tune, a song, haunting and inescapable until written down.

> A love song I had somewhere read,
> An echo from a measured strain,
> Beat time to nothing in my head
> From some odd corner of the brain.
> It haunted me, the morning long,
> With weary sameness in the rhymes,
> The phantom of a silent song,
> That went and came a thousand times.

The 'weary sameness' suggests specific memory of his own 'Mariana'. This was in fact one of the poems he most enjoyed reciting out loud, until it was at long last replaced by *Maud*.

Despite Tennyson's misgivings, Moxon insisted on packing as many as possible of the recent poems into the second volume. Tennyson absolutely refused to include any of the secret 'elegies' for Arthur Hallam, though there were of course deep hidden echoes of his grief in 'Ulysses'. But he did finally admit just one short poem secretly inspired by Hallam's death, 'Break, Break, Break'. Moxon's editorial interventions were largely justified, and the emotional force of all these poems, if not their source of inspiration, was immediately recognised. Even without the elegies, Moxon still believed the collection would have a tremendous impact. FitzGerald had no doubts. 'With all his faults he will publish such a volume as has not been published since the time of Keats: and which, once published, will never be suffered to die. This is my prophesy: for I live before Posterity.'

The first edition of *Poems in Two Volumes* was published on 14 May 1842, and Tennyson received a minimal copyright fee of £150. Eight hundred copies were printed, and five hundred were sold out by the autumn which Moxon assured him was very good. He had 'made a sensation' in the literary world. The collection did indeed mark the decisive declaration of a new post-Romantic generation. It was too early to call it Victorian, but it was clear that some sort of historic transformation was taking place. The poems were different in forms, different in themes, and quite different in tone from almost anything written before 1830. As the *Quarterly* put it, bidding farewell to the previous generation: 'No one but Coleridge among us has ever combined a thoroughly speculative intellect with so restless an abundance of beautiful imagery as we find in Mr Tennyson.'

New echoes of social turmoil were caught in the longer poem 'Locksley Hall'; and in a different way in the historical 'Godiva'. Impressively different kinds of dramatic voice (and metre) took command in such vividly contrasted poems as the heroic 'Ulysses', the sauntering 'Audley Court', the Gothic 'St Simeon Stylites', and the beautiful, sombre 'Morte d'Arthur'. Tennyson's important new turn towards philosophical or 'speculative' meditation, exploring the struggle between science and religious belief, and foreshadowing the central theme of *In Memoriam*, appeared above all in the final version of 'The Two Voices'. Largely rewritten and extended to nearly five hundred lines, it openly debated the question of suicide and intellectual despair, and hinted at the wider issue of total extinction in nature. In one manuscript version Tennyson had ended the whole poem on such a note of profound, almost apocalyptic uncertainty:

Heaven opens inward, chasms yawn,
Vast images in glimmering dawn,
Half shown, and broken and withdrawn.

Ah! Sure within him and without,
Could his dark wisdom find it out,
There must be answer to his doubt.

But must there be such an answer? Such deeply sceptical ideas excited much – often uneasy – comment among reviewers. The back and forth argument seemed to contain 'so many doubts and hopes, and things inscrutable; and thoughts that often present themselves in appalling whispers . . .' So in the last section Tennyson finally rewrote another 120 lines, to reach an apparently reassuring and more traditional conclusion, a sort of aubade or spiritual dawn after the long dark night of intellectual struggle. Moxon printed this extended and finalised version with evident relief.

And forth into the fields I went,
And Nature's living motion lent
The pulse of Hope to discontent.

I wondered, while I paced along
The woods were filled so full with song,
There seemed no room for sense of wrong . . .

There were other kinds of stylistic surprise, such as the plain, blank verse treatment of the traditional folktale of Lady Godiva. Here Tennyson turned towards a genuine historical figure, the Anglo-Saxon heroine who flourished shortly before the Norman Conquest, and whose many charities are recorded in several eleventh-century municipal documents. The beautiful Godiva had attracted several versions of her legendary munificence, the original belonging to a thirteenth-century chronicler, though much embroidered later in the eighteenth century. Tennyson had researched these in some detail, and then added his own distinctive post-Romantic touch.

He presented the well-known story of Godiva's naked ride, but reinterpreted as a political dare or wager, undertaken to protect the oppressed and poor of Coventry. Her bullying husband, the Earl of Mercia, threatens to levy a crushing new tax upon the citizenry. To prevent him, Lady Godiva undertakes to ride naked through the streets of Coventry.

> She told him of their tears,
> And prayed him, 'If they pay this tax, they starve.'

There are obvious echoes here of the kind of social injustice that Chartism was trying to reform. But the wager is a curious one, and there are undertones of sexual jealousy and a marital dispute, which Tennyson subtly exploits in the figure of the Earl.

> He laughed, and swore by Peter and by Paul
> Then filliped at the diamond in her ear;
> 'Oh ay, ay, ay you talk!' . . . and nodding as in scorn,
> He parted, with great strides among his dogs.

In a central passage, in preparation for her ordeal, Tennyson concentrates not on the famous ride itself, but on Lady Godiva's preliminary undressing, a form of diaphanous and gently erotic striptease, no longer watched by the absent Earl, but by what we might now call an invisible male gaze:

> Then fled she to her inmost bower, and there
> Unclasp'd the wedded eagles of her belt,
> The grim Earl's gift; but ever at a breath
> She linger'd, looking like a summer moon
> Half-dipt in cloud: anon she shook her head,
> And shower'd the rippled ringlets to her knee;

Unclad herself in haste; adown the stair
Stole on; and, like a creeping sunbeam, slid
From pillar unto pillar, until she reach'd
The Gateway, there she found her palfrey trapped
In purple blazon'd with armorial gold.

Tennyson evidently set out to rival Keats, in his own tender moonlight undressing scene from *The Eve of St Agnes*. He also added a
touch of humour in the 'creeping sunbeam' of her quick, stealthy
movements. But here are no Keatsian lovers to run away to happiness in the storm, and no marital reconciliation. Lady Godiva wins
her wager with the Earl, and the tax is remitted; but Tennyson ends
his story on a disconcertingly violent note. No one saw her during
her ride, though

    . . . the blind walls
Were full of chinks and holes; and overhead
Fantastic gables, crowding, stared . . .

But there has indeed been one secret male gazer, 'one low churl', who
hopes to glimpse Godiva's nakedness by peering through one of those
many chinks, as she passes through the Coventry streets. He is the
traditional 'Peeping Tom', and he is instantly punished for his effront-
ery (or visual assault) by a kind of symbolic castration:

    . . . his eyes, before they had their will,
Were shrivelled into darkness in his head,
And dropped before him.

Tennyson seems concerned with the notion of sight itself – seeing
and not seeing, of showing and not showing: of what is public and
what is intensely private. It is the display of Godiva's naked body
which appears to be at issue; but it is also her naked feelings. Through
Godiva, Tennyson asks how much can be shown publicly, and how
much must be kept private. Indeed this is a choice that affects his

editing of this whole collection: the problem of whether or not to publish his 'elegies', and of how far he could display naked grief.

What of Godiva's husband the Earl? Tennyson does not relate. We are not told what happened to him, or his marriage, or whether he suffered an equivalent shrivelling fate as the churl. He simply stumps off with his dogs, extinguished from history, while Godiva 'built herself an everlasting name'. Indeed the poem itself proved not only popular, but unexpectedly enduring. In America, Emerson always thought it 'a noble poem that will tell the legend [for] a thousand years'. Yet it is often excluded from modern English editions as inappropriate. However, several lines from it are carved beneath a striking bronze equestrian statue of naked Godiva, in the centre of modern Coventry (1949, reinstalled 2008). Public and private are here strangely reversed.

Volume 2 also contained a few entirely unexpected squibs, such as the long gaudy ballad 'The Vision of Sin'; and the short crack of 'The Skipping Rope', an exercise in teasing metrics but also in Tennysonian sarcasm, with its snapper ending. Here is the complete piece:

Sure never yet was Antelope
Could skip so lightly by.
Stand off, or else my skipping-rope
Will hit you in the eye.
How lightly whirls the skipping-rope!
How fairy-like you fly!
Go, get you gone, you muse and mope—
I hate that silly sigh.
Nay, dearest, teach me how to hope,
Or tell me how to die.
There, take it, take my skipping-rope
And hang yourself thereby.

On examination this little poem reveals itself as a rapid dialogue, or breathless snatch of repartee, between a patronising young man watching a a nimble young woman skipping. His flirtatious advances, 'how fairy-like you fly!', are impatiently repelled by the young woman who will have no such nonsense ('you muse and mope, I hate that silly sigh'). The final exchange sharply and wittily resolves itself in her suggestion that he go hang himself. Here the light badinage turns suddenly and painfully pointed ('hope' rhyming with 'rope', an effect which according to FitzGerald was often the result of Tennyson's mimicry or joking in private conversation).

The book was widely read and favourably reviewed in more than twenty newspapers. Though most of the first reviewers were Cambridge contemporaries, obviously well-disposed: John Sterling in the *Quarterly Review*, Monckton Milnes in the *Westminster Review*, and his old friend James Spedding in the *Edinburgh Review*.

The mixed response of the veteran Leigh Hunt in the *Church of England Quarterly*, representative of the old Romantic generation, was particularly interesting. He felt Tennyson's early poems were 'indolent' and over-refined. 'He grows lazy by the side of his Lincolnshire water-lilies.' But Hunt loved the 'long-drawn musical re-iterations' of 'The Lady of Shalott', though of course he did not 'very well understand' them. He thought 'Mariana' exceptional for its thick and beautifully observed details, picking out 'the blue fly singing in the pane', and quoting eighty-four lines of it. He was amused by the eroticism or 'peculiar exoterical delicacy' of 'Godiva', and what he called 'the ticklishness' of her position. But above all he was immensely impressed by the philosophical severity of 'The Two Voices'. It was his 'favourite poem'. The argument for and against suicide was 'capitally well put on both sides'. He 'admired it so much' that he thought that single poem justified the whole book. He hailed the young Mr Tennyson as the leader of the new generation: 'a kind of philosophical Keats ...'

Monckton Milnes surveyed the imaginative power of 'the rising poets of our time', and concluded: 'It rests with Mr Tennyson to prove that he can place himself at the head of all these his contemporaries.' Sterling declared 'Locksley Hall', in a clarion phrase, to be 'the direct outbirth and reflection of our own age'. While Spedding, knowing something of Tennyson's personal struggles, concentrated more on his friend's moral courage and psychological perception: 'In the Two

Voices we have a history of the agitations, the suggestions and counter-suggestions of a mind sunk in hopeless despondency, and meditating self-destruction; together with the manner of its recovery to a more healthy condition ... Others would have been content to give the bad Voice the worst of the argument ... Mr Tennyson's treatment of the case is more scientific ...'

Spedding summarised grandly but shrewdly: 'Powers are displayed in these volumes, adequate if we do not deceive ourselves, to the production of a great work ...' These brilliant powers were yet displayed 'in fragments and snatches'. But once Mr Tennyson could find 'a subject large enough to take the entire impress of his mind', he would produce a masterpiece.

The note of withheld expectation in these friendly judgements, of faint but careful hesitation, of promise not quite fulfilled, was not lost on Tennyson. He continued to worry about the reception of his work in general, and doubted his own worth. In July he again retreated to Lushington's estate at Park House in Kent, another of his many recuperative summer visits during the 1840s. There he would eventually meet an increasingly eclectic circle of young writers and thinkers. Among them were the painter and future traveller Edward Lear, already author of *Parrots* (1830), but not yet of *Nonsense Songs* (1846). There was also the brilliant young physicist William Thomson, who was still a Fellow at Peterhouse Cambridge but had already written his first astronomical paper 'Essay on the Figure of the Earth' (1840). He was shortly to become, at the age of twenty-two, the youngest Professor of Natural Philosophy at Glasgow, where he got to know Edmund Lushington who was Professor of Greek. Later, ennobled as Lord Kelvin, Thomson would daringly predict the 'heat death' of the sun like any common 'nebular' star. And so, as a burning nebular star, the sun would later appear in Tennyson's poetry. 'There sinks the nebulous star we call the Sun / If that hypothesis of theirs be sound ...'

In stimulating company like this, Tennyson threw off his publication anxieties and attended the summer fair of Maidstone's Mechanics' Institute, the carnival event held each year at Lushington's invitation, in the extensive grounds of Park House. It was enthusiastically reported in the pages of the local press, *The Maidstone and Kentish Advertiser*. He particularly delighted in the various scientific stalls and side shows, partly designed for children. These displayed clockwork boats, model steam engines, electric shock experiments, telescopes, and a hot-air balloon. The whole experience, its bustle and innocent scientific fun, cheered Tennyson greatly and set him thinking about new work. It also encouraged him, in a burst of energy, to move into new and more rural quarters and finally escape the restricting family 'mousetrap' in Tunbridge Wells. He rented a much larger and more comfortable house in the village of Boxley, below the North Downs, two miles through the fields from the Lushington estate, and known rather grandly as Boxley Hall.

Park House gained further festive associations in the autumn when Tennyson's sister Cecilia married Edmund Lushington, a wedding celebrated in the grounds on 10 October 1842. Tennyson wrote to his publisher Moxon: 'My head is yet vertiginous with the champagne I drank yesterday at my sister's marriage.' Five years later, Park House with these dizzy echoes of popular science and family romance, would form the opening to his long-contemplated poem about a women's university, only later to be called *The Princess*.

In an early draft this began as an extension of the remarkable science fiction vision from 'Locksley Hall'. Here it emerged as something rather more industrial, an enormous steam-powered city of the future, tall and thunderous, clanking and belching with manmade energy and force. But also, perhaps, pulsing with menace against the natural world, a potential nightmare that might obscure the sun itself.

> We crossed into a land where mile-high towers
> Puffed out a night of smoke that drowsed the sun;
> Huge pistons rose and fell, and everywhere
> We heard the clank of chains, the creak of cranes,
> Ringing of blocks and throb of hammers mixed
> With water split and spilt on groaning wheels
> Until we reached the court . . .

For the time being, Tennyson took this vision no further. Instead he disappeared on an impromptu summer tour of West Ireland, where he was supposed to meet one of his new admirers, the young (and just published) Irish poet Aubrey Thomas de Vere. De Vere was a great friend of the Irish astronomer Sir William Rowan Hamilton, and Tennyson probably hoped for much starlit conversation. Unfortunately his rushed travel arrangements were so confused that they did not lead to the expected conjunction anywhere on the beautiful Dingle Peninsula, and he never met de Vere at all. Instead Tennyson scrawled a brief note to Edmund Lushington back at Park House, with a hectic survey of his literary hopes and fears, as viewed from a hotel in Dublin. His publisher Moxon had gone on holiday to the Pyrenees but the news was exciting: '500 of my books are sold – according to Moxon's brother I have made a sensation. I wish [Allen's] woodworks would make a sensation! ... I go to Limerick tonight. I hope you are all blooming. What with Ruin in the distance and Hypochondriacs in the foreground I feel very crazy. God help all.'

By 1843 Tennyson's *Poems in Two Volumes* had clearly begun to establish a new reputation, and create a stir especially among the young. Elizabeth Barrett enthused, 'I think such a godship of Tennyson'; and was 'ready to kiss his shoe-ties any day'. Though she had an interesting reservation, rather different from Hunt's. His 'representation of beauty' lacked physicality or sensuality. 'You can no more touch or clasp it, than beauty in a dream. It is not less beautiful for *that*; but less sensual ...' Robert Browning loved it, but thought the revisions to the early poems 'insane'.

Thomas Carlyle was surprised by the revelations of his new friend's poetic energy, which he had not suspected from their early meetings. He told Tennyson that he was taken by storm: 'Truly it is long since in any English book, Poetry or Prose, I have felt the pulse of a real man's heart as I do in this same. A right valiant, true fighting, victorious heart; strong as a lion's, yet gentle, loving and full of music.' Certain lines in 'Ulysses' did not exactly make him weep, 'but there is in me what would fill whole Lachrymatories as I read'. He cited the passage about seeing 'the great Achilles, whom we knew'. When the two volumes reached America, Edgar Allan Poe remarked: 'I am not sure that Tennyson is not the greatest of poets.'

Particular attention was given to Tennyson's attempts at social commentary. Sterling for instance preferred the 'modern and democratic poems', like 'Locksley Hall' and 'Audley Court'. They had a 'clearness, solidity, and certainty of mind'. It was generally accepted that several of the poems were now animated by Tennyson's scientific reflections and speculations, notably the expanded version of 'The Two Voices', with its altered ending. This new material had burst Tennyson out of his old, lyrical Romantic forms, and made him confront the world immediately in front of him. Despite the classical and Arthurian references, the collection as a whole bore witness to shaking social changes and technical innovations.

Yet there was a curious irony in Tennyson's choice of the thunderous railway – 'forward, forward' – to symbolise the modern world. It emerged that his famous line about the swift future progress of civilisation, contained a blatant, if admittedly rather technical, error. 'Let the great world spin for ever down the ringing grooves of change' was a magnificent, dynamic image. But strictly speaking it was an inaccurate one. The wheels of the steam engine (and the carriages) did not of course run in 'ringing *grooves*'. The wheels were flanged, and ran *along the top* of metal rails of very precise dimensions and manufacture. Absurd as it might seem, this engineering error was solemnly pointed out to Tennyson by several reviewers and numerous geeky later readers. One even supplied the original Stephenson standard railway gauge measurements of just over a metre wide, at 1,435 millimetres.

Yet Tennyson was particularly vexed by any kind of scientific inaccuracy, and later he would take immense trouble to explain this particular mistake. He pointed out that he had personally witnessed the historic opening of the first Liverpool to Manchester line in September 1830. 'But it was a black night,' he recalled, 'and there was such a vast crowd round the train at the station that we could not see the wheels.' There was also the simple fact that without his pince-nez he was short-sighted. What he did not mention was that without knowing it, he had also witnessed the fatal injury of William Huskisson. This was quite another kind of symbol of the modern world.

One other response to the scientific themes of the collection came from an unexpected direction, and in an unexpected manner. Charles Babbage, the computer pioneer whose work Tennyson had been diligently reading, sent a curious fan letter. Babbage admired the whole book but decided to express his appreciation with a mathematical witticism, which he enclosed in a private letter. He chose one of Tennyson's odder poems, the burlesque sexual ballad entitled 'The Vision of Sin'. It contained no apparent scientific references at all, but a statistical one. Babbage accordingly picked out the following swinging stanza:

> Fill the cup, and fill the can:
> Have a rouse before the morn.
> Every moment dies a man,
> Every moment one is born.

He commented: 'Sir: In your otherwise beautiful poem "The Vision of Sin" there is a verse which reads – "Every moment dies a man, Every moment one is born." It must be manifest that if this were true, the population of the world would be at a standstill. In truth, the rate of birth is slightly in excess of that of death. I would suggest that in the next edition of your poem you have it read – "Every moment dies a man, Every moment 1 and $1/16^{th}$ is born." The actual figure is so long I cannot get it onto a line, but I believe the figure 1 and $1/16$th will be sufficiently accurate for poetry. I am, Sir, yours, etc., Charles Babbage.'

Babbage had in fact met Tennyson socially, and so felt able to rib him gently, but there was a real 'two cultures' sting in the last mocking phrase, 'sufficiently accurate for poetry'. Years later the physicist and atmospheric scientist John Tyndall came to the poet's defence: 'In regard to metaphors drawn from science, Tennyson made sure of their truth. To secure accuracy, he spared no pains. I found in his room charts of isothermals and isobars intended to ensure the exactitude of certain allusions of his to physical science.'

But Tennyson could always make his own scientific jokes. Once walking out of London on a windy day, he observed the new suspended telegraph wires which were following the new railway lines, steadily branching out in a network of invisible messages across the whole of England. He stopped beneath one of the roadside telegraph poles, and listened intently. Then he said to his companion, possibly FitzGerald, that he could hear 'in the wail of the wires, the souls of dead messages'.

13
Evolution

After the publication of *Poems in Two Volumes* (1842), Tennyson knew he had gained a national reputation, and even perhaps an international one. A long article on his poetry appeared in France, in the summer edition of the *Revue Britannique*. Besides much favourable commentary, it attached complete translations of several of his poems including 'Ulysses' and 'Godiva'. The same two poems, with the addition of 'Audley Court', also appeared in German at this time, an interesting example of the Continental taste for the new 'English' style of semi-dramatised poetry. It was regarded as daringly direct. Curiously, 'Godiva' was also the first of his poems translated into Russian (though not until 1859).

His poems were now being read, reviewed and published (after initial pirating) right along the East Coast of America. They were warmly greeted not only by Emerson and the Transcendentalists in Boston, but also frequently praised in the varied letter exchanges between the fashionable 'Fireside Poets' like Henry Wadsworth Longfellow at Harvard and James Russell Lowell in New York. Edgar Allan Poe in Baltimore reviewed an American edition of the 1842 volume in almost ecstatic terms. 'For Tennyson, as for a man imbued with the richest and rarest poetic impulses, we have an admiration – a reverence unbounded. His "Morte D'Arthur", his "Locksley Hall", his "Sleeping Beauty", his "Lady of Shalott", . . . and many other poems, are not surpassed, in all that gives to Poetry its distinctive value, by the compositions of any one living or dead.' His one reservation was that sometimes Tennyson's sense of beauty could be 'quaint'. The following year Poe's own *The Raven and Other Poems* was accused of plagiarising Tennyson's 'quaintness', as well as his verse forms.

In London his friends, including FitzGerald, began to regard Tennyson not merely as the secret (and sometimes embarrassing) genius in their midst, but also as a public celebrity to be proud of. This also meant that they confidently expected major work to come. Yet paradoxically Tennyson felt little lasting elation or sense of achievement. On the contrary, he felt increasingly ill and depressed. FitzGerald saw this and thought he was suffering essentially from financial worries, though he feared some kind of breakdown. He received one note from Tennyson signed 'Goodbye old Fitz / Lest I lose my Witz'. Fitz had always been sceptical of High Beech, but had hoped it might at least have cured his friend's hypochondria, if little else. But instead it had simply entangled him in the wood-carving scheme, and Tennyson's confidence in Dr Allen and his therapeutic powers had apparently collapsed. He wrote to his erstwhile guru: 'Dear Dr Allen, My spirits have been dejected, my nerves shattered, my health affected by what I have heard ...' Allen himself wrote back shiftily of business debts and warrants of attorney, his own 'proportion of anxiety', and bursting into tears. Carlyle now described the doctor as 'speculative, hopeful, earnest-frothy'.

Shortly after the move to Boxley, in October 1842 Tennyson learned that Allen was attempting to borrow £1,000 from Septimus, still his patient at High Beech; and had approached his sisters and his mother for further loans. Tennyson felt outraged but trapped. 'Dr Allen has already at different times received from our family about £8,000 – the result of which loan has been to my mother and sisters that they are at this time living upon my brother Charles – and to myself that I am a penniless beggar and deeply in debt, besides – which two things have never happened to me before.' Moreover he felt as a result of Allen's influence Septimus had 'estranged himself from his family'.

Once again Tennyson began to suffer from repeated depressive episodes. Years later he described one in London, though this may also have been a memory from a much earlier time, yet the idea of universal extinction always recurs. 'I used to feel moods of misery unutterable. I remember once in London the realization coming over me of the whole of its inhabitants lying horizontal a hundred years hence. The smallness and emptiness of life sometimes overwhelmed me. I used to experience sensations of a state almost impossible to describe in words; it was not exactly a trance but the world seemed dead and myself only alive.' The exact date of Tennyson's revealing entry is not clear, though it is evident from *In Memoriam* that such episodes were recurrent for much of his early life.

The dark terrors of the Kraken were clearly back in the ascendant. Throughout 1843 Tennyson moved restlessly between at least four addresses, besides FitzGerald's: his own rented rooms in Camden, Edmund Lushington's legal chambers at 2 Mitre Court Buildings, the new family house at Boxley, or even back to his old seaside haunts at Mablethorpe. Meanwhile he cast about for a quite different mode of medical cure, and a different kind of guru to replace Allen.

After much research, he came up with the promise of a completely new form of treatment: hydrotherapy – healing by water. He began to investigate a whole series of hydrotherapeutic regimes or 'water cures', visiting a number of special clinics variously in Birmingham, the Malverns and Cheltenham. The Malvern clinic was originally founded by the renowned Dr James Gully and his assistant Dr James Wilson in 1842. It turned out that Dr Gully was rather more commercially astute than Dr Allen. His hugely successful practice eventually made bottled Malvern Water famous, and attracted many celebrities such as Charles Darwin, Charles Dickens, Florence Nightingale and even the sceptical Thomas Carlyle.

Hydrotherapy had become highly fashionable in London circles, and the extension of the Malvern establishment run by Dr Wilson would be the subject of an enthusiastic essay in the *Monthly Magazine* in 1845. It was entitled with a nod to de Quincey and opium, *The Confessions of a Water Patient*. It was written by the 'silver spoon' novelist Edward Bulwer-Lytton, and soon republished as a fifty-page pamphlet. Lytton's description makes the clinic sound like a rehabilitation centre for alcoholics: 'patients accustomed for half a century to

live hard and high, wine drinkers, spirit-bibbers'. But it was also presented as a serious place of mental retreat, of spiritual healing,
eminently suitable for artists and writers.

Lytton promised: 'Men to whom mental labour has been a
necessary – who have existed on the excitement of the passions and
the stir of the intellect – who have felt, these withdrawn, the prostra-
tion of the whole system – the lock to the wheel of the entire
machine – return at once to the careless spirits of the boy in his first
holiday.' In other words it was just the ticket to revive the creative
powers and sparkle of a jaded author, and even of a poet. Or to recover
from what would now be considered a serious mental breakdown.

Tennyson's first extended and rigorous trial of hydrotherapy began
in November 1843, at an offshoot of Dr Gully's flourishing Malvern
practice in Cheltenham. It continued on and off for the next three
years. Cheltenham was already a well-known and well-heeled spa
town, where one of Tennyson's sisters had previously stayed in lodgings, though for recuperation and relaxation – 'taking the waters' – rather
than anything more organised and demanding.

The full and authentic hydrotherapy treatment was considerably
more severe. It involved a demanding regime of cold baths, hot com-
presses, herbal poultices, alternative 'sweatings and coolings', together
with physical exercises, long open-air walks and a strict diet. They
were similar to the modern sauna regimes now imported from Scandinavia. They were originally based on the health teachings of an
Austrian guru, Vincent Priessnitz, who had achieved a dedicated following in Vienna, gave numerous lecture tours, and had published in
England his *Hydropathy; or The Cold Water Cure, as practiced by Vincent
Priessnitz, at Graefenberg, Silesia, Austria* (1842).

Tennyson retreated to a set of rented rooms in Cheltenham, conveniently positioned in a secluded side street at No. 6 Belle Vue, St
James's Square, about ten minutes' walk from the bustle of the Cheltenham Town Hall. Its 'belle vue' consisted of a small garden square
and the sobering sight of the Roman Catholic church of St Gregory
the Great next door, soon to have the addition of a tall spire and bells,
which gave the square a contemplative atmosphere. The cure was

officially run on 'Priessnitzan' lines. Tennyson thought it deeply bene-
ficial, though not at all comfortable. He had a small room at the top
of the house, which he quickly spread with books and papers over all
the chairs and most of the floor. At times he felt almost imprisoned,
and told one visitor that he had two great fantasies of escape: 'one was
to see the West Indies, the other to see the earth from a balloon'.

He recorded the regime with restrained irritation: 'No reading by
candle light, no going near a fire, no tea, no coffee, perpetual wet sheet
and cold bath and alternation from hot to cold. However I have much
faith in it.' One therapy, known as the 'Neptune Girdle', involved
being tightly wrapped from head to toe in a linen sheet, and then
being doused with freezing water at regular intervals for an hour. The
declared aim of hydrotherapy was to 'flush out' or expel physical
'poison' from the bodily system. But it was also intended as a form of
psychological cleansing: to bring to the surface any mental monsters
lying in the deep. Tennyson would indicate to FitzGerald that there
were plenty of those.

Yet part of Tennyson's breakdown, and his 'poison', was evidently
caused by continuing financial anxieties, as FitzGerald had understood. Tennyson once again calculated his loans to Allen as totalling
the hideous sum of £8,000, which now seemed irrecoverable. Sales of
Pyroglyphic furniture were permanently down, and Allen was
declared bankrupt (for the second time in his life) by the end of 1843.
A ruined man, he began suffering from heart problems, withdrew
from directing the asylum, and died in January 1845, still overwhelmed
with debts. Tennyson wondered if a similar fate awaited him.

After receiving several anxious letters from FitzGerald, Tennyson
finally replied to his old friend in February 1844 from Cheltenham
spa. He was evidently in miserable spirits, and the letter was unusually
long, intimate and self-pitying. 'My dear Fitz – it is very kind of you
to think of such a poor forlorn body as myself. The perpetual panic
and horror of the last two years had steeped my nerves in poison.
Now I am left a beggar but I am or shall be shortly somewhat better
off in nerves . . . I have had four crisises . . . Much poison has come
out of me, which no physic ever would have brought to light . . . My

dear Fitz, my nerves were so bad six weeks ago that I could not have written this ... I went through Hell ... No more trips to London and living in lodgings ... You are the only one of my friends who has asked after me ... I shall go over I think to Italy. I certainly cannot live in England and be comfortable.'

FitzGerald sympathised with his brilliant old friend, but would later begin to refer to these dark moods as 'valetudinarian' lapses. Tennyson did not go to Italy, but effectively settled his base at Cheltenham from 1846. Here he was joined by the remaining family – his mother Elizabeth, his two unmarried sisters, Mary and Matilda, and occasionally one or other of his brothers drifting back from Italy or France. They continued to rent lodgings in St James's Square, but in a much larger Regency family house at No. 10, three storeys with elegant ironwork balconies, and plenty of spare rooms for any migrant family member. (There is now a modest Tennyson plaque by one of the ground-floor windows.) It appears that all of them sometimes 'took the waters', but Tennyson also pursued the full hydrotherapy regime elsewhere, once at Birmingham in 1847 and then directly at Malvern under Dr Gully himself.

This was all possible because the financial situation also resolved itself, thanks partly to his poetry sales and partly to his friends. An insurance policy on Allen's life, taken out for Tennyson with great foresight by Edmund Lushington, actually recovered some of the family capital, on Allen's death. The policy eventually paid out £2,000. A further reassurance came in the promise of a Civil List pension of £200 per annum, arranged in part through a kindly intervention by Henry Hallam. But perhaps most reassuring of all, Edward Moxon announced first royalties on no less than three editions of *Poems* of 1842, totalling £746, and with more to come. As FitzGerald had suspected, Tennyson's cure would turn out to be cash as much as water.

So Tennyson gradually emerged from this period of darkness, and once again started writing poetry and visiting his friends in London. One of the first signs of his recovery was the delight he took in a new science book, a work of speculative natural history, the anonymous and sensational *Vestiges of the Natural History of Creation* (1844). It would confirm for him the new scientific theme rising through the elegiac poems that he had been secretly writing, and tentatively starting to collect in a central section, which became *In Memoriam* Nos 50–60. In them he meditated on the conflicting ideas of science and belief, and confronted the increasingly remote evidence for a creative God in the natural world, and the whole idea of extinction.

These dark months revealed what might be called Tennyson's poetry of Evolution, both its promise and its terrors. It was extraordinarily prophetic. Long before he – or anyone else – could have read Darwin, he responded to the metaphysical crisis that Darwinism and the actual theory of evolution would subsequently produce throughout society. It was both an intellectual and an imaginative crisis, the revelation of a whole new world and mankind's transformed place within it. Today it is difficult to appreciate the profound, bewildering impact this would have had. But a modern equivalent might be the discovery of life – intelligent life, not mere bacterial life, but an entire civilisation – on another planet. Our whole view of Creation would have to change.

Tennyson had read a lengthy and favourable review of *Vestiges* in the *Examiner* newspaper in November 1844. Here the book was dramatically praised as 'the first attempt that has been made to connect the natural sciences into a history of creation ... It contains much that at first reading may startle a devout and religious mind; it contains nothing that such a mind will do well to reject on more calm and full reflection.' It was, in effect, a deliberate attempt to describe the whole universe, from star systems to human and animal societies, in evolutionary terms and without the benefit of clergy.

It also put forward a radical vision of humanity itself evolving, both physically and morally, to rise to a new peak, virtually a new kind of *Homo sapiens*. The *Examiner* summarised this second idea in a passage that was to haunt Tennyson. 'Is our race but the initial of a grand crowning type? ... There may be occasion for a nobler type of humanity, which shall complete the zoological circle on this planet, and realise some of the dreams of the purest spirits of the present race.'

Tennyson would revert to this further vision in the very last section of the elegies that became *In Memoriam*. With ideas like this, some brilliant, some slapdash, and some purely fanciful, the range of *Vestiges* was so sweeping, and its conclusions so controversial, that the author would attempt to remain anonymous for several years, to avoid personal attack. This was a lesson not lost on Tennyson, who would continue to keep his own poems concealed for another six years.

The challenging title of the book was drawn from a famous statement made by the Scottish geologist James Hutton, in his classic of late eighteenth-century geology, *The Theory of the Earth* (1788). Hutton had concluded his investigations of rock strata with startling simplicity: 'The result, therefore, of this physical enquiry, is that we find no vestiges of a beginning, no prospect of an end.' He proposed that geological time was of a different order of magnitude from anything in human history, or indeed anything in biblical history. *Vestiges* explored in detail the implications of this idea.

Tennyson promptly ordered a copy from Moxon. 'I want you to get me a book which I see advertised in the *Examiner*. It seems to contain many speculations with which I have been familiar for years, and on which I have written more than one poem. The book is called *Vestiges of the Natural History of Creation . . .*'

When *Vestiges* arrived at Cheltenham, Tennyson's hands shook with anticipation as he opened the package. He recalled the moment years afterwards, explaining to the young poet William Allingham that all his life he had been anxious 'to get some insight into the nature and prospects of the human race'. He had gathered from all the talk about the new book 'that it came nearer to an explanation than anything before it'. 'I trembled as I cut the leaves,' Tennyson told Allingham with a smile.

The evolutionary arguments of *Vestiges*, so daring and controversial as they were, did not satisfy Tennyson, but they encouraged him to take risks. They would have a powerful influence over his writing and editing of the scientific sections of *In Memoriam* over the next five years. They gave him renewed courage to voice his philosophic doubts about religious belief, and to expand the poem far beyond the limits of personal elegy and private grief. In a similar mood of radical thinking, Tennyson also ordered Ruskin's *Modern Painters* 'on the greatness of Turner' and his later painting. He was fascinated by Turner's own grappling with the modern world, with geology and meteorology, and even the railways, as in his famous and daring picture *Rain, Steam and Speed – The Great Western Railway* exhibited in that same year, 1844.

Vestiges became one of the most influential and controversial popular science books of the nineteenth century. It outsold Darwin's *On the Origin of Species* (1859) for a generation. It promoted an early theory of evolution, described in reckless and sweeping generalisations, across all the sciences, including astronomy, geology and biology. Whewell called it 'bold, speculative – and false', and Darwin himself thought its 'geology bad . . . zoology far worse'. Yet he would later praise its 'powerful and brilliant style' and believe it had done 'excellent service' in 'removing prejudice and preparing the ground' for his own great work on evolution.

Its sensational impact was due partly to its apparent rejection of any concept of a biblical creation, and its aggressive tone. 'How can we suppose that the august Being who brought all these countless worlds into form by the simple establishment of a natural principle flowing from his mind, was to interfere personally and specially on every occasion when a new shell-fish or reptile was to be ushered into existence on *one* of these worlds? Surely this idea is too ridiculous to be for a moment entertained.' It constantly threw out provocations: 'Man is seen to be an enigma only as an individual; in the mass he is a mathematical problem.' It suggested evolutionary views on subjects as diverse as canine loyalty, photography and slavery.

But it was also partly due to its anonymity which provoked frantic speculation as to its author's identity, and even their gender. Charles Babbage, with his usual mischievous wit, suggested that Byron's brilliant daughter, the computer expert Ada Lovelace, known for her freaks, could have written it. The twenty-eight-year-old Countess of Lovelace was in fact secretly assisting Babbage at the time, and had just written – not *Vestiges* – but a brilliant series of Notes (1843) on the 'creative' possibilities of Babbage's latest Analytical Machine. She speculated that it might be capable of composing music, and many other forms of creative activities, if not actual original thought. (Alan Turing would later famously call this 'Lady Lovelace's Objection' to the claims of artificial intelligence.) Lovelace even suggested that a new composite discipline was emerging, combining mathematics and imagination, which she called 'Poetical Science'.

She deeply impressed many of the leading scientific men of her day, including John Herschel and the physicist Michael Faraday. Babbage himself described Ada unforgettably to Faraday in a letter of September 1843: 'That Enchantress who has thrown her magical spell around the most abstract of Sciences and has grasped it with a force that few masculine intellects (in our own country at least) could have exerted over it.' So her anonymous authorship of *Vestiges* was not such a wild suggestion.

Moreover *Vestiges* did specifically refer to Babbage's own theory that natural 'procedures' of evolution might occasionally be interrupted by a 'higher law'. This might operate less like a divine intervention (which Babbage doubted) and more like the override command of a computer program, and could also be 'creative' in its effect. The *Examiner* reviewer was quick to pick out this novel, possibly blasphemous, concept of the sudden, unorthodox development of the human 'brain and heart', implying the special creation of the human soul quite distinct from ordinary animal intelligence. '[The writer] gives a remarkable suggestion, from some data of Mr Babbage's calculating machine, that this ordinary procedure may be subordinate to a higher law which only permits it for a time, and in proper season interrupts and changes it.'

Its actual author, a daring Scottish journalist and popular science writer, Robert Chambers, carefully remained under cover in Edinburgh until the initial storm had died down, and a second edition could be prepared with corrections and explanatory notes. His replies were seriously argued, and his success eventually prompted him to go into business with his brother, and found *Chambers's Encyclopaedia*. Meanwhile, right across society, extensive and excited comments began to appear in letters or articles by the novelist Benjamin Disraeli, by Charles Lyell and Charles Darwin, by Sir John Herschel, and even among the intellectuals and businessmen of Prince Albert's courtly circle. Darwin was provoked to write his 'Second Preliminary Essay'(1844) on evolution, a marvel of compression compared to the eventual *Origins* of 1859, running to a mere 230 pages.

Vestiges aroused extraordinarily passionate feelings, both in favour and against its arguments. But above all it revealed the particular prejudices of its commentators. One hostile reviewer, Adam Sedgwick, the Professor of Geology at Tennyson's old Cambridge college, Trinity, gave full rein to his horror and then his misogyny. Sedgwick initially turned down several invitations to review *Vestiges*, pleading lack of time. But in March he read it closely and on 6 April discussed with other leading clergymen the 'rank materialism' of the book, 'against which work he & all other scientific men are indignant'. He thought the background reading 'very shallow'. Moreover, the hasty jumping to conclusions and the 'shooting ahead' of facts, all indicated one shameful truth: the author was a woman. This would not have impressed Mary Somerville.

In a letter to Charles Lyell about 'the foul book', Sedgwick's intellectual disgust unfurled itself: 'If the book be true, the labours of sober induction are in vain; religion is a lie; human law is a mass of folly, and a base injustice; morality is moonshine; our labours for the black people of Africa were works of madmen; and man and woman are only better beasts! ... I cannot but think the work is from a woman's pen, it is so well dressed and so graceful in its externals. I do not think the "beast man" could have done this part so well.'

Sedgwick covered his edition with furious pencilled notes, some of which have survived. Against Chambers's speculation about the evolution of 'mutually adapted' language powers throughout the animal kingdom, Sedgwick blazed two asterisks and scrawled angrily below the margin: 'But why do not monkeys talk?' His misogyny also became more and more explicit as he bent to his task. 'Most women have by nature a distaste for the dull realities of physical truth, and above all for the labour pains by which they are produced.' On 10 April he finally contacted the editor of the *Edinburgh Review*, and volunteered his long, angry article, which was quickly accepted. Its publication, and the storm aroused, would have alerted Tennyson to the risks he was taking in his own poetry.

Yet Tennyson had written directly and bleakly about the challenge posed to the notion of a beneficent and creative Divinity by these nascent theories of evolution. In doing so, he lit upon phrases that have been burned into the English language ever since, and which were directly inspired by his scientific reading. He gave a whole new dimension to these fearful revelations, by considering their impact not only on traditional religious dogma, but much more on the human imagination itself. They altered the old innocent Romantic trust in nature, but equally questioned the new robust Victorian hope in the future. If the source of Creation was itself doubtful, what about its end? Must mankind itself become extinct? How to bear such a crushing philosophy? The first of what can be called Tennyson's central evolution sequence, No. 50, though not specifically about science, begins in such a desperate, yet defiant mood.

> Be near me when my light is low,
> When the blood creeps, and the nerves prick
> And tingle; and the heart is sick,
> And all the wheels of Being slow.
>
> Be near me when the sensuous frame
> Is rack'd with pangs that conquer trust;
> And Time, a maniac scattering dust,
> And Life, a Fury slinging flame.

The exact sequence or dates of the poems Tennyson gathered in this mid-section of *In Memoriam* is not clear. Tennyson later claimed some were conceived 'some years' before he read Chambers; yet they might in fact have been written any time in the mid-1840s. But in the key sequence of three poems, No. 54, No. 55 and No. 56, which he placed at the almost exact halfway point of the entire collection, he formulated a series of terrible doubts and bitter questions which seem to respond directly to *Vestiges*. Chambers had proposed: 'It is clear from the whole scope of Natural Law, that the Individual, as far as the present sphere of being is concerned, is to the Author of Nature, a consideration of inferior moment. Everywhere we see the arrangements for the species perfect; the Individual is left, as it were, to take his chance amidst the *melée* ...' This is also typical of Chambers's casual manner of dealing with the deity's apparent lack of concern with humanity.

In Memoriam No. 54 expresses by contrast the desperate hope that this might not be true. Surely the Individual was precious, no single life form would be casually 'destroyed'? It is a groan of early Victorian anguish in the face of a new and monstrous perception that the whole of creation might be ultimately – in a strikingly plain, bleak and domestic word – simply 'rubbish'.

> Oh, yet we trust that somehow good
> Will be the final end of ill,
> To pangs of nature, sins of will,
> Defects of doubt, and taints of blood;
>
> That nothing walks with aimless feet;
> That not one life shall be destroy'd,
> Or cast as rubbish to the void,
> When God hath made the pile complete;
>
> That not a worm is cloven in vain;
> That not a moth with vain desire
> Is shrivell'd in a fruitless fire,
> Or but subserves another's gain.

If *Vestiges* gave Tennyson courage to voice these fears, much of his specific imagery for the next two poems No. 55 and No. 56 was drawn from Lyell's masterwork, *The Principles of Geology*. Lyell had written of the inevitable extinction of whole species or 'types' in Volume 2 of *Geology*. Two whole chapters (Volume 2, chapters 8 and 9) were given

over to 'Changes in the Animate World which Tend to the Extinction of Species'. A third (chapter 13) is entitled: 'Embedding of
Organic Bodies and Human Remains in Blown Sand', which would
clearly inspire Tennyson's line 'Be blown about the desert dust'. A
fourth (chapter 17) spreads out in a vast, disturbing panorama entitled, 'How the Remains of Man and his Works are becoming Fossil
beneath the Waters'.

Babbage had concluded the same about the inevitable 'extinction of
every race' in his *Ninth Bridgewater Treatise*. But Lyell assembled a
mass of detailed evidence for global change and extinction, drawn from
sites as far apart as the alluvial riverbanks of the Nile and the delta of
the Ganges to the eroded sea coasts of Norfolk. He would continually
throw out suggestive mysteries, like the 'so-called submarine forest of
Happisburgh', a village many miles south of Mablethorpe but on Tennyson's own North Sea coast. Such details were riveting to Tennyson.

In No. 55 Tennyson responded with the idea that nature, far from
demonstrating the benign power of God, as in traditional Natural
Theology, might actually be at war with Him. Here he used Lyell's
precise and contemporary scientific term 'type', to denote any defined
species in animal or plant life:

> Are God and Nature then at strife,
> That Nature lends such evil dreams?
> So careful of the type she seems,
> So careless of the single life;
>
> That I, considering everywhere
> Her secret meaning in her deeds,
> And finding that of fifty seeds
> She often brings but one to bear,
>
> I falter where I firmly trod,
> And falling with my weight of cares
> Upon the great world's altar-stairs
> That slope thro' darkness up to God,

I stretch lame hands of faith, and grope,
And gather dust and chaff, and call
To what I feel is Lord of all,
And faintly trust the larger hope.

Lyell concluded *Geology* with a chilling philosophic overview. 'Amidst
the vicissitude of the earth's surface, species cannot be immortal, but
must perish, one after another, like the individuals which compose them.
There is no possibility of escaping from this conclusion.' Even the dead
would become extinct. 'None of the works of a mortal being can be
eternal ... And even when they are included in rocky strata, when they
have been made to enter as it were into the solid framework of the globe
itself, they must nevertheless eventually perish, for every year some portion of the Earth's crust is shattered by earthquakes or melted by volcanic
fire, or ground to dust by the moving waters on the surface.'

In response to this Tennyson produced a magnificent answering
poem, one of the finest in his life. He repeated his own question, 'So
careful of the type?', and followed through these terrors with pitiless
and unshaken logic, to the last trembling line: the possibility of some
'veiled' answer or redemption for mankind. So much of Tennyson's
scientific reading, his doubts and reflections, were now brilliantly
compressed into *In Memoriam* No. 56.

'So careful of the type?' but no.
From scarped cliff and quarried stone
She cries, 'A thousand types are gone:
I care for nothing, all shall go.

'Thou makest thine appeal to me:
I bring to life, I bring to death:
The spirit does but mean the breath:
I know no more.' And he, shall he,

Man, her last work, who seem'd so fair,
Such splendid purpose in his eyes,

Who roll'd the psalm to wintry skies,
Who built him fanes of fruitless prayer,

Who trusted God was love indeed
And love Creation's final law—
Tho' Nature, red in tooth and claw
With ravine, shriek'd against his creed—

Who loved, who suffer'd countless ills,
Who battled for the True, the Just,
Be blown about the desert dust,
Or seal'd within the iron hills?

No more? A monster then, a dream,
A discord. Dragons of the prime,
That tear each other in their slime,
Were mellow music match'd with him.

O life as futile, then, as frail!
O for thy voice to soothe and bless!
What hope of answer, or redress?
Behind the veil, behind the veil.

Here Lyell's apocalyptic scientific language has become the grim geological vision of Tennyson's 'desert dust' and 'iron hills'. Chambers's
remote evolutionary deity stands back from the human 'melée' and
the 'countless' suffering of individual lives. Herschel's astronomy and
beautiful stars disappear behind 'wintry skies'. Churches generate
'fruitless' prayers and theology. Babbage's Creator is revealed as a calculating machine, whose 'law' is mathematical. William Buckland's
dinosaurs become Tennyson's 'dragons of the prime', who in their
thrashings and slime, have a distant undersea echo of the Kraken. Yet
who or what exactly is the 'monster, the dream, the discord' of the penultimate stanza? Tennyson's masterful surging poem, in the very
desperate urgent rush of its syntax, its questions and exclamations,

leaves this disturbingly unclear. Is it Mankind, or Nature, or God, or even the Absence of God? He would leave this enigma, and its veiled answer, with an entire generation of Victorian readers.

But most immediately disturbing, and unavoidable, was Tennyson's recognition of the extreme, ruthless and inescapable physical violence in nature. Creation had no 'final law' of love at all. It had only a universal violence, that scientific study witnessed throughout the natural world. This was most particularly evident, and horrifying, in the animal and marine kingdoms. The phrase 'red in tooth and claw', now merely worn and proverbial, would then carry a profound emotional shock – dismay, even revulsion – to any reader in the 1840s and 1850s. It was a truly savage vision. Especially when specifically set against the assumed benevolence of God's creation. The famous stanza bears reconsideration:

Who trusted God was love indeed
And love Creation's final law—
Tho' Nature, red in tooth and claw
With ravine, shriek'd against his creed—

There was nothing remotely like this in *Vestiges*, or in the seventh edition of Mary Somerville's *On the Connexion of the Physical Sciences* (1846), or in the great Victorian naturalist P.H. Gosse's hugely popular *Glimpses of the Wonderful* (1845). Nothing, that is, until the promise of lucid violence contained in Darwin's calm phrase 'natural selection', enshrined in the subtitle of *On the Origin of Species by Means of Natural Selection* of 1859.

Yet there is a curious possibility that Tennyson may have been inspired by a phrase, not from any scientific study, but from Carlyle's latest political tract, *Past and Present* (1843), his further meditations on the Condition of England Question. In his introduction, Carlyle had invoked the myth of the Sphinx, and her unanswerable riddles. He suggested that nature revealed herself to modern man as being like the Sphinx, and was full of hidden cruelties. The passage is striking, not least for its unconscious misogyny. 'Nature, like the Sphinx, is of womanly celestial loveliness and tenderness; the face and bosom of a goddess, but ending in claws and the body of a lioness.' Carlyle then sharpened this idea in an additional passage. 'Answer her riddle, it is well with thee. Answer it not, pass on regarding it not, it will answer itself; the solution for thee is *a thing of teeth and claws*; Nature is a dumb lioness, deaf to thy pleadings, fiercely devouring.'

Tennyson can hardly have ignored this haunting phrase, though he stamped it absolutely as his own with the unforgettable blood-mark addition of the adjective 'red'. His line also runs on: 'red in tooth and claw *with ravine*'. The deliberate archaic word, meaning any predator's 'prey', but also echoing the idea of 'rapine', is given add-itional force and disturbance by its alliteration. Nature's inherent violence becomes vicious, relentless and inescapable.

But there was in fact one similarly grim, but intensely private rec-ognition of the truth, dating from some thirty years before Tennyson, and forty before Darwin. In March 1818 John Keats spent a day sit-ting on the rocks in Teignmouth bay, watching the breakers bursting among the 'green sea-weed'. That evening, from his lodgings, he wrote a verse letter to his old friend J.H. Reynolds, describing what he had seen and the shock of the experience. This remarkable letter was not published until Monckton Milnes's edition of Keats in 1848, but had Tennyson read it then, he would perhaps have recognised himself on the beach at Mablethorpe. Keats set the scene with decep-tive simplicity.

> ...'twas a quiet eve,
> The rocks were silent, the wide sea did weave
> An untumultous fringe of silver foam
> Along the flat brown sand; I was at home
> And should have been most happy, – but I saw
> Too far into the sea, where every maw
> The greater on the less feeds evermore. –
> But I saw too distinct into the core
> Of an eternal fierce destruction,
> And so from happiness I far was gone.

Keats's vision haunted and depressed him throughout the following day. It might be described as the pre-shock of evolution theory: a pre-monition of Darwin's 'natural selection' and Herbert Spencer's 'survival of the fittest' (1864). For Keats it took on specific images of the unexpected savagery of individual creatures – both on land and under the sea. Even the small companionable garden robin, in the act of hunting a worm, became like a huge ravenous snow leopard or 'ounce'. Keats was forced to reject the whole horror of it, dismiss it as a mere bad 'mood', and return to drafting *The Eve of St Agnes*.

> Still am I sick of it, and tho', to-day,
> I've gather'd young spring-leaves, and flowers gay
> Of periwinkle and wild strawberry,
> Still do I that most fierce destruction see, –
> The Shark at savage prey, – the Hawk at pounce, –
> The gentle Robin, like a Pard or Ounce,
> Ravening a worm, – Away, ye horrid moods!

But Tennyson could not turn away, could not dismiss the vision. His final hope of finding an answer 'behind the veil', was the fragile ges-ture slipped into his last line. But together these seven devastating stanzas of interrogation, half furious accusation and half desperate prayer, became one of Tennyson's most powerful and openly agnostic poems. It managed to voice the scientific and spiritual experience – fierce discovery and terrible doubt – of an entire generation. It is especially remarkable that Tennyson carefully omitted it from the first private printing of *In Memoriam*. Like Darwin struggling with early drafts of *Origin of Species*, for a moment he considered it too challenging. His dilemma was revealed in an earlier letter to Emily Sellwood in the 1839 exchange: 'God cannot be cruel. If he were, the heart could only find relief in the wildest blasphemies, which would cease to be blasphemies ... I dare say my own progress is impeded by holding this hope however dimly.'

14

Recognition

It becomes clear in retrospect that these central Evolution poems were the consequence of a long line of increasingly agnostic thinking that had been silently emerging from the late 1830s. They had gathered powerful and contemporary scientific imagery, steadily forming in Tennyson's 'butcher's notebook'. They had slipped in among the purely personal elegies, and frequently burst out of them. An early glimpse of this process appeared in this passage, which Tennyson finally placed in *In Memoriam* No. 3:

> 'The stars,' she whispers, 'blindly run;
> A web is wov'n across the sky;
> From out waste places comes a cry,
> And murmurs from the dying sun:

> 'And all the phantom, Nature, stands—
> With all the music in her tone,
> A hollow echo of my own,—
> A hollow form with empty hands.'

This was Herschel's enlarged astronomy of 'island universes', with another strange premonition of William Thomson's theory of 'the heat death of the sun', much discussed at this time. Gradually the true implications of Hallam's own death, and whether it signified immortality or extinction, had become a dominating question about the true meaning of the universe itself. These poems now gravitated towards the central sections of *In Memoriam*, and transformed its elegiac significance. Not only 'red in tooth and claw', but such phrases as 'the dust of continents to be' (No. 35) or 'Life, a fury slinging flame' (No. 50); or 'the secular abyss to come' (No. 76), would spring out upon the reader suggesting a new kind of intellectual demand, and a new kind of philosophical crisis.

Similarly the fascination with being buried alive, which had haunted Tennyson at High Beech, made a significant appearance in another early group (Nos 31–34). This circled uneasily round the celebrated New Testament story of Christ raising Lazarus from the dead. 'When Lazarus left his charnel-cave . . .' (No. 31). The miracle of raising the dead – either literally or metaphorically – obviously bore on Tennyson's thoughts about Hallam. But he took no easy comfort from it, and began by retelling the story from an unexpected and sceptical angle. Again Tennyson used the form of a question. What did Lazarus's sister Mary feel about it?

> When Lazarus left his charnel-cave,
> And home to Mary's house return'd,
> Was this demanded—if he yearn'd
> To hear her weeping by his grave?

Tennyson was fascinated by the detailed way in which the miracle was reported in chapter 11 of Saint John's Gospel, and also by what was left out. (For example, his other sister Martha says she fears Lazarus's corpse would be decomposed – 'Lord, by this time there will be an odour, for he has been dead four days' (St John, 11.40). Tennyson still longed to believe the resurrection story, but instinctively questioned it with a kind of scientific logic. The poem dramatises this questioning in four vividly compact stanzas, just sketching in the visual setting of the Mount of Olives with its 'purple brows'. If Lazarus was dead and buried in a tomb for four days, *what exactly happened* to his body or to his spirit, during the interval? The unflinching question is put into his sister Mary's mouth, with just a hint of domestic impatience:

'Where wert thou, brother, those four days?'
There lives no record of reply,
Which telling what it is to die
Had surely added praise to praise.

Neither Lazarus, nor Christ, are allowed to 'tell', or give any explanation in Tennyson's poem. Even though such an explanation would 'surely' have enhanced the wonder of the miracle. This blank silence, this absolute lack of reassurance, let alone any rationalisation, becomes the driving force of the poem. There is no statement of belief, just the presentation of a blank fact. This prompts Tennyson to conclude with another of his suspended final lines, but with a hint of irony. He unexpectedly turns attention away from the problematic miracle to the reporter of it, to the Gospel writer. St John may himself have doubts:

Behold a man raised up by Christ!
The rest remaineth unrevealed;
He told it not, or something sealed
The lips of that Evangelist.

Tennyson would much later say confidently that 'the cardinal point of Christianity is the life after death'. But there is no such confident faith in this poem. There is a hesitation, there is 'something sealed'. The very word 'sealed', used of the writer's mouth, is noticeable because it is a word also associated with the tomb, with closing the 'charnel-cave'. It is not mere doubt, it is a form of agnosticism that Tennyson implies. The Gospel writer, the Evangelist (the 'bringer of Good News'), is refusing to commit himself beyond the evidence.

Tennyson's evolving agnosticism in *In Memoriam* would eventually have a profound impact. It was T.H. Huxley, Darwin's bulldog, the secretary of the Royal Society, who formally defined the idea of Victorian agnosticism in 1869, and then in several later essays. 'It simply means that a man shall not say he knows or believes that which he has no scientific grounds for professing to know or believe.' Huxley also pointed out the unique representative power of *In*

Memoriam. 'Tennyson is the only modern poet, in fact the only poet since Lucretius, who has taken the trouble to understand the work and tendency of the men of science.' Tennyson himself defined his own agnosticism in another phrase that became famous, 'honest doubt'. It appears in No. 96:

> Perplext in faith, but pure in deeds,
> At last he beat his music out.
> There lives more faith in honest doubt,
> Believe me, than in half the creeds.

The topics of materialism, scepticism and evolution always remained with Tennyson, and would lead to many later discussions in the 1860s with the physicist and dauntless Alpine climber John Tyndall. Tyndall climbed both physical and intellectual mountains, with great daring and directness. One of his most famous revelations was an early explanation of why an unclouded sky looks blue. He also agreed with Tennyson's view: 'No evolutionist is able to explain the mind of Man or how any possible physiological change of tissue can produce conscious thought.'

In 1868 Tyndall would invite Tennyson to become a founder member of the Metaphysical Society, a 'semi secret' scientific group meeting in London dedicated to debating the claims of science and religion. He would also quote Tennyson's poetry in his famous 'Belfast Address' of 1873, on the scientific status of the Book of Genesis. Although a Darwinian, Tyndall worried away at the problem of human consciousness, as Tennyson had done. 'But the passage from the physics of the brain to the corresponding facts of consciousness is unthinkable, granted that a definite thought and a definite molecular action in the brain occur simultaneously. We do not possess the intellectual organ, nor apparently any rudiment of the organ, which would enable us to pass, by a process of reasoning, from the one to the other.'

Tennyson still did not reveal any of this new challenging poetry, even to close friends like FitzGerald. So when a first, popular assessment of his career appeared in the summer of 1844, it was positive but conventional. *The New Spirit of the Age* was a collection of contemporary biographical essays by a young satirical journalist Richard Horne and by Elizabeth Barrett (still Tennyson's fan). Their lively group portrait of the literary movers and shakers of the day was inspired by Hazlitt's *Spirit of the Age* of 1825. Horne had previously published an epic poem *Orion* (though it sold for a farthing, it was not a success) and shortly afterwards emigrated to Australia. Elizabeth Barrett had just published her debut volume of *Poems* (1844) and was still living with her family in Wimpole Street, and was not to elope to Italy with Robert Browning for another two years. Only then would she become 'EBB'.

Their collection was eclectic, covering both poetry and prose, including short studies of Dickens and Browning, as well as Harriet Martineau and Mary Shelley. Though unlike Hazlitt's broad sweep of poets and philosophers, there was nothing on such contemporary scientific thinkers as Lyell, Herschel, or Mary Somerville. A thoughtful, but distinctly post-Romantic chapter was given to Tennyson, described as 'the master of many spells'. He was a poet 'slowly, calmly' achieving public recognition, and gradually distinguishing himself from Keats, Coleridge and Shelley. 'In music and colour he [is] equalled by Shelley; but in *form*, clearly defined ... Tennyson stands unrivalled.'

Horne shrewdly suggested that Tennyson had been through a period of personal upheaval and turmoil since the 1830s. In consequence, in his latest poetry Tennyson had become 'awake to the actual world'. He was 'awake after the storms, after the wrecks, the deepest experiences of life'. In the ten years' interval he had known and suffered, Horne particularly praised the 'preternatural spirit' of enchantment that Tennyson could throw over his exact observations of 'simple incidents and objects of reality'.

With a keen journalist's eye, Horne was another of the early critics to appreciate the vivid symbolic life-force of physical objects, in the otherwise 'dim and desolate' poem 'Mariana'. Like Leigh Hunt, he marvelled at the descriptions of the tinkling latch, the buzzing fly and the shrieking mouse in the wainscot. He also identified Tennyson's power to evoke 'the fantasy of things', and his instinct to create dream or mythological figures, going right back to the Kraken. 'His Muse was his own Lady of Shalott – she was metamorphosed into the Merman and the Mermaid; they reunited at the bottom of the sea in the form of the Kraken, and lay swelling with the sense of ages beneath enormous growths floating upon the surface.' Referring to the 1842 volume, and giving an opinion obviously shared with Elizabeth Barrett, Horne added: 'Why this "Kraken" should have been omitted in the present collection puzzles and annoys us . . .'

Tennyson was evidently aware of the forthcoming essay, and in August invited Horne to 'come and drink tea some night' to discuss it. He rather hoped to 'talk him out' of including his actual portrait in the book, and suggested they meet at Lushington's rooms at 2 Mitre Court Buildings. It was during this tête-à-tête over the teacups – which did not include the book's other editor, Elizabeth Barrett – that Tennyson must have revealed something of his troubled years to the sympathetic Horne. But he still did not speak of the elegies.

In January 1845 Tennyson finally left his secret medical retreat at Cheltenham spa and descended upon London for some serious dining out. He flitted between Spedding's apartment in Lincoln's Inn, Henry Lushington's chambers at Mitre Court, and FitzGerald's new Bloomsbury lodgings in Charlotte Street. Once again he seemed oddly homeless in London. When no friend was conveniently to be found, he resorted to the Hummums Turkish Bath establishment off Covent Garden, where beds could be hired by the night.

Turning up unannounced one night at FitzGerald's, he was in better spirits about his financial affairs and at last began to talk about his new poems for *In Memoriam*. 'Tennyson suddenly came up fresh from Water Cure, and drinking a bottle of wine daily. The man who swindled his money died suddenly – and AT is come in (I hope) for a Life Insurance, or Policy, or whatever the Devil they call it, so that Apollo seems to have directly interfered, and slain the offending Doctor with one of those sudden painless darts which Homer tells of. AT has near a volume of poems – elegiac – in memory of Arthur Hallam.' Fitz had his usual tender but half-grudging response. 'Don't you think the world wants other notes than Elegiac now? [Milton's]

Lycidas is the utmost length an elegiac should reach. But Spedding praises: and I suppose the elegies will see daylight – public daylight – one day.' It was clear from this that Fitz still understood the poems to be pure, traditional literary elegies. Tennyson had evidently not spoken of the 'other notes', of the new controversial, agnostic themes.

His old Cambridge friend Brookfield, thinking he needed feeding up and cheering up (though no longer in need of opium), took Tennyson to dine at the fashionable Reform Club, recently restored and rebuilt (1841) at No. 104 Pall Mall, Piccadilly. 'After dinner, in contempt of all formality, Tennyson persisted in resting his feet upon the table. His friends remonstrated, in vain, until one of them said, "Take care, Alfred, they will think you are Longfellow". Down went the feet.' The sly suggestion that he might be taken for the popular American poet, who had recently achieved huge success with his collection of backwoods *Ballads* including 'The Village Blacksmith' (1842), was an effective reproach. In fact, unknown to the Trinity friends, Longfellow was one of Tennyson's most eager Transcendental readers at Harvard.

Tennyson was also talking more freely about his poetry in general, and his critical standing, to other acquaintances. One was the young Irish poet Aubrey de Vere, whom he had failed to meet on his family estate in Limerick. De Vere's friendship with the astronomer Sir William Rowan Hamilton, and his studies at Trinity College, Dublin (where he had read 'Kant and Coleridge'), and his newly published verse *The Search after Proserpine: Memories of Greece* (1843), all interested Tennyson, who was already something of a legend to the thirty-year-old de Vere. The sparky Irishman kept a lively diary, recording political and literary gossip, the 'lions' of the London salons, and what he considered the best speeches at the House of Commons. Among these highlights he included some memorable recitals from Tennyson.

On 17 April 1845 he noted: 'I called on Alfred Tennyson and found him at first much out of spirits. He cheered up soon, and read me some beautiful Elegies, complaining much of some writer in *Fraser's Magazine* who had spoken of the "foolish facility" of Tennysonian poetry.' Tennyson said 'he was dreadfully cut up by all he had gone through'. Much impressed by Tennyson's complaints and confidences, de Vere returned the next evening when Tennyson '*crooned* out his magnificent Elegies till one in the morning'. The following day de Vere also listened to readings from what he called 'the University of Women' – evidently drafts of *The Princess*.

But de Vere was never sure of Tennyson's mood. Only a few days later Tennyson burst out that 'he could no longer bear to be knocked about the world, and that he must marry and find love and peace or die'. But then he added that 'he was very angry about a very *favourable* review of him. Said he could not stand the chattering and conceit of clever men . . .'

In fact all this indicated a subtle but growing self-confidence during this spring of 1845. Tennyson's taste for reciting his work in public (not merely to close friends or admiring sisters, as before); his willingness to talk to literary journalists like Horne; or his eagerness to discuss his critical reviews, and even his romantic frustrations, with new acquaintances like de Vere; all suggested his new sense of himself as a recognised poet with (as he hoped) an eager potential audience. No longer a private, or even an entirely withdrawn figure, Tennyson was now prepared to recruit a new circle of supporters, and de Vere had become one of the early enthusiasts.

Intimate records of Tennyson in London, except for Carlyle's prognostications and FitzGerald's candid notes, had been a comparative rarity. Yet now they were becoming more frequent, especially as Tennyson's physical appearance conformed more than ever to the imposing, post-Romantic archetype; or rather perhaps, invented it. This was quite different from Byron's beautiful, slight satanic figure, so neat and stylish, which had held Regency society in thrall to the end of the 1820s. By contrast Tennyson was large, chaotic, ill-dressed and physically imposing when he entered a room. He struck most observers as tall, dreamy, dusky, distracted and aloof; with his 'almost Spanish complexion', and above all his superb mane, his 'massive abundance', of fine dark curling hair. There was also an element of shambling, and a distinct whiff of tobacco. Over the next decade, as fame approached, such records would become a positive flood of Victorian anecdotes, journal entries and diary records, particularly by younger disciples such as de Vere, William Allingham, Francis Pal-grave, and Edmund Gosse (and eventually Henry James).

Yet such accounts varied wildly. At the very time that de Vere found Tennyson complaining and much out of spirits, FitzGerald dined with him in Bloomsbury, 'looking well and in good spirits'; and enthusing about 'two hundred lines of a new poem [*The Princess*] in a butcher's book'. He was also quite frisky after Cheltenham. 'He and I made a plan to go to the coast of Cornwall or Wales this summer, but I suppose we shall never manage to do it . . .'

In Mitre Court, Edmund Lushington was also read parts of this new poem about a women's university, of which 'he had heard noth-ing before'. He was amazed to learn that it had been inspired by the Mechanics' Institute's summer fair held at Park House four years pre-viously, as Tennyson had never mentioned it at the time.

Tennyson also began accepting more formal invitations in London. One was a grand literary summons to visit Samuel Rogers, regarded as the ancient doyen of the Romantic poets, now aged eighty-two. His address at 22 St James's Place was referred to with respect as the Hal-lowed Temple of the Muses. Tennyson was supported by Spedding and Lushington in this ordeal. At dinner they met the diarist Crabb Robinson, himself seventy, friend of Wordsworth and Coleridge from thirty years before. Robinson was impressed by all three youthful tal-ents of the new generation, 'belonging to Young England'. He felt Tennyson was 'by far the most eminent of the young poets. His poems are full of genius, but he is fond of the enigmatical, and many of his most celebrated pieces are really poetic riddles.' These riddles perhaps referred to Tennyson's scientific speculations. They had a long tête-à-tête about Goethe, but were interrupted by a surprise lady guest who had come 'on purpose to see Tennyson'. This turned out to be 'the much eulogised and calumniated' poet Caroline Norton, wrongly accused of adultery with Lord Melbourne, and later campaigner for married women's rights. Tennyson was clearly moving up in society.

Tennyson paid another of his surprise visits to Thomas Carlyle in Chelsea, but finding him out, spent the whole evening alone with Carlyle's wife, an unusual breach of Victorian social etiquette, though perhaps allowable for a poet. Jane Carlyle, in her mid-forties, hand-some and witty, and already famous for her mischievous letter-writing, took full advantage of the occasion. Welcoming him into the parlour, she sat him firmly down and 'got out pipes and tobacco – and brandy and water – with a deluge of tea over and above'. She had heard that Tennyson was 'dreadfully embarrassed with women alone', but to her relief she found him delightful company, attentive and intimate, though he did smoke without stopping 'for three mortal hours'. He was 'talking like an angel – only exactly as if he were talking with a clever <u>man</u> – which being a thing I am not used to ... strained me to a terrible pitch of intellectuality'. Nonetheless this was far better than 'men always *adapting* their conversation to *what they take to be* a wom-an's taste'. The visit was a success, and on his return, Carlyle was impressed to find the empty parlour thick with tobacco smoke, and his wife triumphant. He later boasted to FitzGerald of this happy solo visitation by 'the fiery Son of Gloom'.

Tennyson's new importance as an author was such that in August 1846 he was taken on a leisurely tour of the Swiss Alps by his pub-lisher Edward Moxon. They sailed down the Rhine, travelled by coach and steamer from Lucerne to Geneva, and walked together through the beautiful Lauterbrunnen valley. It was the first such extended European journey he had made since his Rhine trip with Arthur Hallam thirteen years previously, and it was full of sup-pressed echoes from the past. He was also travelling in the footsteps of Shelley, as Hallam would no doubt have reminded him. Tenny-son kept a sketchy journal, packed with droll enough domestic details, like the man who snored in the hotel room above him, 'enough to shake the walls of Jericho'. Yet he was thrilled by the wonderful Alpine light. 'Far off Jungfrau looking as if delicately pencilled ... strange aspect of hill, cloud and snow as if the moun-tains were on fire ... We watch the clouds opening and shutting as we go down and making framed pictures of the lake ...' On a more earthly note, there was also much fine dining: 'souper, eels, chops, eau de vie, 2 flasks Liebfraumilsch'. The portly Moxon was a jolly travel companion, but evidently no Hallam. Together they could at least celebrate many of the stock complaints of English men abroad: 'bad beer ... sour ill looking maid ... bad dinner ... infamous beds swarming with fleas'.

Despite these disenchantments, they discussed Tennyson's thoughts on the higher education of women at length. There were also some encouraging encounters bearing on this subject: 'Agreeable Swiss young lady to whom I quoted Goethe and she spouted William Tell – sorry to lose her.' Tennyson was even inspired to write a new lyric poem at Grunewald, placing an ideal woman in the magic Alpine landscape above them: 'Come down, O maid, from yonder mountain height'. Moxon was delighted to hear that this might eventually find its place in the new – and otherwise rather severe-sounding – 'Women's University' poem.

The new poem was presented as a simple 'Shepherd's Song', celebrating the Alps in all their wild magnificence, untouched by modern tourism. The shepherd begs his love to descend from the dizzy heights and 'monstrous ledges', and dwell safely as his wife in the valley below, 'For Love is of the valley, come thou down ...' But the song is far from simple, and with its rich operatic language, it sets out a kind of metaphysics of altitude. The Alpine landscape, in its extreme geography, its inhuman combination of absolute Depth and perilous Height, expresses for Tennyson some fundamental division of the heart. It is as if human love belongs to the lush valley below, and not to the 'cold splendour' of the peaks above. These terrible bare icy heights are nightmare places, zones of spiritual bleakness, with

> Their thousand wreathes of dangling water smoke
> That like a broken purpose waste in air.

These indeed are some of the places visited in the bleaker sections of *In Memoriam*. But here the Shepherd's Song closes with an evocation of the safe and sweet harmonies of the Alpine valley. A magic soundscape is conjured up in what became three of Tennyson's most famous onomatopoeic lines:

Myriads of rivulets hurrying through the lawn,
The moan of doves in immemorial elms,
And murmuring of innumerable bees.

Those elms and those 'innumerable bees' somehow sound more English than Swiss, as if some ideal country park is being evoked. But the poem has a classical source in the lovelorn shepherds of Theocritus, and the last line was inspired by a passage in Virgil's pastoral *Eclogues*, where the Latin for bees is *apes*. This produced a rare joke in Tennyson's journal on 10 August. Moxon had remarked on their fellow tourists *swarming* noisily round them in the evening at their inn. Recalling his 'innumerable bees', Tennyson replied whimsically with a Latin pun: 'Very feeble sunset – tea – infernal chatter of innumerous *apes*.'

Later they ran into Charles Dickens at a hotel in Lausanne and the joke recurred in a different form: Dickens gave them 'some fine Rhine wine, and cigars *innumerable*'. Tennyson perhaps thought the great Dickens was being mildly patronising. He was at the height of his popular success as a novelist, and announced to all that he was working hard – 'racking my brains' – on *Dombey and Son*. He seemed amused to find Tennyson was holidaying with his own publisher. He noted that Moxon sported 'a very limp and melancholy straw hat', and was 'an odd companion for a man of genius'. Tennyson himself seemed 'quite at home' with this arrangement, but kept producing from his pocket the shredded remains of a five-pound note, which he said was reserved solely for 'emergency' travel expenses. The implication was that Moxon, his publisher, was financing the entire trip. In Dickens's company Tennyson was careful not to play the poet, and made no reference to his Alpine visions. When he did talk about the fantastic Mer de Glace, Dickens thought his plain descriptions were 'that of the man who had been to Niagara and said it was nothing but water'. In fact Tennyson knew a great deal about Niagara from his reading of Lyell's *Geology*, but evidently chose to say nothing about this to Dickens.

Back in London he was reassured to learn that the Civil List pension of £200 was confirmed for life. He understood this was partly due to the lobbying of his old Cambridge friend Monckton Milnes, now embarked on a successful career in Parliament. Anxious to support the Arts in Britain, Milnes read the poem 'Ulysses' out loud to Sir Robert Peel, the prime minister, as an example of the best in contemporary poetry. Peel (who had recently founded the Metropolitan Police) claimed to know nothing about Poets, but said he knew a 'good speech' when he heard one, and approved the pension. This seemed a further hint that Victorian taste might be steadily coming round to Tennyson's work. There seemed no doubt that since the 1842 *Poems in Two Volumes* public recognition was growing.

Yet he felt mixed about this early glimpse of fame. 'People fete and dine me every day,' he wrote that autumn, 'yet I am somewhat unwell and out of spirits.' There was inevitably less time for some of the old original circle of faithful friends, who had had less worldly success (so far). Edward FitzGerald felt this particularly. He had heard absolutely nothing from Tennyson since rumours of the Swiss summer jaunt had reached him, apparently with Dickens and other glamorous literary figures. Tennyson waited until November to write him a long and surprisingly passionate explanation, clearly thinking that with Fitz, attack was the best line of defence.

'Why should you say that it is in vain to expect a line from me? Don't you know that I esteem you as one of those few friends who would still stick to me though the whole polite world with its great idiot mouth (wider than ever was a clown's at a fair, staring at a show) howled at me. If I write not, it is not because I do not love and remember, but from some small absurd cause of not having pen, or paper, or ink, or a Queen's Head [stamp] in reach ... Yet I write to you, old Fitz: after that say I won't write.'

He added for good measure that Switzerland was not so exciting after all: 'mountains, great mountains, disappoint me'. The Swiss people were 'miserable-looking poor devils', and the most notable thing about Dickens was that he was 'very hospitable' with biscuits, cigars and Liebfraumilch 'which is being interpreted *Virginis Lac* as I dare say you know'. He signed off rather briskly, 'Ever thine.' It was a gallant if not entirely convincing performance, and Fitz was appeased for the time being. But he must have worried when he read that word 'esteem' (until it became 'love and remember'); and been puzzled by the very un-Alfred-like image of public recognition being 'a great idiot mouth'.

Yet his high Alpine adventures, with such wonderful unclouded views of the stars (when Mont Blanc had not 'kept his nightcap on'), had indeed been inspiring for Tennyson. That winter he returned again to his study of the night skies. He had been excited to read about a new giant telescope, named the 'Leviathan', far bigger even than Herschel's instrument at Slough. It had been constructed by the engineer and philanthropist Lord Rosse on his large estate at Parsonstown, just south of Dublin. Begun in 1840, the long series of trials had lasted over five years, and had obviously been followed by de Vere, with his astronomical connections in Ireland. The Leviathan's dimensions were astonishing: the main tube was seventy-two feet long, with a reflector mirror six feet in diameter, consisting of a speculum metal lens five inches thick and weighing three tons. The whole was supported in a movable lattice iron framework, a marvel of early Victorian engineering, which was cradled in a huge open casing of brickwork turrets. At a distance it looked more like an enormous artillery piece, its barrel aimed at the heavens. Now at last its tremendous magnification powers were beginning to yield fantastic discoveries among distant and hitherto unknown star clusters or nebulae.

William Rosse's observations went far beyond the Milky Way that Herschel had identified a generation before. Rosse gradually filled this new Deep space with strange new concepts and names, and made detailed drawings of what he observed. The first new object he discovered was a star cluster shaped like a spider, which quickly became known as the Crab Nebula. Rosse then found that many very distant star clusters were actually individual galaxies, and proposed that most had a dynamic 'spiral' structure, revolving steadily upon themselves through unimaginable periods of Deep time. He named the first of these spirals he had located as the 'Whirlpool Galaxy', now known to modern astronomy as Messier 51. His Leviathan remained the biggest and most powerful telescope in the world for the rest of the century (until 1917). Finally completed in February 1845, its discoveries were just becoming known to the general public, and catching the imagination of many writers like Tennyson (and not least the young Marian Evans before she became George Eliot).

Among the most responsive was Thomas de Quincey who, like Tennyson, was fascinated by these latest developments in astronomy and cosmology. In 1846, while a poverty-stricken freelance journalist, and still dependent on opium, he lodged for three months in the attics of Glasgow Observatory thanks to its kindly director John Pringle Nichol. This led to an extraordinary encounter between scientific and literary minds. Nichol was then Regius Professor of Astronomy at the city's university, where Edmund Lushington was also Professor of Greek. Nichol was already a widely read popular science writer, whose early books were generally known by Tennyson and the Lushingtons. But it was Nichol's new cosmological study, *Thoughts on Some Important Points Relating to the System of the World* published in 1846, which first popularised Lord Rosse's work. The book was immediately purchased by Tennyson, and soon after reviewed by de Quincey in *Tait's Edinburgh Magazine*. FitzGerald also saw de Quincey's review, which he thought humorous but very odd.

De Quincey's long article was entitled 'System of the Heavens as Revealed by Lord Rosse's Telescope' and took the dramatic form of an imaginary journey into an 'infinite universe' at 'unutterable' speeds through blazing suns and glimmering stars, accompanied by an angel, but using all Rosse's scientific observations. These marked 'a new era for the human intellect'.

De Quincey first considered the new metaphysical visions of Time and Space opened up by Kant's cosmology. He then described Herschel's astronomy and compared it with the latest discoveries by Rosse. In increasingly apocalyptic style he concentrated on one of Rosse's most dramatic astronomical objects: not the Crab, but the huge Orion nebula, as magnified by the Leviathan, and as drawn by Herschel. It was 'famous for its frightful Magnitude and for the frightful Depth to which it is sunk in the Abyss of the heavenly wilderness'. Here de Quincey moved from pure science to science fiction. He claimed to discern in it, not merely a new star cluster, but the outlines of *a monstrous head*. 'The reader must look to Dr Nichol's book, at page 51, for the picture of the abominable apparition.' He compared it to the Egyptian statue of Memnon II in the British Museum (the same that Shelley had evoked in 'Ozymandias'), and imagined it to be some huge creature of mysterious evil power, approaching earth out of the depth of space. 'You see a head thrown back, and raising its face, (or eyes, if eyes it had,) in the very anguish of hatred, for some unknown heaven . . .' What de Quincey had found was, in effect, a menacing version of Tennyson's Kraken in deep space.

Cosmology, benign or malignant, fantastical or empirical, always led Tennyson back to the eternal questions. They would resurface continuously in the later sections he added to *In Memoriam* throughout the remainder of the 1840s; and sometimes obsessively in his talk. So in May 1846, just before the froideur provoked by the Swiss tour, FitzGerald had witnessed Tennyson in a fierce debate about science and the possibility of an afterlife with an aroused and implacably sceptical Carlyle. The argument evidently went on at stunning length, and certainly exhausted FitzGerald. 'I met Carlyle last night at Tennyson's, and they two discussed the merits of this world and the next, till I wished myself out of <u>this</u>, at any rate. Carlyle gets more wild, savage and unreasonable every day . . .' FitzGerald reported Carlyle as relentlessly mocking Tennyson for having notions of immortality still based on the Bible. 'Eh! old Jewish rags!' Carlyle jeered. 'You must clear your mind of all that. Why should we expect a hereafter?'

They finished the argument by telling each other two contradictory versions of the Tale of the Traveller at the Inn, a parable of what happens after death. In Carlyle's version, each traveller spends only one night at the Earthly Inn, then the next morning gets up and disappears into oblivion without a word. The following night the exactly same thing happens and so on in endless, meaningless repetition. Carlyle implied by this bleak tale a purely materialist view of life, and even an atheistical conclusion. But Tennyson responded more hopefully, even if agnostically. He imagined a traveller who 'leaves the inn

in the morning, and goes on his way rejoicing, with the sure and certain hope and belief that he is going *somewhere*, where he will sleep
the next night'. FitzGerald was delighted by this rare glimpse of Tennysonian optimism, with his liturgical 'sure and certain hope', and
even more by the end of the whole endless argument. 'You have him
there,' he concluded encouragingly.

The next day Fitz briskly took Tennyson 'to get a pint or two of
fresh air at Richmond'. But perhaps his impatience with so much
metaphysics explains something of the uneasy silence that followed
this comradely conclusion, in the long autumn after the Swiss tour.

15

University

What Tennyson published next was not a first selection from his *In Memoriam* elegies, as FitzGerald and other friends expected and half-dreaded, because it menaced a return to mournful themes. It was something quite different and much more forward-looking. It was his long-promised ideological poem about the education of women. Its experimental form also took them by surprise: a long semi-dramatic work in seven parts, mixing modern with medieval settings. This most controversial, and in some ways least typical, of all Tennyson's longer poems – and which provoked the most varied responses – was published by Moxon in December 1847 under the innocent and apparently anodyne title, *The Princess: A Medley*.

Tennyson announced this in conciliatory style to FitzGerald after these months of silence. 'My dear Fitz. Ain't I a beast for not answering you before? Not that I am going to write to you now . . . I am going to bolt as soon as ever I can . . . I would go to Italy if I could get anyone to go with me which I can't . . . Here I am alone and wish you were with me . . . My Book is out and I hate it and so no doubt will you. Never mind you will like me none the worse and now goodnight . . . Ever yours, A. Tennyson.'

Of course the book had originally been conceived as long ago as 1839, with various changing titles over the intervening eight years, such as 'The University', 'The New University', 'The Commonwealth of Women', or 'The University of Women'. It had been initially influenced at that time by the example of his sisters at High Beech, and

their all-women reading and discussion group, the 'Husks'. His sister
Cecilia's correspondence had steadily expanded the group to include
Emily Sellwood, Julia Hallam (Arthur's sister) and several other
friends and cousins. One of the poem's leading characters, a tall and
outspoken teenager called Lilia, may even have been directly mod-
elled on a specific member of the group, Mary Neville, who happened
to be six foot and particularly vehement in her feminist opinions.

But its literary roots certainly go back as far as a passage in
Samuel Johnson's philosophical romance of 1755, *The History of Rasselas, Prince of Abyssinia*. This was a gently satirical study of dreams
and delusions, and mankind's perennial search for happiness. In
the concluding chapter 49 ('In which Nothing is Concluded'), the
Prince and his sister Princess Nekayah, together with the poet
Imlac, having sought advice from many savants (including a philosopher, an engineer and an astronomer), see only one obvious
hope for future progress. 'The Princess thought that of all sublun-
ary things Knowledge was the best; she desired, first, to learn all
the sciences, and then proposed to found a College of learned
Women in which she would preside . . .' But of course this too may
be delusory.

Rumours about Tennyson's fascination with the subject had circulated around the London literary world for several years. Edmund
Lushington had heard about it in 1845, and Elizabeth Barrett in 1846.
She wrote wonderingly to Browning in January that year: 'It is in
blank verse and a fairy tale and called *The University*, the university
members being all female . . . It makes me open my eyes. Now isn't
the world too old & fond of steam for blank verse poems in ever so
many books to be written on the fairies?' But of course she had not
actually read it at that date.

By August 1847 Barrett's friend Mary Russell Mitford had heard
more intriguing, but still second-hand accounts, based on first corrected proofs of Tennyson's work going to Moxon. 'His new poem is
a "Commonwealth of Women". – A man gets in, and you can imagine
the *dénouement*. It is said to be very beautiful, but not favourable to
female intellect or character.' Even more suggestive, Tennyson was 'a

great torment to Mr Moxon, keeping proofs a fortnight to alter, and then sending revises'.

FitzGerald said Tennyson had read him 'the first three books' in May 1847, which presumably included the Prologue based on the modern Science Fair at Lushington's estate. He was initially pleased that Tennyson was at last engaging with a contemporary issue, but far from shocking he found it 'rather monotonous' and claimed to have fallen asleep during Tennyson's reading. He passed this off as his own fault, rather than a criticism: 'I was knocked up: and also I may be fast growing out of my poetical age. Everyone likes the poem.' It is also possible that FitzGerald did not really approve of education for women in the first place. Yet he knew Tennyson had returned to the topic again and again over eight years, and was never quite certain how to present it himself. That uncertainty remained with the final version, given its decorative title and apologetic subtitle: *A Medley*.

In reality women's education was a deeply divisive subject, uneasily debated throughout these years, including in the House of Commons in May 1836 (where speeches were 'indecent ... flippant ... unmanly'); and inevitably bearing on the nature of Victorian marriage. In Carlyle's sense it was a Condition of England Question. No full university courses were open to women until a generation later, with the foundation of Girton College, Cambridge in 1869. This was followed by University College, London in 1878; and Somerville College and Lady Margaret Hall, Oxford in 1879. The philosopher Mill, who had earlier praised Tennyson's poems, was now himself working on the question with his beloved feminist partner Harriet Taylor. But their views were held back as too controversial to publish, though Taylor first submitted an essay 'The Enfranchisement of Women' to the *Westminster Review* in July 1851.

Mill did not dare to publish his controversial polemic *On the Subjection of Women* until 1869. In chapter 1 he would set out his fundamental position. 'Until conditions of equality exist, no one can possibly assess the natural differences between women and men, distorted as they have been. What is natural to the two sexes can only be found out by allowing both to develop and use their faculties freely.' Such free

development of faculties naturally implied higher education. Mary
Somerville expressed exactly the same view in her *Recollections*, but
again did not publish until 1873.

By contrast Tennyson's proto-feminist character, the teenage Lilia,
leaps straight into the fray, declaring the embattled position early on
in the Prologue, in response to her host's patronising and teasing:

> Quick answered Lilia: 'There are thousands now
> Such women, but convention beats them down;
> It is but bringing up; no more than that.
> You men have done it – how I hate you all!
> Ah, were I something great! I wish I were
> Some mighty poetess, I would shame you then,
> That love to keep us children! O I wish
> That I were some great Princess, I would build
> Far off from men a college like a man's,
> And I would teach them all that men are taught;
> And twice as quick! . . .'

Lilia's dream of building just such a college 'far off from men' and
directed by 'some great Princess' becomes the central drama of Tennyson's poem. The fundamental question at issue was the rights of
women to a true higher education ('all that men are taught'), not
merely to conventional lessons in music, drawing and botany. The real
challenge was to give a full curriculum of adult lectures to women-
only audiences: to take them on field trips, to teach them the new
astronomy, mathematics and geology, and encourage genuine intellectual independence and debate. This challenge led directly to the
foundation of Queen's College, the first all-woman institute for higher
education in England, at 45 Harley Street, London in this very same
spring of 1847. It was 'empowered to issue Certificates of Education' to
women in all subjects.

Its founder and first director was the idealistic Frederick Denison
Maurice, a Christian Socialist and champion of the liberal arts, who
had been to Trinity College a few years before Tennyson, a member

of the Apostles, and who now openly courted controversy in aca-
demia. He later became a great friend of Tennyson's, and when in 1854
he was dismissed by London University for his unorthodox and liberal views on education, expressed in his *Theological Essays* (1853),
Tennyson dedicated an amusing poem to him, praising Maurice as
one of the 'honest few / Who give the Fiend himself his due'.

Maurice insisted on teaching science and maths, ridiculing the
objections that it would be mentally 'damaging' for women. Instead
he insisted the professors of Queen's College should be highly motivated, and their pupils above any fashionable ideas of success. 'They
are not to study … arithmetic, or language or literature or history, in
order to shine or be admired; that if these are their ends, they will not
be sincere in their work or do it well.' Among many outstanding
young women who later attended the college, each thirsting for their
own kind of independence, would be Gertrude Bell (1894), Katherine
Mansfield (1903) and Jacqueline du Pré (1959).

The idea of Queen's College was perhaps one model for Tennyson's poem, and Maurice in turn would refer to the poem itself in his
opening lecture on 'The Objects and Methods' of the college. It is also
possible that the female professors of *The Princess* may reflect what
Tennyson knew of the new generation of contemporary 'bluestockings'. Among them was the mathematician Ada Lovelace; the poet
Caroline Norton, who had come expressly to meet him at Samuel
Rogers' dinner party; Harriet Martineau whose essays he had read in
the *Athenaeum*, describing them to his Aunt Russell as 'very *wonderful*'; and of course Mary Somerville whose scientific work he already
knew in considerable detail.

But Tennyson's inspiration also went further back into feminist
history. He wrote in a headnote to one later draft of the poem: 'I
believe the *Vindication of the Rights of Woman* (1792) by Mary Wollstonecraft first turned the attention of the people of England to the
"wrongs of women".' Wollstonecraft's other books, notably *Thoughts
on the Education of Daughters* (1787), had specifically raised the question of higher education. So her popular supporter, the poet and
actress Mary 'Perdita' Robinson, had concluded her influential tract,

Letter to the Women of England (1799), with this ambition, printed in capital letters: 'Had fortune enabled me, I would build a UNIVERSITY FOR WOMEN.'

After many false starts Tennyson determined to avoid any direct preaching or pedagogy, but explore the whole subject in the form of a poetic drama, partly comic in intention. He dropped his previous apocalyptic vision of the industrial city of the future, powerful as it was. Instead, he set the opening scene as a lively description of the Festival of the Mechanics' Institute at Lushington's Park House. This was a light-hearted vision of Victorian science at play, highly original in itself (including both boys and girls), a carnival of carefully observed scientific demonstrations, games and experiments. Tennyson's medley style here, quite unlike Lilia's declamatory one, is sprightly and amused, adopting an almost novelistic manner that might even be compared to Dickens in its precise and busy details.

> A man with knobs and wires and vials fired
> A cannon: Echo answer'd in her sleep
> From hollow fields: and here were telescopes
> For azure views; and there a group of girls
> In circle waited, whom the electric shock
> Dislink'd with shrieks and laughter: round the lake
> A little clock-work steamer paddling plied
> And shook the lilies: perch'd about the knolls
> A dozen angry models jetted steam:
> A petty railway ran: a fire balloon
> Rose gem-like up before the dusky groves
> And dropt a fairy parachute and past:
> And there thro' twenty posts of telegraph
> They flash'd a saucy message to and fro
> Between the mimic stations; so that sport
> Went hand in hand with Science . . .

Into this scene Tennyson introduced his cast of seven argumentative 'college' friends, to dramatise a 'seven-fold story'. They are shades of the Cambridge Apostles, but now a mixed company, and including the young and fiery Lilia. Urged on by the festival's genial host, Sir Walter Vivian (a version of Lushington), they begin to debate the Woman Question, and 'talk of College and of Ladies' rights'. Incited by Lilia's vehemence, they decide to set up their own kind of scientific experiment in the form of a speculative game. What would happen if a women's university really existed? They imagine a university that has some resemblance to Cambridge, but built as a medieval fortified castle, with an inscription over its gate: 'LET NO MAN ENTER IN ON PAIN OF DEATH'.

It will be ruled by a headstrong female vice-chancellor, the Princess Ida, who staffs it with a brilliant group of female dons. With deliberate provocation, the passionate little Lilia is given the leading role of the tall and commanding Princess Ida to act out.

'Take Lilia, then for heroine,' clamoured he,
'And make her some great princess, six feet high,
Grand, epic, homicidal . . .'

Ida is learned in all the sciences, a passionate polymath, but also on occasions something of a pedant. Characteristically, she greets a romantic sunset with a glancing reference to the theories of Laplace:

'There sinks the nebulous star we call the sun,
If that hypothesis of theirs be sound.'

Tennyson develops the rest of the poem in this form of an intricate chivalric game, an improvised drama, partly serious and partly satirical. The friends each contribute, taking on the various leading roles. Among them are Princess Ida's fellow female dons, the fearsome and competitive Lady Blanche; the beautiful and intellectual Lady Psyche; and the romantic Lady Melissa ('a rosy blonde' who looks like a daffodil in her yellow academic robes). Meanwhile, outside the university, other characters gather: a glamorous Prince who wishes to marry Princess Ida; and her conventional Victorian father who wishes to bully her into returning home. Both wish to disrupt the university experiment. So an increasingly wild plot spins into action, always rapid but frequently confusing. As FitzGerald afterwards pointed out, it is often impossible to tell exactly who among the male characters is speaking. Yet sometimes this barely matters, as speech after speech wittily engages with the newly emerging feminism of the 1840s.

The Prince decides to infiltrate the university secretly, with two members of his court, Cyril and Florian. The three men put on female disguise (or frankly drag) which is perhaps another extension of the notion of 'medley'. They prepare to mock at what they find in the quadrangles and university theatres, assuming it will be ridiculous. But what they witness on their tour, to their astonishment and discomfort, is a series of packed lecture halls, and a number of challenging lectures on the widest range of historical, cultural and scientific subjects.

The bantering tone and dialogue of the early sections steadily gives way to passion and polemic. The teaching curriculum includes feminist history, and the new scientific ideas of pre-Darwinian evolution, emphasising the latest discoveries in astronomy and geology. The lecturers draw on an impressive general syllabus, covering an ambitious sweep of scientific and literary subjects.

> ... Then we dipt in all
> That treats of whatsoever is, the state,
> The total chronicles of man, the mind,
> The morals, something of the frame, the rock,
> The star, the bird, the fish, the shell, the flower,
> Electric, chemic laws, and all the rest,
> And whatsoever can be taught and known ...

A high point arrives with a star lecture given by Princess Ida's twenty-year-old assistant, Lady Psyche. Psyche is Tennyson's model of a young female university don, described as 'a quick brunette, well-moulded, falcon eyed' standing very erect at her satinwood lecture desk. She is easily capable of holding the attention of a packed lecture hall of lively and restless young women.

Psyche's lecture is more specifically scientific, and revolves round the comprehensive idea of evolutionary development, much as described by Robert Chambers in his *Vestiges of Creation*, and with no holds barred: 'Here might they learn whatever men were taught.' This includes the 'nebular hypothesis' as originally outlined in 'The Palace of Art', and now specifically footnoted by Tennyson 'as formulated by Laplace'. In her brisk lecture-hall manner, Psyche begins with her own complete Natural History of creation, summarised in a mere eight lines. It dashes from the emergence of galaxies and our own planetary solar system, to the appearance of first life on earth, with the 'monster' dinosaurs and followed by early *Homo sapiens*. This includes a passing swipe at the 'crushing' patriarchy of primitive societies, which still persists in the 'lowest' part of the modern world.

'The world was once a fluid haze of light,
Till toward the centre set the starry tides
And eddied into suns, that wheeling cast
The planets: then the monster, then the man
Tattooed or woaded, winter-clad in skins,
Raw from the prime, and crushing down his mate
As yet we find in barbarous isles, and here
Among the lowest.'

She then sketches the historical emergence of female power, from 'the legendary Amazon, as emblematic of a nobler age' to those 'who first had dared / To leap the rotten pales of prejudice'. These include heroic figures like the poet Sappho, the political genius Queen Elizabeth the First, and the inspired military leader 'peasant' Joan of Arc. In the future Psyche sees women distinguishing themselves in every sphere and profession. They will sound 'the abyss of science and the secrets of the mind', and especially make their mark in the arts: 'Musician, painter, sculptor, critic, more ...' They will even become 'Poets, whose thoughts enrich the blood of the world'.

The men stagger out of these lectures, 'gorged with knowledge', but arguing privately among themselves about the authenticity of such female pedagogy. '"Why Sirs, they do all this as well as we . . . But when did women ever yet invent?" . . . "Ungracious!" answered Florian.' So their debate continues, until they are joined by Princess Ida herself, and walk with her round the park in deep and serious discussion. Abruptly the mood changes again. Ida asks them for a song 'that gives the manners of your country women'. Perversely Cyril breaks into a tavern song, thoroughly 'unmeet for ladies', and all their disguises are discovered.

Burlesque takes over, as Princess Ida and her whole court take to horse and gallop back into the safety of the university buildings. To top it all, Ida's steed misses the entrance plank, and with an unexpected descent into complete farce, 'blind with rage' she is thrown into the moat. The men gallantly plunge in and rescue her.

It is at this point (by the end of Part 4) that any reader – and especially faithful doubters like Elizabeth Barrett or Edward FitzGerald – might wonder what sort of poem Tennyson was writing. His insistent reply (to be extended in his Conclusion section) would be: a Medley, in which he moved in a zigzag or 'strange diagonal' between varied styles: 'mock-heroic gigantesque . . . burlesque . . . banter . . . true-sublime'.

Some sort of order is eventually restored to the plot, and the issue of sexual equality and misogyny in marriage is addressed in Parts 5 and 6 of the poem. The grossness of patriarchal attitudes are effectively satirised by Ida's unreformed father:

'Man is the hunter, woman is the game:
The sleek and shining creatures of the chase,
We hunt them for the beauty of their skins;
They love us for it, and we ride them down.'

While various gender arguments are gently explored:

'For woman is not undeveloped man,
But Diverse: could we make her as the man,
Sweet Love were slain: his dearest bond in this,
Not like to like, but like in difference . . .'

Yet from here on *The Princess* also expends itself in romantic complications. There is a declaration of war, and a medieval joust between various disputants and rivals. The Prince and his two friends are wounded. Finally Princess Ida and her fellow academics revert to the traditional womanly profession of nurses and carers. Love and harmony is declared, but by the end the woman-only status of the university is dissolved.

> So was their sanctuary violated,
> So was their fair College turned to Hospital,
> At first with all confusion; by and by
> Sweet order lived again with other laws,
> A kindlier influence reigned, and everywhere
> Low voices with the ministering hand
> Hung round the sick . . .

This fantastical plot, a mixture of pedagogic earnestness, social satire, romantic melodrama and distinctly arch mockery, is combined with several interwoven love stories. Indeed some of the female dons – or at least the festival visitors acting them – seem to be seduced. There are also intense emotional rivalries between the women themselves. (The older Lady Blanche reveals a throat 'like a vulture'.) These amorous subplots, with overtones of *Love's Labour's Lost*, gradually overwhelm Tennyson's original and apparently serious intentions. His critics were soon to point this out. As the *Athenaeum* noted: 'The grand error of the story is inconsistency . . . its different parts refuse to amalgamate . . . what might else have been appreciated as genial satire loses its force from its juxtaposition to tragic emotion.'

Yet *The Princess* remains strong in isolated virtuoso passages. Among these are many striking sections drawing on Tennyson's scientific reading. Notable is the description of a geological field trip led by Princess Ida and Lady Psyche together:

> . . . we wound
> About the cliff, the copses, out and in,

Hammering and clinking, chattering stony names,
Of shale and hornblende, rag and trap and tuff,
Amygdaloid and trachyte, till the Sun
Grew broader towards his death, and fell, and all
The rosy heights came out above the lawns.

Visions of the geological past also filter unexpectedly into the debate
about relations between the sexes, and the problem of male violence.
Princess Ida believes that war – an exclusively male pursuit – itself
should be relegated to the past, in scientific images of extinction that
would recur in *In Memoriam*:

'I would the old God of war himself were dead,
Forgotten, rusting in his iron hills,
Rotting on some wild shore with ribs of wreck,
Or like an old-world mammoth bulked in ice,
Not to be molten out.'

The images of the iron hills, and the mammoth frozen in ice, were
one of the startling revelations of Lyell's geology, which referred to
the remains of a huge elephant 'found in a mass of ice on the shore of
the North Sea'. Similar images of the frozen monster appeared later
in *Vestiges*, and Hugh Miller's popular study of fossils *Old Red Sandstone* (1845).

Yet overall Tennyson's own views on the ultimate viability of a
women's university remain deeply ambiguous. Did he lose faith in it,
even as he wrote? The university experiment certainly seems to have
been a failure as far as Princess Ida is concerned. She appears to end
her career as a senior hospital matron, perhaps a premonition of Florence Nightingale rather than a memory of Mary Wollstonecraft. Even
the feminist Lilia seems disappointed with the outcome: afterwards
she 'took no part in our dispute', though she was 'crammed with theories out of books'. Much later the comic potential of all this confusion
would be brilliantly exploited in Gilbert and Sullivan's opera *Princess
Ida*, first produced in 1870.

Initially *The Princess* received a very mixed press reception, though the first printing sold out two thousand copies, and eventually it ran to no fewer than seventeen editions within Tennyson's lifetime. General readers seemed fascinated by its theme, but deeply puzzled by its treatment, just as Elizabeth Barrett had been. Her own long poem *Aurora Leigh* (1856) was to some degree her attempt, a decade later, to handle the same theme of women's intellectual independence, but in a radically modern manner, as a sort of novel in verse.

When Moxon proposed a first reprint as early as February 1848, Tennyson responded immediately. He began to adapt, alter and add new material, and this process continued over several years. *The Princess* would become the most revised of all his longer poems. Tennyson's view of the 'woman question', and its thesis of female independence, was felt to be brave but mismanaged and frequently absurd. His half-humorous treatment, his shift between modern and medieval settings, his uncertain attitude to Princess Ida herself, seemed to lack clear design. The characters of Lady Blanche and Lady Psyche were thought unsympathetic, and the sudden introduction of Psyche's lost (and apparently illegitimate) child regarded as confusing. None of this shape-shifting – especially from Tennyson – struck reviewers as entirely convincing. Dickens's friend John Forster, writing in the *Examiner*, more in sorrow than in anger, asked: 'Why should Mr Tennyson have thrown all this into a medley? He had something serious to say – why graft it on burlesque? ... Eminently he is worthy to be the poet of our time. Why does he not assume his mission?'

The *Athenaeum* repeated its observation that despite a worthy theme, and occasional brilliance, the so-called Medley lacked 'harmony of design'. It noted drily that Princess Ida herself began as a modern bluestocking to be 'quizzed and bantered'; but ended rather alarmingly as a remote and 'stately Amazon'. It concluded: 'we lose all faith in her individuality'.

For once FitzGerald agreed with the *Athenaeum*'s views, and his initial reservations suddenly deepened into active dislike of the whole concept. 'The criticism of the *Athenaeum* is candid and true, I think,' he privately told the young scholar Edward Cowell. 'I would so gladly have persuaded Tennyson not to publish it, but I should have had all his friends about my ears if I had done so.' He recognised the defence made by Spedding and others that 'the Public cannot understand an original thing at once, and at first etc'. Yet his own 'intuition' told him the treatment was flawed.

FitzGerald's evident disappointment suddenly provoked further disenchanted reflections about Tennyson's entire career. He quoted grimly from the *Inferno* about the crisis of mid-life. 'I despair now of Tennyson doing anything great; and mourn over this grotesque abortion at the *mezzo del cammino* of his life when he should have been conceiving a poem like Dante's.' This was savage enough, but then he added an unusually personal outburst against his old friend. 'I know nothing which would now restore him to his native and abdicated powers, but such an event as the invasion of England! That would shake him up from his inglorious pipe, petty digestive solicitudes, and make him burst the whole network of selfishness twined about him by so many years of self-indulgence and laziness.' This impatient explosion, sent privately to Cowell, was never of course meant for Tennyson's own ears. Yet paradoxically it revealed Fitz's passionate concern about the direction of Tennyson's future work, which would become more and more outspoken.

Whatever his own uncertainties, Tennyson sent a first printed copy to Emily Sellwood, though signing it simply 'A. Tennyson'. But he also composed – and wrote out for her – a new and separate lyric, the beautiful lullaby, 'Sweet and Low'. She was delighted with it, and immediately set it to music.

> Sweet and low, sweet and low,
> Wind of the western sea,
> Low, low, breathe and blow,
> Wind of the western sea!
> Over the rolling waters go,
> Come from the dying moon, and blow,
> Blow him again to me;
> While my little one, while my pretty one, sleeps.

He decided to give this traditional lullaby to Ida herself, at the end of Part 2, where the six hundred women undergraduates are assembled for evensong. It seems an interesting shift of mode for the fiery vice-chancellor. After Moxon's promise of a second edition of *The Princess* for 1848, Tennyson also began inserting a series of other purely lyric 'Songs'. These he later said were the 'best interpreters' of his text, softening its argumentative structure, just as he had once softened his 'preaching' letters to Emily. Once again, one wonders if he was having second (and third) thoughts about the meaning of *The Princess*. Over the next three years he added seven other lyrics, which included some of his most beautiful fugitive pieces, all appearing in the third edition of 1850.

Triumphantly among these later songs, Tennyson included perhaps his most perfect love sonnet. It was on the theme suggested so long ago by Arthur Hallam, when he wrote his last letters from the Vienna art museum in 1833: Titian's *Danaë*. Accordingly, Tennyson brilliantly combines two traditions: that of modern astronomical observation, and the classical myth of the Greek princess Danaë, future mother of Perseus who became an entire constellation. But now, fifteen years later, the poem has gained extraordinary enigmatic qualities. Though recited by Princess Ida in Part 7 of *The Princess*, the identity of the lover becomes mysterious: it might be either a man or a woman. Moreover its haunting musical quality also effortlessly disguises the fact that it is a sonnet that is dependent on subtle half-rhymes and subliminal repetitions.

> Now sleeps the crimson petal, now the white;
> Nor waves the cypress in the palace walk;
> Nor winks the gold fin in the porphyry font:
> The fire-fly wakens: waken thou with me.
>
> Now droops the milk white peacock like a ghost,
> And like a ghost she glimmers on to me.
>
> Now lies the Earth all Danaë to the stars,
> And all thy heart lies open unto me.

Now slides the silent meteor on, and leaves
A shining furrow, as thy thoughts in me.

Now folds the lily all her sweetness up,
And slips into the bosom of the lake:
So fold thyself, my dearest, thou, and slip
Into my bosom and be lost in me.

In addition, it incorporates images of a beautiful slumbering night garden – the porphyry pools, the peacock, the fireflies – which seem to come from a quite different tradition, that of Persia and the poetry of Hafiz, which was just being discovered in England at this time. That may have particularly delighted FitzGerald – or reconciled him – to elements in *The Princess*, as he would shortly explore the Persian tradition very fully for himself.

But having urged Tennyson to publish for so long, FitzGerald continued to be critical of the overall result. In May 1848 he would tell friends openly that he was more interested in Thackeray's *Vanity Fair* or Dickens's *Dombey and Son*, both just published, than Tennyson's *Princess*. The latter was 'elaborate trifling'. Perhaps this reflected the increasing power of Victorian prose over Victorian poetry. But it was also a personal judgement. 'Tennyson is now in Ireland, I think, adding [Songs] to his new poem the *Princess* ... I am considered a great heretic for abusing it; it seems to me a wretched waste of power at a time of life when a man ought to be doing his best; and I almost feel hopeless about Alfred now. I mean about his doing what he was born to do ...'

These were severe judgements indeed, though evidently marked by FitzGerald's genuinely frustrated sense of what his friend, now nearly forty, could still achieve. To Tennyson himself he passed off his lack of enthusiasm about the poem as just 'one of Fitz's crotchets'. But what exactly did FitzGerald think his great friend was born to do? For the first time he hints at his idea that Tennyson might produce a great contemporary scientific poem. This was revealed by his throwaway suggestion in June, that Tennyson might translate Lucretius's epic *De Rerum Natura*. In his opinion, Tennyson was still the only 'great poet' alive in England who could 'naturalize' the science of Lucretius into English.

Meanwhile Tennyson was leading a new kind of social life. He had taken upmarket rooms at 2 Ebury Street, Belgravia, and Carlyle noted, 'Tennyson has been here for three weeks; dining daily till he is near dead . . .' Tennyson for once issued his own invitations to dinner parties, though they were usually still held at Lushington's rooms at No. 2 Mitre Court. On one single evening he invited Thackeray, Leigh Hunt, John Forster, 'etc etc', to join him, an unusually crowded session. Though such strictly literary invitations did not seem to include FitzGerald at this time.

Tennyson's old science tutor William Whewell, impressed – or perhaps provoked – by *The Princess* and its subject of university education, wrote from Trinity College requesting Tennyson write an 'Installation Ode' to mark Whewell's election as Master. But Tennyson felt overwhelmed by the idea, and gently refused. 'You will understand that I decline from no unwillingness to oblige you but from a sheer dread of breaking down.' Even at this time of increasing recognition and success, Tennyson still felt vulnerable, the shadow of the Kraken still upon him. Whewell eventually persuaded Wordsworth to write the Installation verses.

But after his long silence, and this uncertain reception, Tennyson was at least eager to talk about his work with younger writers. The twenty-four-year-old Francis Palgrave, a classicist just down from Oxford (and the future anthologist of the *Golden Treasury*), met him walking on Hampstead Heath the following spring, 1849. 'Walked towards Hampstead with A. Tennyson. Conversed on Universities, the "Princess", his plans etc; he very open and friendly: a noble, solid mind bearing the look of one who had suffered greatly; – strength and sensitiveness blended . . .' Later, on another walk along the Hampstead Road, Tennyson read Palgrave some of the beautiful 'Song' sections he was adding to the poem. Palgrave felt honoured by these confidences. 'Long conversation with him: he read me songs to be inserted in the "Princess", and poems on A. Hallam, some exquisite.'

Besides 'Now sleeps the crimson petal', these songs made an extraordinary anthology of Tennyson's best lyrical work. They included the heady mountain aria, 'Come down, O Maid', brought back from his Swiss tour; and a beautiful 'Swallow' song, which seems to reflect something of Tennyson's continuously repeated – but never fulfilled – intentions to depart for Italy. Then there was the lullaby 'Sweet and low' with its invocation to the 'western seas', perhaps these seas off Cornwall; and the famous 'Bugle Song' to be brought back from Ireland with its rousing opening 'The splendour falls on castle walls'. This memorably exploited the distancing effect of acoustics with the multiple 'dying, dying, dying'. Tennyson now added with a certain scientific deliberation that when he had heard a boatman's bugle blown across a lake in Killarney, he had counted 'eight distinct echoes'.

Many of these songs, only added in the 1850 edition, with their airy inspiration from 'far, far away' (that is, from Tennyson's real or imagined trips out of England), have no obvious connection to the plot of *The Princess* but were intended as breathing interludes ('linnets songs'). All subsequently became disassociated from the main poem, and were separately set to music by late Victorian composers (such as Charles Villiers Stanford, and Arthur Sullivan), who exploited and indulged their sentimental overtones to the full.

They also powerfully influenced Palgrave in his essentially lyrical selection of pieces for the first *Golden Treasury* of 1861, explicitly subtitled *English Songs and Lyrics*. This became the most popular 'people's' poetry anthology for the rest of the nineteenth century, though increasingly traditional, even staid, in its selection. But Tennyson's songs have quite other qualities, magical and hypnotic elements, wonderfully brought to life in Gustav Holst's settings of 1905, carefully scored for an unaccompanied (*a cappella*) female chorus. In an extraordinary way this setting, explicitly for a group of women singing in close harmony, returned the songs to the original idea for a women's university.

16

Epic

Tennyson's way of dealing with first reviews of *The Princess* was to escape the 'fevered' dinners of literary London entirely, and plunge into the pastoral anonymity of Irish country house life. This Irish retreat came after repeated invitations from Aubrey de Vere, who now 'carried him off bodily' to the west of Ireland to stay in the large family estate at Curragh Chase, thirty miles from Limerick, in the spring of 1848. Here he was lapped for five weeks in all the landed comforts of the Protestant ascendancy. For a moment there was a glimpse of the way Tennyson might respond to real celebrity, if it ever came.

On the whole he was a grateful if demanding guest. He was taken on tours to all the 'old Castles and Abbeys' in the neighbourhood; but was much more impressed by the eight-hundred-foot cliffs at Slieve League, whose tremendous breaking Atlantic surf and huge upsurging waves added to his catalogue of marine wonders. He was shocked 'by the poverty of the peasantry' and the 'marks of havoc wrought through the country by the great potato famine', but did not wish to enquire further. Instead he would insist on breakfasting upstairs in solitary state; smoke his pipe in the house's 'most comfortable bedroom' especially set aside for him; wander in the local woods on his own; and 'sit alone half the day, musing', and (it was understood) preparing for the new edition of *The Princess*. De Vere, a patient and admiring host, found him 'strong, dark and impressive', and his behaviour 'as simple as a child' – and perhaps as wayward.

Tennyson could prove particularly challenging in the evenings, unless he was encouraged to give recitals of poetry, 'commonly choosing pathetic pieces'. On one occasion after such a reading, he glanced round reproachfully and said, 'I do not see that any of you are weeping!' On another, when Lady de Vere hosted a dance in his honour, he sat by himself in a corner, grumbling and refusing to join in such a 'stupid thing'. De Vere was surprised by Tennyson's sudden shift of mood, so familiar to FitzGerald. Stubbornly sitting there in his poetical sulk, Tennyson was confronted by 'a certain Lady G', a formidable woman 'who was accustomed to speak her mind to all alike' in the Irish manner. She swept across the room and scolded the distinguished author of *The Princess*. 'How would the world get on if others went about it growling at its amusements in a voice as deep as a lion's? I request that you will go upstairs, put on an evening coat, and ask my daughter Sophie to dance.' Without a word Tennyson went meekly up and changed, came down and danced with Sophie and all the other Limerick girls with the greatest enthusiasm for the rest of the evening. He turned out to everyone's surprise (and de Vere's evident relief), to be 'an excellent dancer'.

Back in London, Thomas Carlyle too had become fascinated by the surprising contradictions of Tennyson's character. They were evidently all the more interesting as he became more widely known as a poet. Carlyle, the author of *On Heroes, Hero-Worship, and the Heroic in History* (1841), regarded himself as an acknowledged master of biography, and would find Tennyson an increasingly intriguing biographical case to write about. He thought *The Princess* was a blind alley professionally, and even more than FitzGerald he doubted the whole future direction of Tennyson's work. He felt he should better 'apply his Genius to Prose' and told FitzGerald that Tennyson was 'a lifeguardsman spoilt by making poetry'. To which Tennyson would reply evenly that Carlyle was 'a Poet to whom Nature has denied the faculty of Verse'.

Carlyle had written a long, melodramatic and rather sceptical reassessment in December 1847 to Emerson. 'A truly interesting Son of Earth, and Son of Heaven – who has almost lost his way, among the will-o'wisps, I doubt . . .' (This was perhaps the world of 'The Princess'.)

But worse, Tennyson 'may flounder ever deeper, over neck and nose at last, among the quagmires that abound ... He wants a *task*; and alas, that of spinning rhymes, and naming it "Art" and "high Art" in a Time like ours, will never furnish him.'

Yet Carlyle remembered fondly the long night walks they took together at this time, through deserted London streets, around Chelsea and along the Thames, just as Boswell and Johnson had once done along Fleet Street and the Strand. They had held 'long and free discussions on every conceivable subject'. Striding along in huge clouds of tobacco smoke, they had argued and laughed each in their own style. Carlyle raged against the 'acrid putrescence' of commercial London, while Tennyson agreed but felt this was not the concern of poetry. This was the only time they 'almost quarrelled'. Carlyle again accused him of claiming poetry to be 'High Art', and privately called him 'indolent, somnolent, almost imbecile'. Tennyson calmly replied: 'I never in my whole life spoke of High Art.'

Public assessments of Tennyson became more frequent. He had already appeared in William Howitt's *Homes and Haunts of the Most Eminent British Poets* Vol. 2, of 1847, the only living poet so included. It was full of praise, but also of suggestive misinformation. Though Tennyson is correctly described as born in Somersby Lincolnshire, his social life in London or therapeutic spells in Cheltenham are barely mentioned. Instead Tennyson is presented as a mysterious, rootless, provincial figure drifting along southern English sea coasts, Hastings, Eastbourne, the Isle of Wight and Cornwall. This wandering poet is only to be glimpsed, unrecognised and unaccompanied, in remote country inns, contentedly reading and smoking his meerschaum pipe. Such was his image until *The Princess* was published.

His greatest poems were given as 'Locksley Hall' and 'The Two Voices'. Both of these were now seen as representative of the new, self-questioning spirit of the age. Tennyson had moved beyond the 'spiritual' assurance of the Romantics (specifically of Wordsworth), and was grappling with all the 'doubts and fears' of the new generation, torn between belief and disbelief. He was confronting this, in Howitt's striking phrase, with 'graphic power and logical force'.

Perhaps, however, the crown of all Tennyson's verse is 'The Two Voices' ... In this poem there is no person who has passed through the searching, withering ordeal of religious doubts and fears as to the spiritual permanence of our existence – and who has not? – but will find in these simple stanzas the map and history of their own experience. The clearness, the graphic power, and logical force and acumen which distinguish this poem are of the highest order. There is nothing in the poems of Wordsworth which can surpass, if it can equal it ...

Now the author of *The Princess* suddenly had a more urban presence. Other ambitious young poets gathered round him, besides Aubrey de Vere. Among them was the twenty-five-year-old Coventry Patmore, as yet an unknown junior assistant at the British Museum, but soon to become the author of the long, bestselling narrative Victorian love poem, *The Angel in the House* (1854). Its ideal of domestic affections and family loyalties could not have been further from the world of Lady Psyche and Lady Blanche. But Patmore's name appears on an impressive memo of Tennyson's dining engagements for May 1848, which runs in an unbroken list between the third and the twenty-third of that sociable springtime. Among them were a couple of his old Apostle friends from Cambridge, including William Brookfield, and at least one literary lioness, the glamorous young Lady Duff Gordon. But again there was no sign of Edward FitzGerald's name.

Over these evenings, one of the subjects much discussed apart from poetry was the latest political developments in France. *The Princess* had ended with a prophetic glance southwards, across the peaceful English countryside to the Channel and the restless Continent:

A red sail, or a white; and far beyond
Imagined more than seen, the skirts of France.

The skirts were ruffled. Eighteen forty-eight became the Year of Revolutions across the whole of Europe. Tennyson had his own source of information about events abroad. His beloved sister Emily, married six years previously but still exorcising her memories of Hallam in her own way, was adventuring in the 'tumult' of insurrectionary Paris with her sailor husband, the handsome – and as it would prove – garrulous Captain Richard Jesse. They witnessed the terrifying street-fighting, the burning down of the Palais Royal, and the revolutionary crowds singing the Marseillaise. Emily wrote long accounts to her brother, describing the chaos, the failure of the provisional government, the state of 'constant alarm'. They became involuntary witnesses of social change. 'Provisions are growing very scarce, and the cry for bread is now strong. Yesterday half the mob were drunk.'

Tennyson was much struck by the vivid fear of Emily's accounts, and though usually silent on politics, now wondered how the universal search for 'social truth . . . and justice' had somehow been subverted in France. The insurrectionary Paris mob had become, in his angry phrase put into No. 127 of *In Memoriam*, 'the red fool-fury of the Seine'; and the city riots in the name of freedom appeared to him merely to 'pile her barricades with dead'.

The unusual vehemence of Tennyson's reaction was due to the revival of the equivalent Chartist movement in England. In spring 1848 this produced enormous and alarming gatherings in London, and riots in Manchester, Glasgow and elsewhere. Buildings were damaged, windows smashed, iron railings were pulled down, and hundreds of demonstrators were arrested. But the Charter (in its third version since 1829) was still only demanding constitutional changes, like annual parliaments and universal manhood suffrage. So these British street demonstrations were peaceful by comparison with the sustained revolutionary violence of those in Paris. There thousands were killed, and the July monarchy of Louis-Philippe completely overthrown.

The great open-air meeting that took place in Kennington Green, south London, on 10 April, was recorded in one of the earliest panoramic photographs. By contrast it showed a vast and orderly gathering, peacefully listening to speeches from a flag-draped wooden stage, while many of the listeners are well-dressed in frock coats and wearing the distinctive Victorian half-top hat. Carlyle continued to walk the London streets, fascinated to observe these 'Signs of the Times' at first hand, but finally diagnosing a deeper evil. 'Our true

Deity is Mechanism. It has subdued external Nature for us, and we think it will do all other things.' The essential political problem was the increasingly inhuman, mechanical nature of industrial society. Carlyle now turned all his literary powers on this prophecy.

By contrast Tennyson, apparently turning his back on political upheavals, spent much of this summer wandering through the remotest parts of Cornwall. His route took him along the wildest coasts and harbours, especially Morwenstow, Tintagel and Penzance. After *The Princess*, he was thinking about returning to his old idea of an Arthurian epic, developed from his much earlier fragments such as 'Sir Launcelot and Queen Guinevere' (1829), and particularly 'The Morte d'Arthur' (1833). This idea for a chivalric epic dated right back to his childhood at Somersby, and came as an evident relief in the present turmoil.

But after *The Princess* it represented an apparent retreat from contemporary themes. With its promise of Arthurian magic and Celtic folklore, he seemed to be again rejecting the whole modern world of social reform or scientific ideas. Yet as he adventured through this wild country, threading the lonely exposed coastal paths, encountering sheer cliffs beneath and towering outcrops of ancient rock above ('Upper Devonian'), he was constantly reminded of Lyell's new vision of deep geological time, and all nature's pitiless erosions. The heaving seas, the craggy coves and the dark caves stirred familiar shapes.

Tennyson intended to base himself at Bude in north Cornwall, where he dramatically informed Aubrey de Vere he might find a match for the mighty Atlantic breakers of Limerick, and the west coast island of Valentia. 'I hear that there are larger waves there than in any other part of the British coast: and must go thither and be alone with God!'

He set off on 29 May, but had a 'very rough introduction to Cornwall', and his impetuosity almost led him to a closer encounter with the Almighty than he intended. On his first night at Bude, staying at the inn, he unpacked and impatiently demanded to see 'the sea! the sea!' The maid opened a back door from which came the thump of waves. Dashing out into the dark, he forgot his short-sightedness, and plunged over a sheer embankment just outside, landing heavily on the pavement seven feet below, as he put it viciously 'fanged with cobbles'. He was cut, bruised and shaken, and his wounded leg troubled him for the next fortnight. But soon he pressed on through Boscastle and as far as Tintagel in the rain, where he sat in the ruins of King Arthur's legendary castle, 'cliff all black and red and yellow, weird looking thing'.

From then on he found Cornwall full of dark poetical mysteries as well as frequent physical discomforts. He loved its dramatic layerings of ancient rock exposed by the cliffs, and its deep and apocalyptic caverns, one of which he visited by candlelight. He was carried into another through the surf, on the back of a Mr Stokes, who announced he was the local poet. 'Coast looked gray and grand in the fading light. Went into cave. Rembrandt-like light through the opening.'

He soaked up the Arthurian atmosphere and wrote a travel journal, a series of snapshot entries, mixing picturesque tourism with careful geology. Meticulously kept throughout June and July 1848, the style was spare but strangely evocative. No poems, not a word of politics or social troubles, but everywhere glimpses and gleams of epic shapes and unsettling images, caught in single words or short phrases. He rejoiced especially in the bad weather, the gloom, and the contrast between the towering Cornish cliffs and the flickering appearance of local children on the beaches.

> June 5th: Went through the sea-tunnel cavern over great blocks. Walls lined with shells, pink or puce jellies. Girls playing about the rocks as in a theatre. June 6th: Slate quarries, one great pillar left standing; ship under the cliff loading; went into a cavern all polished with the waves like dark marble with veins of pink and white. Followed up little stream falling through the worn slate, smoked a pipe at a little inn, dined, walked once more to the old castle darkening in the gloom.

At Slaughter Bridge, Camelford, he found King Arthur's reputed stone, and sat by it for a long time in the rain, 'under two or three sycamores'. Then he walked seaward to a little churchyard. 'Large rich crimson clover; sea purple and green like a peacock's neck.'

He stayed at small inns, talked and drank with the weatherbeaten fishermen and picked up their accents: 'supped on *sas-ages*'. He visited the many ivied towers and Gothic follies built by the local Cornish landowners, many in 'splendid polished porphyry and jasper'. He spent a long time 'talking geology' with a local savant, Charles Peach; and borrowed a microscope to study fossils extracted from the Cornish clay, various minerals, and also 'zoophytes, corallines and a spider'. His learned lists of specimens became rather like those of Princess Ida's on one of her field expeditions. But as ever he was especially observant of fleeting effects of light and moving water. 'Turf – fires on the hills; jewel-fires in the waves from the oar which Cornish people call *bryming*.'

The walking tour, after much heroic 'hobbling', was eventually upgraded to a series of horse-drawn gigs and lumbering pony carts. These took him over Bodmin to south Cornwall, as far as Penzance and the Lizard Point. Here he went down to the beach, joined a group of workmen to smoke his pipe, and sat transfixed watching the enormous seas until sunset. 'Glorious glass-green monsters of waves . . . sat watching wave-rainbows.' At Polperro bay he actually sea-bathed. At Tavistock he was surprised to hear the music of Ethiopian singers, 'the sole hope of the evening', perhaps bringing memories of Coleridge's Ethiopian maid from 'Kubla Khan'. His tour ended in late July, when he reached Plymouth, and civilisation reasserted itself at the Royal Hotel with the presence of fleas. He rode back gratefully on a 'pretty railway by the sea' as far as Exeter, then immediately went off to recover for several weeks at Dr Gully's hydrotherapy establishment at Malvern.

His expedition was especially enlivened by his meetings with the eccentric folklore writer and vicar of Morwenstow, the Reverend Robert Stephen Hawker. 'June 2ⁿᵈ: Took a gig to Reverend Hawker at Morwenstow. Walk on cliff with him, told of shipwreck.' After initial suspicions on both sides, they became walking companions, with a shared fascination for Cornish legends and flamboyant hiking gear. Hawker had recently published his *Echoes from Old Cornwall* (1846), and took Tennyson to visit his famous look-out hut perched on the

clifftops. From here he catalogued storms, observed smuggling and
wrecking operations, and organised rescues for the crews of ship-
wrecked boats, together with generous charities to support them. A
popular local figure, but still unknown in London, Hawker recited his
'Song of the Western Men' to Tennyson, with the chorus line that
eventually brought him celebrity (thanks to an article by Dickens):
'And shall Trelawny die?/ Here's twenty thousand Cornishmen/ Will
know the reason why!'

All this greatly appealed to Tennyson, who quickly warmed to
Hawker, and unusually confided to him his future plans, which deeply
impressed the scholarly vicar. 'Seated on the brow of the cliff . . . He
revealed to me the purpose of his journey to the West. He is about to
conceive a Poem – the Hero King Arthur – the Scenery in part the
vanished Land of Lyonnesse, between the Mainland and the Scilly
Isles.'This was indeed Tennyson's initial conception of the *Idylls of the
King*, the long multifaceted saga which would begin to appear over a
decade later. Hawker himself was later to publish the first part of an
Arthurian epic, *The Quest of the Sangraal* (1864), so their conversations
must have been animated.

They made a spectacular and outlandish pair as they patrolled the
windswept cliffs together. Hawker observed 'a tall swarthy Spanish
looking man', in a long, flapping black cloak and wide-brimmed
black hat, with an eye 'like a sword' gleaming beneath. While Tennyson noted with approval a large, rumpled figure in a blue fisherman's
jersey, with no sign of clerical collar, but broad shoulders draped with
what appeared to be an old ragged, yellow horse blanket, and big feet
encased in sea-boots. Best of all, there was something like a pink
Turkish fez clamped on his head. Only one thing spoiled this unexpected encounter for Tennyson: 'they put me into the Cornish
newspapers'.

On Tennyson's return to London from Dr Gully's care, FitzGerald found him in November back at his old, unglamorous lodgings at
25 Mornington Place, 'just emerged from the water-process at Malvern'. He was more than ever sceptical of such therapy, worried about
Tennyson's poor health and what seemed to him Tennyson's

directionless life. He wrote a series of anxious letters to friends, again remarking on Tennyson's 'bottle of wine a day', his continuous smoking, and his squandering all the benefits after 'many weeks of privation and penance'. Yet beneath FitzGerald's genuine concern, there may by now have been some subtle element of jealousy lurking in the deep. He mocked his old friend's hypochondria with a new, angry impatience: 'Tennyson has emerged half-cured or half-destroyed from a Water Establishment ... This really great man thinks more about his bowels and nerves than about the Laureate wreath he was born to inherit ... Now *the Princess* is done, he turns to King Arthur – a worthy subject indeed ... and spent some time visiting his traditional haunts in Cornwall. But I believe the trumpet can wake Tennyson no longer to do great deeds ... How are we to expect heroic poems from a valetudinary? I have told him he should fly from England and go among Savages.'

What exactly FitzGerald implied by this last, ringing admonition is not quite clear. He cannot literally have meant the Savage islanders, as discovered by Captain Cook in 1774, deep in the South Pacific. Yet the idea of the Noble Savage, and the notion that more 'primitive' societies could heal the corruption of 'civilization', was of course current since Rousseau, and by 1853 was sufficiently popular to be mocked in an essay by Dickens, 'The Noble Savage'. But FitzGerald seems to have been making some more specific reference to Tennyson's own lines from 'Locksley Hall', whose soldier-narrator had fantasised about fleeing Western civilisation for a tropical paradise, those 'summer isles of Eden', and finding there an alternative life with 'some savage woman'. If so, this was surely ironic, and its real significance was the new sarcasm in FitzGerald's confidences to Cowell. Tennyson did indeed consider going abroad around this time, as he frequently told FitzGerald, leaving his rooms with dramatic farewells, only to reappear in Charlotte Street after a couple of days. But his plan was not exactly 'to go among savages'. It was to join his brother Frederick in Italy.

Other cultural influences were at work at the end of the 1840s, more subtle perhaps than the political ones that enflamed all Europe.

These profoundly affected the way Tennyson was thinking about the developing themes of *In Memoriam,* and how the poem might be structured. Similar literary responses spread far beyond Thomas Carlyle's polemical essays. The young would-be poet of Charles Kingsley's autobiographical novel *Alton Locke* (1850), though attracted by Chartism, was told that politics was not enough. 'And be sure, if you intend to be a poet for these days (and I really think you have some faculty for it), you must become a scientific man. Science has made vast strides, and introduced entirely new modes of looking at nature, and poets must live up to the age.' In the same vein, in the autumn of 1848 Robert Hunt, a young Cornish mineralogist based in Falmouth, published *The Poetry of Science*, which attracted considerable interest in London, and led to his election to the Royal Society. The book quickly ran to a second edition, with much discussion in the press.

Hunt was forty-one, a man of ranging interests and energy. Already a Fellow of the Royal Statistical Society, he was also a locally published poet, the author of *The Mount's Bay: a Descriptive Poem* (Penzance, 1829). An amateur photographer, who sent his pioneering pictures to John Herschel, he had a natural feeling for light and landscape. Like the German poet Novalis a generation earlier, he eventually became a Professor at the Government School of Mines, a remarkable combination of talents.

Throughout his life Hunt was fascinated by the possible interaction between poetic and scientific knowledge. As he put it in the general introduction to his book: 'In science we find the elements of the most exalted poetry; and in the mysterious workings of the physical forces we discover connections with the illimitable world of thought.' In his text he gave passing quotations from Goethe, Lucretius, Milton, Wordsworth, Keats, Shelley and Coleridge, and many other poets. These illustrated extensive passages of scientific analysis from Newton, John Herschel, Charles Lyell, Michael Faraday and Mary Somerville.

In chapter 12, 'On Chemistry', he asked what exactly did Shakespeare mean by the haunting lines from *The Tempest?*

Full fathom five thy father lies;
Of his bones are coral made;
Those are pearls that were his eyes; . . .

Did Shakespeare have any grasp of the actual chemical process by which coral, or pearls, was formed? And did this matter? Yes, it did thought Hunt, and this was not the whole meaning of Ariel's song. Shakespeare 'painted, with considerable correctness, the chemical changes by which decomposing animal matter is replaced by a siliceous or calcareous formation. But the gifted have the power of looking through the veil of nature, and they have revelations more wonderful than even those of the philosopher, who evokes them by perpetual toil and brain-racking struggle with the ever-changing elements around him.'

Hunt observed that poets and scientists frequently shared particular words and concepts, especially those connected with 'rhythm' and 'undulatory motion' throughout nature. For example, in water ripples or sound waves, or pulses of light, as Mary Somerville had observed. Hunt himself used this 'ripple' analogy in his introduction, to describe the necessary co-existence of 'bare' scientific truth with universal poetic truth.

'To rest content with the bare enunciation of a truth, is to perform but one half of a task. As each atom of matter is involved in an atmosphere of properties and powers, which unites it to every mass of the universe, so each truth, however common it may be, is surrounded by impulses which, being awakened, pass from soul to soul like musical undulations, and which will be repeated through the echoes of space, and prolonged for all eternity.'

Hunt's prose could be over-elaborate, and ultimately committed to a confident kind of Natural Theology, quite alien to Tennyson's subtle agnostic doubts. 'The task of wielding the wand of science,—of standing a scientific evocator within the charmed circle of its powers, is one which leads the mind through nature up to nature's God.'

Nevertheless Hunt was keen to argue that the same language was shared by poets and by scientists, that they were not fundamentally opposed in their appreciation of nature, or in their contrasting ideas of 'the material and the ideal'. His book touched on many of such ongoing themes of science and poetry, belief and unbelief, that Tennyson was now gathering together for *In Memoriam*. Tennyson found even Charles Dickens, stepping beyond his usual literary remit, choosing to review Hunt's book enthusiastically and at length for the *Examiner* in December 1848. Dickens asserted that Hunt's aim was:

'To show that Science, truly expounding nature, can, like nature herself, restore in some new form whatever she destroys; that, instead of binding us, as some would have it, in stern utilitarian chains, when she has freed us from a harmless superstition, she offers to our contemplation something better and more beautiful, something which, rightly considered, is more elevating to the soul, nobler and more stimulating to the soaring fancy; is a sound, wise, wholesome object.'

It was no coincidence that FitzGerald was also considering the role of science in poetry at this time. He had been rereading Homer in his cottage at Boulge, and reflecting on the possibilities of a modern epic. He wrote to his friend Edward Cowell, now more than ever his confidant on Tennysonian matters, a long, groundbreaking letter dated 24 July 1847. 'I often think it is not the poetical imagination, but bare Science that every day more and more unrolls a greater Epic than the Iliad – the history of the World, the infinitudes of Space and Time! I never take up a book of Geology or Astronomy but this strikes me.' Though he did not say so, FitzGerald surely had some such unwritten epic, some 'heroic poem' by Tennyson in mind. He did not yet realise that his friend's *In Memoriam* might fulfil, in a quite different way, the role of such a modern epic.

In the same letter, FitzGerald turned specifically to Charles Lyell's own epic descriptions of Niagara, and its vast timescales, which had so influenced Tennyson himself on first reading chapter 10, 'Aqueous Causes', in *The Principles of Geology*. Lyell had demonstrated that the enormous falls, so magnificent and monumental in appearance, and looking to the human eye like a feature eternally fixed in nature, had always – in geological fact – been unfixed and ever-changing. The 'eternal' falls were deeply unstable and mutable. Eroding their way through the lip of the shale and limestone rock 'southwards', the falls had steadily moved backwards up the riverbed over thousands of years. 'The falls of Niagara afford a magnificent example of the progressive excavation of a deep valley in solid rock ... It must have required nearly ten thousand years for the excavation of the whole ravine ... Should the erosive action not be accelerated in the future, it will require upwards of thirty-thousand years for the falls to reach Lake Erie (twenty five miles distant) ...'

FitzGerald observed that Lyell could calculate this astonishing, relentless movement southwards in units of thousands of years, to which all human lives and history were 'the twinkling of an eye'. Moreover the 'soft strata' of rock beneath the falls contained 'evidences of a race of animals', and of an inland sea washing over them, that had existed long before the falls themselves had even begun to flow. FitzGerald summarised: 'So that, as Lyell says, the Geologist looking at Niagara forgets even the roar of its waters in the contemplation of the awful processes of Time that it suggests.' Fitz concluded pointedly: 'It is not only that this vision of Time must wither the Poet's hope of immortality; but it is in itself more wonderful than all the conceptions of Dante and Milton.'

Still apparently thinking of Tennyson, he pursued this theme in his next letter of 30 July. 'As to my Epic theory, I do not say the duration of the bodily fabric of the world is a greater interest than the *Soul* of man, but that it is of greater interest than most other subjects *in the* Soul of man ... Therefore I think that Geology which has certainly discovered to us so much of the Past – and the Being of this Earth *when we were not*; is a more wonderful, grand and awful, and therefore *Poetical* idea, than any we can find in our poetry ...' And so far, perhaps, in Tennyson's poetry.

To FitzGerald himself, not usually given to big metaphysical statements, these truths were epic and existential. He sat alone in his Boulge cottage garden, among the flowers and pigeons, and contemplated them. 'The Facts of Man's history ... recorded in granite rock ... which we can just fathom so much as to know it is unfathomable, this really fills, and immeasurably over-fills, the human Soul with Wonder and Awe and Sadness! There is no end to looking into this Vista.'

Unknown to him, during these late 1840s, Tennyson was adding more to his great gathering of evolution poems for *In Memoriam*. After the crucial midway group of dramatic questioning, Nos 50–56, came a later reflective group, Nos 118–123. Together these would form what was in effect the fragments of a great Science Epic, though FitzGerald barely suspected their existence among the many elegies he had glimpsed in the famous but now elusive butcher's book. These poems recognised, more vividly than ever, how contemporary science – especially geology and astronomy – had shown that human civilisation had grown out of a dark and violent past, infinitely older and more savage than one conceived by contemporary religion and belief in a benign – or specifically Christian – Creator. But they also admitted that the consequence of this knowledge was terrifying and monstrous, and almost impossibly hard to accept. It was in articulating and holding steady the tension between these two kinds of fundamental beliefs, both essentially Victorian, that the greatness of Tennyson's unknown epic lay.

In one of these late evolution poems, No. 123, Tennyson returned to the memorable image of the dark blank city street (Wimpole Street), where he had once stood mourning for Arthur Hallam (No. 7). But now he accepted that the death of an individual, however beloved, counted for nothing within the vast and pitiless scale of geological death and extinction. Science had revealed that nothing abides, the land itself changes like mist, like clouds. Within the depth of the 'central sea' still lurks the Kraken of this terrible knowledge. And yet, and yet, he cannot finally abandon the other kind of knowledge, the previous kind of belief. Both remain, painfully and beautifully, suspended in this brief but masterly poem:

> There rolls the deep where grew the tree.
> O earth, what changes hast thou seen!
> There where the long street roars, hath been
> The stillness of the central sea.

The hills are shadows, and they flow
From form to form, and nothing stands;
They melt like mist, the solid lands,
Like clouds they shape themselves and go.

But in my spirit will I dwell,
And dream my dream, and hold it true;
For tho' my lips may breathe adieu,
I cannot think the thing farewell.

Tennyson's poem directly reflected not only the gathering observations of Victorian geologists, but also naturalists and explorers like his exact contemporary Charles Darwin. Long before the *Origin of Species* (1859) Darwin had recently published a second edition of *The Voyage of the Beagle: Journal of Researches* (1845), which was widely read. Compiled from his daily journals, the book overflowed with the same sense of the relentlessly changing earth that had haunted Lyell and now haunted Tennyson. 'Where on the face of the earth can we find a spot on which close investigation will not discover signs of that endless cycle of change, to which this earth has been, is, and will be subjected?' Or again: 'Daily it is forced home on the mind of the geologist that nothing, not even the wind that blows, is so unstable as the level of the crust of this earth.'

Indeed some of Darwin's detailed descriptions of the bleak Pacific coast of Chile, in chapter 15 of *The Voyage of the Beagle*, might be specifically captured in Tennyson's poem. His beautifully compressed and suggestive first line, 'There rolls the deep where grew the tree ...' seems almost a summary of one whole paragraph of Darwin's journal. 'It required little geological practice to interpret the marvellous story which this scene at once unfolded; though I confess I was at first so much astonished, that I confess I could scarcely believe the plainest evidence. I *saw* the spot where a cluster of fine trees once waved their branches on the shores of the Atlantic, when that ocean (now driven back 700 miles) came to the foot of the Andes. I *saw* that they had sprung from a volcanic soil which had been raised above the level of the sea, and that subsequently this dry land, with its upright trees, had been let down into the depths of the ocean.'

Both the scientist and the poet consider this idea of an ancient 'central' sea or ocean to be 'a marvellous story' of the earth's formation. But where Darwin calmly accepts the astonishing facts as the 'plainest evidence' with the greatest tranquillity, Tennyson was forced back into his own 'spirit', challenged and dismayed by the loss of the old Christian idea of creation, or the divine order itself, the indescribable 'thing' of faith. (Or perhaps, unspoken, a more specific 'thing': the belief in Hallam's immortality.) He acknowledges this older, reassuring idea to be nothing but a 'dream' now, not to be rationally accepted; though he cannot finally find it in his heart to dismiss it. This state of hovering, or trembling, between science and religion, between empirical evidence and traditional faith, is exactly what Darwin does not describe, but which Tennyson so painfully and brilliantly captures.

17

Private

During 1849 Tennyson spent much time revising *In Memoriam*, still sometimes at his retreat in Cheltenham, but also during trips to Scotland and further solitary visits to Mablethorpe beach – which he now described as 'this desolate sea-coast'. He still sought out old Lincolnshire friendships, notably the Rawnsleys at Halton Holegate, with whom he stayed for several days in November. His old flame Sophie Rawnsley was long since married and become (very happily) Mrs Elmhirst, but her brother Drummond and his wife Catherine were increasingly important to him. Tennyson confided more than ever in the Rawnsleys, both about Emily Sellwood and the possible publication of his poem, two things more than ever inextricably linked in his mind.

But also visiting Halton at this time was Weston Cracroft, one of the old Harrington Hall set, who had known Rosa Baring. This talkative figure from the past, now MP and lieutenant colonel in the North Lincoln Militia, evidently made Tennyson uneasy. Cracroft kept a diary, and reminisced about the strange Tennyson children, 'in days of yore', who used to visit the hall from the nearby rectory at Somersby. 'We were all children together when my Father and Mother lived at Harrington and they at Somersby.' He remembered that the Tennyson sisters brought over some carefully cut-out 'little paper dolls', which he and his brother Peter threw into the Harrington Hall fireplace. Apparently they were 'well punished' for this disdainful act, though Cracroft clearly still thought of it as rather a good joke. Yet for Tennyson, the thought of burning paper dolls at Harrington may have had a more painful symbolism.

Cracroft, in his mid-thirties and still feeling his social superiority, continued to patronise the distinguished author of *The Princess*. 'Alfred Tennyson the poet dined with us. When he lights up his face is almost fine, but his expression coarse when unexcited.' He added: 'Complexion dirty and sallow – he wears spectacles – conversation agreeable ... a scrubby looking fellow by daylight – dress in no way neat – in short a poet.' It is interesting to compare this with the exuberant contemporary descriptions by Carlyle and FitzGerald. Most typical of a poet, Tennyson had to be told off by his hostess for boorish behaviour. 'I was amused,' recalled the Lt-Colonel, 'when the Poet put his feet up on the sofa and Mrs Rawnsley [senior], careful soul, whom I had observed to grow fidgety, all of a sudden quietly warned him off.' Cracroft might have been even more amused had he known the same sort of poetical behaviour had occurred in the Reform Club, Piccadilly.

But Tennyson sometimes seemed barely aware of his surroundings this winter. Concentrating on inner matters and decisions, his outward movements were uncertain, and his plans shifting and disorganised. Back in London, he yet again planned to abscond to Italy to stay with his brother Frederick. Carlyle also heard rumours of Tennyson's drifting: 'Tennyson, it seems, has returned to Town: a glimpse of him was got the other day, "walking with large strides into Regent Street", in a northerly direction; and then he went over the horizon again, and has not re-emerged since.' FitzGerald continued to 'despair of his ever doing what he was born to do', though thought he was 'still the same noble and droll fellow he used to be'; and in his absence commissioned half a dozen lithograph copies of the Laurence portrait to circulate among friends.

Yet it was probably during these restless months that the other late group of evolution poems were drafted or at least revised, *In Memoriam* Nos 118–123. They opened with a grand traditional overview:

Contemplate all this work of Time,
The giant labouring in his youth ...

But in the course of six poems they marked a sudden shift in his views, and radically altered his interpretation of evolution itself. Tennyson revisited the essentially hopeful astronomy of William Herschel, with his ideas of deep time and space, and a 'constructive' or developing cosmology of 'island universes', or separate galaxies under continuous evolutionary development. But Tennyson now explored the idea further, and suggested a quite radical new possibility: that cosmic evolution among the stars might produce a *moral* or spiritual evolution among mankind. We might evolve into a 'nobler' race, or 'higher' type of humanity.

He put this with formidable power and sweep. Out of the great swirling, gaseous 'tracts of fluent heat', the planets condensed and began their orbits; and on planet earth the 'seeming-random' but marvellous forms of life started to evolve, eventually producing mankind with his 'ever nobler ends'. According to this view:

> The solid earth whereon we tread
> In tracts of fluent heat began,
> And grew to seeming-random forms,
> The seeming prey of cyclic storms,
> Till at the last arose the man;
>
> Who throve and branch'd from clime to clime,
> The herald of a higher race,
> And of himself in higher place,
> If so he type this work of time
>
> Within himself, from more to more;
> Or, crown'd with attributes of woe
> Like glories, move his course, and show
> That life is not as idle ore,
>
> But iron dug from central gloom,
> And heated hot with burning fears,
> And dipt in baths of hissing tears,
> And batter'd with the shocks of doom

To shape and use. Arise and fly
The reeling Faun, the sensual feast;
Move upward, working out the beast,
And let the ape and tiger die.

Here Tennyson summoned an overview of the emergence of life on the planet, triumphantly arising out of volcanic eruptions and animal savagery. The magnetic iron core at the centre of the earth would drive all life forward, although with great pain and violence. Tennyson's strange hope, a wholly unscientific and unDarwinian speculation, was that in this process mankind itself would evolve upwards, no longer 'the greater ape', but a 'higher race' altogether. Paradoxically it was in this idea of higher *human* evolution that he now found scientific evidence for the existence of a divine purpose guiding the universe. The human race would 'move upwards', working out the violent 'beast' in its lower nature, the ape or tiger: or say the Kraken. Such a belief moved him beyond science itself, as he confided in *In Memoriam* No. 120.

I trust I have not wasted breath:
I think we are not wholly brain,
Magnetic mockeries; not in vain,
Like Paul with Beast, I fought with Death;

Not only cunning casts in clay;
Let Science prove we are, and then
What matters Science unto men,
At least to me? I would not stay.

Let him, the wiser man who springs
Hereafter, up from childhood shape
His action like the greater ape,
But I was *born* to other things . . .

Where did this new idea, and new confidence, come from? This kind of evolutionary 'social optimism' would soon become widespread in the work of Herbert Spencer, Thomas Huxley, Alfred Russel Wallace and others. But the initial concept was certainly inspired by Robert Chambers in his *Vestiges*, which spoke explicitly of the human species being 'but the initial of a grand crowning type'. In his final chapter, 'Purpose and General Condition of the Animated Creation', Chambers had expanded on this hopeful version of evolution: 'The exertions of the present race, I conceive as at once preparations for, and causes of, the possible development of a higher type of humanity – beings less strong in the impulsive parts of our nature ... more fitted for the delights of social life, because society will then present less to dread and more to love.'

At the very end of *In Memoriam*, Tennyson briefly convinced himself that it was Arthur Hallam who had somehow represented this evolutionary development. It was Hallam himself who became the symbolic forerunner of this 'higher type':

Whereof the man, that with me trod
This planet, was a noble type,
Appearing ere the time was ripe,
The friend of mine who lives in God.

It is interesting to contrast Tennyson's wild moment of evolutionary wish-fulfilment, with John Ruskin's unappeased fears at this time. He was about to publish his great critical work, *The Stones of Venice* (1851), but his personal beliefs were increasingly undermined. Writing from Cheltenham in May 1851 he confided to his friend, the physician Henry Acland, that his 'beloved science' geology was destroying his faith and his peace of spirit: 'You speak of the Flimsiness of your own faith. Mine, which was never strong, is being beaten into mere gold leaf, and flutters in weak rags from the letter of its old forms; but the only letters it can hold by at all are the old Evangelical formulae. If only the Geologists would let me alone, I could do very well, but those dreadful Hammers! I hear the clink of them at the end of every cadence of the Bible verses.'

Tennyson was now confiding in some of his friends, trying out chosen passages on various listeners, including Spedding, Brookfield, Patmore and young Francis Palgrave. Charles Kingsley also seemed to have glimpsed some manuscript pages. But they still appeared to think the themes were entirely elegiac, rather than scientific: they were 'poems on A. Hallam, some exquisite'. Even FitzGerald had apparently seen none of the latest controversial poems. In June he wrote with puzzlement: 'I heard Alfred had been seen flying through town to the Lushingtons: but I did not see him. He is said to be still busy about that accursed *Princess*.'

He indicated his impatience by advising Tennyson's brother Frederick to 'beg, borrow, steal or buy' an entirely different collection just published by Moxon: John Keats's *Life, Letters and Literary Remains*, edited by Tennyson's friend Monckton Milnes. These two volumes contained 'wonderful bits of poems' and unlike Tennyson's years of labouring on *The Princess*, they were written 'off-hand at a sitting, most of them'. FitzGerald particularly admired this spontaneity, and added: 'I only wonder that they do not make a noise in the world.'

FitzGerald was right to spot this sudden revival of Keats's reputation, and quick to realise his poetic influence on Tennyson's earlier work. Edward Moxon's publication was effectively the first collected edition of Keats's poems, and added to the various editions of Tennyson's poems would effectively establish Moxon as the leading – and indeed almost the sole – mainstream publisher of major Victorian poetry from 1840 to 1870. Now based at Dover Street, Piccadilly, and rivalling John Murray's nearby at Albemarle Street, Moxon's publishing house would eventually include on its poetry list works by Shelley, Wordsworth, Browning and Swinburne. But Moxon's particular connection with Tennyson would soon become as famous as Murray's with Lord Byron.

Tennyson was naturally most concerned about the response to the sceptical and scientific sections of *In Memoriam*, and wondered if they would be attacked in the press as anti-Christian, or even possibly atheist. There was also the question of whether the elegiac theme, passionately mourning a young man who had died some seventeen years earlier, would be thought excessive, or even perverse. But most especially Tennyson worried about Emily Sellwood's opinion of the entire collection. What would it be? They had remained out of touch for so long, despite the occasional smuggled letters, the go-between diplomacy of the Tennyson sisters, or the sympathetic interventions of Rawnsleys, it was impossible to know.

Because so much of their private correspondence before 1850 was lost or destroyed, much about this final courtship – if that is what it was – remains obscure. He and Emily had met again, apparently completely by chance, at Park House sometime in September 1847, when Tennyson called 'unannounced before breakfast' at the Lushingtons' and found Emily visiting Cecilia. Astonishingly it was their first meeting for seven years. Little is known about this emotional reunion, except that it lasted for a single weekend, and Tennyson left abruptly on the Monday. The Rawnsleys, who had now moved south from Lincolnshire to Shiplake on the Thames, saw more of him and were soon taken into his confidence. They believed that Tennyson had seized the chance – if chance it was – to propose marriage again, but Emily 'had even definitely refused him'.

They believed that she was still in love with Tennyson, but felt she could not marry him because of continuing Sellwood family disapproval. Part of this was Tennyson's chronic lack of financial security in Henry Sellwood's eyes, despite the state pension, and Moxon's increasing sales of Tennyson's poetry. Yet friends considered the fundamental reason was her own distrust of his sceptical religious views and doubts, and how they might appear in any future published poems. As Catherine Rawnsley put it: 'She had grown to feel that they too moved in worlds of religious thought so different that the two would not "make one music" . . .' The Rawnsleys took the phrase 'make one music' from a late stanza added to *In Memoriam*.

But there must also have been more practical reservations on Emily's part: Tennyson's ill-health, his black moods, his circle of old Cambridge cronies (as she saw them) and his rootless bachelor life, forever shuttling around various London addresses, or dashing off to Cheltenham or Malvern; or still disappearing to the wild Lincolnshire coast to commune alone with the sea. This was to mention nothing of his dirty clothes, his ceaseless tobacco smoking and his late-night drinking and reciting, which even FitzGerald had begun to find oppressive.

Tennyson's own friends suspected the real reasons were deeper reservations, and ambiguities, on Tennyson's side. Had the emotional turmoil of the 1830s – his father's dark influence, his passionate feelings for Arthur Hallam, his deep frustrations with Rosa Baring, his visions of madness at High Beech – really been resolved? His uncertain commitment to Emily had been held suspended for so many years. His sense of achieving professional independence, and his ambition to be a great and 'acknowledged' national poet (in the Shelleyan sense), were yet to be fulfilled. Whatever the reasons on either side, obviously a great deal – two whole lifetimes – depended on the reception of *In Memoriam*.

At last, in the autumn of 1849, Tennyson took decisive action. He decided to add a special Prologue to the poem. It turned out to be astonishingly – even shockingly – *orthodox*. It was evidently intended to reach out to a wider public and more conventional readership. And also perhaps to reassure – or at least appease – Emily. At any rate this forty-four-line Prologue was an extraordinary document. It appeared to be a recantation of all his previous agnosticism, and much of his previous emotions. It seemed to reject all his subtle, agonised enquiries into the nature of belief; and his years of dauntless confrontations with the terrors of science; and the submerged psychological menace of the Kraken. Instead it opened with a direct appeal to God, and a great organ peal of faith, in the deeply traditional rhythms of an Anglican hymn:

Strong Son of God, Immortal Love,
Whom we, who have not seen Thy face,
By faith, and faith alone, embrace,
Believing where we cannot prove ...

This abrupt, thunderous and surprising assertion of pure religious faith ('faith *alone*'); together with this dismissal of the very notion or relevance of scientific proof ('where we *cannot* prove'), gave the whole

Prologue its sense of a public recantation. It seemed like a return to orthodoxy and the old nostrums of Natural Theology. This tone of profound apology continues throughout all eleven stanzas of the Prologue. It is perhaps the most bewildering document that Young Tennyson ever wrote. Unless one considers it from an entirely different viewpoint: as the first document written by the Old Tennyson, now suddenly anxious for acceptance.

In stanzas 5 and 6 Tennyson added what appears to be a disclaimer about the reliability or permanence of scientific knowledge.

> Our little systems have their day;
> They have their day and cease to be:
> They are but broken lights of thee,
> And thou, O Lord, art more than they.

> We have but faith: we cannot know;
> For knowledge is of things we see;
> And yet we trust it comes from thee,
> A beam in darkness: let it grow.

The implication was that the 'little systems' of science might change just like the rest of nature; and scientific theories would inevitably themselves become extinct. The 'little systems' phrase was both sweeping and dismissive. It might refer to the recent science books and speculations by such as Robert Chambers, John Pringle Nichol and Lord Rosse. But by implication it also swept aside ('broken lights') the 'grand' systems of Lyell, Laplace and Herschel. 'We have but faith: we cannot know'. So God alone supplied the permanent light, the reliable 'beam in darkness'. If nothing could stop the growth of scientific knowledge itself, its essential purpose was 'reverence' for the divine.

> Let knowledge grow from more to more,
> But more of reverence in us dwell;
> That mind and soul, according well,
> May make one music as before ...

In that last line alone there was still a glimmer, a hint, of secular hope, at least metaphorically. Science and religion might eventually 'make one music': the very matrimonial phrase the Rawnsleys would later pick out. Thus a kind of harmony – both intellectual and spiritual – might ultimately be achieved.

In the final stanzas of the Prologue Tennyson makes a show of humbly begging forgiveness for the 'Confusions of a wasted youth'. These confusions – in the plural – now included not only the recent years of dangerous scientific scepticism, but (even more painfully) his earlier time of passionate mourning for Arthur Hallam.

> Forgive my grief for one removed
> Thy creature whom I found so fair.
> I trust he lives in thee, and there
> I find him worthier to be loved.

It is painful to accept this whole *apologia* at face value. After all, expressing and exploring such overwhelming grief was the initial motive for the entire *In Memoriam*. Yet Tennyson surely meant it all. So what the Prologue most reveals is his sense of risk and exposure at the moment of publication. The word 'forgive' appears four times in these last three stanzas. In them Tennyson confessed to his 'sin in me' (without saying exactly what this was), and to his failures in 'truth' (without saying exactly what these were). But perhaps the final stanza of the Prologue seemed especially appropriate for Emily Sellwood, and was a final appeal directed to her wisdom (as well as God's):

> Forgive the wild and wondering cries
> Confusions of a wasted youth;
> Forgive them where they fail in truth,
> And in thy wisdom make me wise.

Would Emily believe such a recantation? Would his old friends like FitzGerald? Would new ones like Carlyle? Would anyone who knew him well? Perhaps not. However, this was evidently intended to have a reassuring effect on all Victorian readers. And the astonishing thing is that it very largely did so. Charles Kingsley, for example, would respond to this Prologue by calling Tennyson, 'the willing and deliberate champion of vital Christianity and of an orthodoxy the more sincere because it has worked upward through the abyss of doubt'.

Having written this Prologue last of all, Tennyson placed it first in his trial edition, at the very beginning of *In Memoriam*, and specifically dated it 1849. It was an attempt to prepare – or at least forearm – his readers for everything that followed. The question was, would it be enough for them; or for Emily Sellwood? By the end of the year he was still uncertain about publishing, or even what his title should be. As late as December FitzGerald was still in the dark too, having heard nothing from him except a single note saying he had 'taken chambers in Lincoln's Inn Fields', a much smarter address. He thought anxiously that his old friend was 'about to print but (I think) not to publish those Elegiacs on Hallam'. He knew nothing about the Prologue, which in the old days they would have discussed at length. Instead he drew attention to quite another kind of autobiographical success, that of Dickens's *David Copperfield*.

There was one other thing that FitzGerald did not know. Tennyson had written abruptly to Catherine Rawnsley at Shiplake in a letter postmarked Christmas Day 1849. 'I have made up my mind to marry in about a month. I have much to do and settle in the mean time. Pray keep this thing a secret. I do not mean even my own family to know.' One wonders if even Emily Sellwood knew at this stage; or how much still depended on *In Memoriam*.

For several weeks during the New Year of 1850 Tennyson seems to have laid low in London, avoiding his mother and sisters in Cheltenham, but staying with Spedding in Lincoln's Inn, and making arrangements about his future publication rather than his future marriage. It was here that FitzGerald finally caught up with him. He wrote triumphantly on 17 January: 'I found A. Tennyson in chambers in Lincoln's Inn: and recreated myself with the sight of his fine old mug, and got out of him all his dear old stories, and many new ones. He is republishing his Poems – the *Princess* with Songs interposed.'

It was a surprisingly fond reunion, relieved and affectionate after months of apparent coldness, with lively talk and banter on both sides. Yet, significantly, nothing seems to have been mentioned about either Emily Sellwood or even the imminent publication of *In Memoriam*. From what FitzGerald wrote cheerfully to his Cambridge friend W.B. Donne in late January, it seems that FitzGerald had not himself realised Tennyson's new reticence with his old comrade.

But there may have been reticence now on both sides. When FitzGerald was shown the new Songs in manuscript by Tennyson, including the beautiful 'Now sleeps the crimson petal', he evidently reacted favourably. But later, to Donne, he was distinctly reserved about them. 'I cannot say I thought them like the old vintage of his earlier days, though perhaps better than other people's. But even to you such opinions appear blasphemies.' Better 'than other people's' was hardly a Fitz recommendation. But there was no chance to discuss this further, as Tennyson was off again; 'A.T. is gone on a visit into Leicestershire: and I miss him greatly.'

There was one straw in the wind, suggesting that Tennyson's poetry might soon appeal to a much wider class of readers than hitherto. It might even catch a national mood. The straw had come in December, in the form of an unexpected request for a single signed copy of his poetry from an anonymous working man in Lancashire. This turned out to be none other than Samuel Bamford, the notorious radical reformer and activist, who had once been arrested and tried for treason after the great Manchester demonstration of Peterloo in 1819. It was the event celebrated thirty years previously by Shelley in *The Mask of Anarchy*. Bamford had settled down in the weaving business, and just published his memoirs in a book that would become a classic of working-class history, his *Passages in the Life of a Radical* (1844).

Perhaps for this reason the approach to Tennyson could not be direct, but was tactfully relayed through the intermediary of the rising young Manchester novelist Elizabeth Gaskell. Gaskell was the same age as Tennyson, and had just published her own first novel, *Mary Barton* (1848). She had written to John Forster, now editor of the

Examiner, on her own and Bamford's behalf, both of them being in effect the poet's northern fans. She presented Bamford ('a great, gaunt, stalwart Lancashire man'), with his literary leanings and his straitened situation in Manchester, using all her novelistic flair.

Bamford was '*the* most hearty admirer of Tennyson I know (and that's saying a good deal) ... who repeats some of Tennyson's poems in so rapt, and yet so simple a manner, utterly forgetting anyone is [near]by, in the delight of their music and exquisite thoughts ...' She had heard Bamford once 'blaze up' in his 'beautiful broad Lancashire' when Tennyson's work was criticised in an argument. But he was so poor he could not afford actually to buy his own copy of the poems, either new or (what, significantly, he could not find) second-hand. So far he had simply had to memorise them, yet he loved doing just this. Bamford had told Gaskell proudly: 'Thank God he had a good memory, and whenever he got into a house where there were Tennyson's poems, he learnt as many as he could off by heart; and he thought he knew better than twelve.' So, in short, could Tennyson do Bamford the honour of sending him a signed copy via Forster?

Tennyson was greatly impressed by this whole story, and swiftly agreed to organise a personalised copy through Moxon. He wrote back thoughtfully to Forster from Mablethorpe: 'Of course I will give him a copy ... All that account of Samuel Bamford is very interesting. I reckon his admiration as the highest honour I have yet received.' It was a particularly interesting honour, since not only was Bamford a working-class reader from the north, but he found Tennyson's the kind of poetry he could easily *learn by heart*, and took pleasure in reciting it out loud in public. Altogether this suggested a quite new role for Victorian poetry, quite different from the old intimate privacy of the Romantic lyric.

But the delicate question of the essential privacy of *In Memoriam* remained. In early March 1850 Tennyson asked Moxon to print off just half a dozen copies of the poem, with the Prologue, but still untitled and still anonymous. He was to send them directly and confidentially to Tennyson alone: 'Give none away and retain none yourself.' He specifically omitted the controversial evolution poem

No. 56. One of these special copies was sent to the Rawnsleys' house on the Thames at Shiplake where Emily Sellwood was staying, so for the first time she could read the whole work.

It is likely that Emily had seen a few very early drafts in manuscript, possibly ten years previously, before the break of 1840, when she had copied as many as seventeen sections for him. These early drafts still exist in her handwriting. But she still had absolutely no idea of the complete work of 131 sections, or its overall impact. Nevertheless Tennyson asked the Rawnsleys to return the complete printed copy within a week, apparently giving her (and them) very little time to read. So he awaited Emily's response impatiently: upon it depended their future together.

Emily's hurried reply is strange and awkward, and exists only in a partial copy. It appears to be dated as late as 1 April 1850, but may have been written earlier. It was enclosed in a covering letter to Catherine Rawnsley, where Emily appears very uncertain in her response. 'My dearest Katie ... Do you really think I should write a line with the Elegies, that is in a separate note to say I have returned them? I am almost afraid, but since you say I ought to do so I will, only I cannot say what I feel ...' Yet she was evidently overwhelmed, and could not 'willingly part from what is so precious'.

Emily then took the curious step of copying in her letter a quite 'separate note', in truly magnificent praise of the poems. This note was not from herself at all, but from someone quite else, unsigned and unidentified, but whom she called 'a true seer'. In effect, she was supplying two replies in one: her own guarded thanks ('a sort of label only'), and the mysterious passage of unrestrained praise.

This third party endorsement, as it were, turned out to be from one of the Rawnsleys' confidants, the young radical clergyman Charles Kingsley, author of *Alton Locke* and later of *The Water Babies* (1863), who had secretly pressed to see the printed copy of the Elegies in the little time available. (Curiously enough, Kingsley, a future Darwinian, would later manage to obtain a pre-publication copy of *On the Origin of Species*.) Without admitting its source, his assessment was to be passed on directly to Tennyson, as part of Emily's reply. It is evident why: Kingsley's note was euphoric. 'I have read the poems through and through and through and to me they were and they are evermore and more a spirit monument grand and beautiful, in whose presence I feel admiration and delight, not unmixed with awe. The happiest possible end to this labour of love!'

So Kingsley spoke for Emily, and said what perhaps she was too shy to say explicitly herself. Yet her approval was touchingly manifest in her hesitations, and she seemed to show no further concerns about his excessive mourning for Hallam, his religious doubts, or even his scientific scepticism (though she had not seen No. 56). Instead Emily added a loving if painfully modest postscript. 'After such big words shall I put anything about my own little I? That I am the happier for having seen these poems and that I hope I shall be the better too.'

So among many other things, *In Memoriam* effectively became Tennyson's long-awaited marriage proposal; which in this oblique way, Emily Sellwood lovingly and gratefully accepted.

18

Public

Tennyson's masterpiece was finally published by Moxon in late May 1850, still anonymous, and still with its full marmoreal title: *In Memoriam A.H.H. Obit MDCCCXXXIII*. It was dedicated with due formality, not of course to Emily Sellwood, but to his old friend and faithful host at Park House, Edmund Lushington. Reviews followed immediately, and suddenly it was clear that Tennyson had found a quite new kind of popularity. Moxon's initial cautious print run of five thousand copies, quickly went into two more editions that summer and was going through fifty thousand sales by mid-autumn. For the first time (but not the last) Tennyson had something like a bestseller on his hands. He was becoming a national phenomenon: a poet who might soon replace not only Byron but even Wordsworth in the nation's cultural identity, its sense of itself. As Bamford's example had shown, he was also a poet whose work might be *learned by heart* too.

It was Bamford's champion John Forster who led off the extraordinary reviewing in the Saturday *Examiner*, explicitly comparing the poem to Milton's *Lycidas* and also, more daringly, to Dante's *Purgatorio*. It was a mixture of subtle thought and delicate imagery with a quality of 'massive grandeur'. The 'appalling shock' of a tragic bereavement led to 'varied and profound reflections on individual man, on society, and their mutual relations'. A host of other enthusiastic reviews followed in the *Spectator*, the *Morning Post*, the *Guardian*, *Tait's Edinburgh Magazine* and *Fraser's Magazine*. In October the

Westminster Review dedicated a nineteen-page eulogy to Tennyson. The *North British Review* announced definitively that its publication was one of 'the great epochs in the history of poetry'.

G.H. Lewes, the radical voice in literary London (and very soon to fall in love with George Eliot), agreed, and wrote more reflectively in *The Leader* that the poem would become 'the solace and delight of every house where poetry is loved'. For this reason he thought it superior even to *Lycidas*. This idea of solace would be echoed by thousands of Victorian readers, until it eventually received a kind of royal seal of approval, when Queen Victoria announced that she was keeping a copy by her bed at Osborne. Since Wordsworth had died on 23 April, Lewes was able to call Tennyson without reservation 'our greatest living poet'. There was a sense of triumph, even the possibility of the Laureateship. Elizabeth Barrett Browning, now far away in Italy, believed Tennyson 'stood at last on a pedestal and was recognized as a master spirit'.

By the end of the year the book had run to four editions and sold no less than sixty thousand copies. This was a huge sale for the period, comparable to what both Byron and Walter Scott had achieved in the previous Romantic generation. Tennyson himself, with growing confidence, described his poem as 'The Way of the Soul'. But he did not forget its uncertain beginnings. He sent a special copy, '*handsomely* bound' by Moxon's printers, to his faithful Aunt Russell, whose 'bounty' had never failed.

There was surprisingly little concern, or awkwardness, about *In Memoriam* being manifestly addressed to the memory of an idealised young man, 'the creature whom I found so fair', only mildly disguised by the initials AHH. Its themes of bereavement, and spiritual doubts, seemed so obviously universal. No critic yet raised any unease about the male object of all this mourning, even though his identity soon became widely known as Arthur Henry Hallam, son of the distinguished barrister and historian. Instead, favourable comparisons continued to be made with *Lycidas*. This, after all, commemorated another intense Cambridge undergraduate friendship, John Milton's with Edward King, who had drowned aged twenty-five off the coast of Wales in 1637.

Charles Kingsley faithfully transformed his private praise into public approval. He put aside any worries about 'the private histories connected with the book'. He was 'utterly lost in admiration of it', and thought 'the love passing the love of woman' was a theme 'ennobled by its own humility'. He wrote a superb long assessment of Tennyson's whole development as a poet since 'Mariana', for *Fraser's Magazine* in September 1850. He concluded on an exalted note. 'Blessed it is to find the most cunning poet of our day able to combine the complicated rhythm and melody of modern times with the old truths ... to see in the science and the history of the nineteenth century new and living fulfilments of the words which we learnt at our mothers' knee.'

He followed this up with a personal salute in *Alton Locke*, where the young narrator 'fell on Alfred Tennyson's poetry, and found there, astonished and delighted, the embodiment of thoughts about the earth around me which I had concealed, because I fancied them peculiar to myself'.

Yet over the next months there were a few dissenting voices, most notably about the perceived 'monotony' of Tennyson's verses. It was FitzGerald, writing in private to Tennyson's brother Frederick on 31 December, who put this criticism most shrewdly. 'In Memoriam is full of the finest things, but it is monotonous, and has that air of being evolved by a Poetical Machine of the highest order ...' This idea that Tennyson had become a brilliant kind of poetry *engine* was not just a throwaway remark by Fitz. He was evidently thinking of Charles Babbage's early computers: the famous No. 1 Difference Engine of 1832; and even more aptly of his subsequent Analytical Engine of 1844, a computer of a 'higher order'. This was precisely the machine which Ada Lovelace had claimed was potentially capable of composing poetry (as is now claimed for AI). So FitzGerald implied that Tennyson's genius had taken on the repetitive and perhaps unfeeling quality of a wonderfully programmed computer, a superb 'mourning Machine', admittedly of the highest order, but fatally churning out manufactured lines. Or rather, 'evolving' them, another loaded scientific term.

Was FitzGerald justified in making this damaging criticism (if only to Tennyson's brother)? It might well seem true that *In Memoriam*, if read – and especially if *recited* out loud (as Fitz could well imagine) – continuously, section after section, with the same lyrical metre and the same rhyme scheme, might have some such stunning and mechanical effect on its audience. Yet the same of course could be said of 'The Lady of Shalott', one of FitzGerald's 'champagne' favourites. But perhaps he recognised the great paradox in so much of Tennyson's work, that its most brilliant effects would always be lyrical and local, rather than narrative and extended.

Surprisingly FitzGerald did not comment on the paradox of the Prologue. But he added a subtler, and stranger, observation. 'So it seems to be with him now, at least to me, the Impetus, the Lyrical *oestrus* is gone ...' This was a curiously biological term for FitzGerald to use, as the *oestrus* (from the Latin) means the period of sexual receptivity in specifically female creatures. It is the time when they are capable of being made pregnant. For FitzGerald, Tennyson's real lyrical gift was in some profound sense feminine. It was a brilliant imaginative fertility. Accordingly, he went on in this same, long letter to repeat his longstanding regret about Tennyson's lack of active poetical ambition, as he saw it. 'It is the cursed inactivity (very pleasant to me whom am no Hero) of the nineteenth century which has spoiled Alfred, I mean spoiled him for the great work he ought now to be entering on, the lovely and noble things he has done must remain.'

This seems like genuine disappointment for his old friend, with perhaps some ironic element of deflected jealousy. But it might also have been the truth according to Carlyle's recently proclaimed and essentially Victorian concept of 'The Heroic in History' (1841), in which the national myth must be forged by outstanding individual figures – whether prophets, statesmen, soldiers or poets – who must not refuse their destiny. Had Tennyson refused his? It remained to be seen.

In fact the possible 'monotony' of Tennyson's four-line stanza form was the only criticism generally voiced. As the *Inquirer* put it: 'The measure is of too obvious facility, and in a less masterly hand, might break into the very false gallop of verses.' After all, Tennyson had raised it himself: 'The sad mechanic exercise / Like dull narcotics, numbing pain.' The critic of the *Atlas* admitted the poem was sometimes 'too mournfully monotonous'. Even John Forster, who had encouraged the link to working-class readers like Bamford, thought the music of the verse could be repetitive, though 'exquisitely adapted' to its solemn burden and 'its slow, yet not imposing march'. Yet as Dickens's future biographer, he thought Tennyson would soon absolutely dominate the world of popular Victorian poetry, just as Dickens already did the world of popular fiction.

Remarkably, the one criticism that Tennyson feared, that the poem had too much scientific scepticism, contained unorthodox religious speculations, or reflected 'materialist' or evolutionary views or doubts, did not yet appear anywhere. In this sense the Prologue had done its precautionary work. Tennyson was saved from the kind of outraged response that greeted J.A. Froude's autobiographical novel *The Nemesis of Faith* (1849), which was universally decried in the press, and publicly burned during a furious debate at Oxford. The shadowy shape of the Kraken was not recognised rising beneath the surface. It was not until the next century, and the shrewd criticism of T.S. Eliot, that the poem's dark doubts were acknowledged as one of its most powerful elements. As Eliot wrote in 1932: '*In Memoriam* can, I think, justly be called a religious poem, but for another reason than that which made it seem religious to his contemporaries. It is not religious because of the quality of its faith, but because of the quality of its doubt. Its faith is a poor thing, but its doubt is a very intense experience.'

In fact *In Memoriam* was universally praised for its religious qualities. The *Guardian* thought it was 'full of religious feeling', and *Fraser's Magazine* praised its 'pure Christianity'. The failure to see the vast body of new and controversial science that Tennyson had so skilfully evoked, and his constant grappling with a new kind of Victorian agnosticism, is puzzling. Possibly the reviewers were themselves nervous about pointing out unorthodox views. Only very occasionally were suspicions raised, as in the *Britannia* magazine which observed regretfully that there was 'an almost total absence of those higher consolations which religion should suggest'. While the conservative *English Review* did suspect agnosticism, or possibly worse in particular passages. 'It is most falsely, and we may add, offensively assumed that the unbeliever in Christianity can possess a faith of his own, quite as real and stable as that of the believer!' It picked out the famous agnostic lines from No. 96:

> There lives more faith in honest doubt,
> Believe me, than in half the creeds.

These lines were 'infinitely mischievous', both spiritually and intellectually. The *English Review* also suspected the poet of believing in evolutionary ideas, but disguising these in beautiful images and subtle speculations. 'We remain undecided as to Mr Tennyson's faith, though we opine that, strictly speaking, *he has none*, whether negative or affirmative, and advise him, for his soul's good, to try to get one ! ...'

But many of Tennyson's scientific readers were enormously grateful for what he had achieved, using poetry to find a subtle line of intelligence and feeling between science and atheism, a thing perhaps impossible at that time in prose. Yet it often took them years to admit this in public. Professor Henry Sidgwick, the Cambridge philosopher, could only write long after Tennyson's death: 'I remember being struck by a note in *Nature* ... which regarded him pre-eminently as the Poet of Science. I always felt this characteristic important in estimating his effect on his generation.' Unlike Wordsworth, who had interpreted nature 'by religious and sympathetic observation', Sidgwick believed that for Tennyson 'the physical world is always the world as known to us through physical science: the scientific view of it dominates his thoughts about it; and his general acceptance of this view is real and sincere, even when he utters the intensest feeling of its inadequacy to satisfy our deepest needs'.

Remarkably it was a passage from the Prologue, or at least its last four and a half lines beginning 'Let knowledge grow from more to more ...', which eventually became the official motto of the second-biggest new Victorian science magazine, *Knowledge*. Launched by the science writer and lecturer Richard Proctor in November 1881, *Knowledge* was the direct rival to *Nature*, and addressed a huge, non-specialist readership. With Tennyson's cited words from *In Memoriam* blazoned on the title page, Proctor placed a symbolic rising sun image on its masthead, with the proclamation: 'KNOWLEDGE – An illustrated Magazine of Science – Plainly Worded – Exactly Described'.

Proctor distinguished the magazine's aims from the specialist *Nature*, in terms that echo Tennyson's own sense of writing for an enquiring, if anxious general readership. 'Knowledge is a weekly magazine intended to bring the truths, discoveries, and inventions of Science before the public in simple but correct terms.' It was notable for its articles on astronomy, geology, electricity and microscopes. It also considered the more general impact of science on society: the effect of telegraphy, railways and even a series 'On the Evolution of the Sense of Beauty'. All this was promoted under the Tennysonian strapline.

As time went on, and *In Memoriam* found its place among Victorian readers, Tennyson himself was increasingly inclined to underplay the uniquely personal aspect of its original, elegiac inspiration. Perhaps the poem was not fundamentally about Arthur Hallam at all, but some much larger loss; and it was certainly not about Tennyson's own autobiography. 'It must be remembered,' he wrote much later, 'that this is a poem, <u>not</u> an actual biography … the "I" is not always the author speaking of himself, but the voice of the human race speaking through him … as to the localities in which the poems were written, some were written in Lincolnshire, some in London, in Essex, Gloucestershire, Wales, anywhere I happened to be.'

By the time he talked to James Knowles, the architect and co-founder of the Metaphysical Society in 1869, whose aim was precisely to reconcile science and religion, Tennyson was keen to emphasise this more general relevance. His real theme had been universal and timeless. 'It is rather the cry of the whole human race than mine. In the poem altogether private grief swells out into thought of, and hope for, the whole world. It begins with a funeral and ends with a

marriage – begins with deaths and ends in the promise of a new life –
a sort of Divine Comedy, cheerful at the end. It is a very impersonal
poem as well as personal. There is more about myself in [my poem]
"Ulysses" . . .' So much, at any rate, he assured Knowles and the members of the Metaphysical Society. The marriage in the poem was of
course his sister Cecilia's to Edmund Lushington, back in 1842. But
should it have been Arthur Hallam's to his sister Emily, long ago? Or
would it come to be understood as his own in the near future?

For while the public debate about the exact meaning and significance of *In Memoriam* developed, Tennyson had finally married
Emily Sellwood very privately, and with sudden precipitation, after
their ten-year delayed courtship. The ceremony took place on 13 June
1850, just a fortnight after publication. It was held at the parish church
of Shiplake-on-Thames, so swiftly that Catherine Rawnsley did not
have proper time to prepare either the bride's wedding dress, or the
wedding cake, both of which arrived at the last moment. Tennyson
did not even forewarn his mother, let alone old friends like Spedding
and FitzGerald, or new ones like John Forster and de Vere. But he
made no apology, writing crisply to Forster: 'I told nobody, not even
her who had most right to be told, my own mother. She was not
angry, why should you be? It is nothing but the shyness of my nature
in everything that regards myself.'

On the marriage licence he described himself as 'Alfred Tennyson
of Lincoln Inn Fields'. The wedding itself was a quiet, country affair
at the Shiplake church of St Peter and St Paul, largely organised by
Catherine Rawnsley. Her husband Drummond Rawnsley was the
officiating vicar, and the church bells were briefly rung. The reception
took place at the vicarage garden next door, though being mid-June
it rained steadily. The ceremony was witnessed by only a very small
circle: Emily's old father Henry Sellwood, her brother-in-law Charles
Weld, Tennyson's sister Cecilia and her husband Edmund Lushington, and some of the Rawnsleys' friends. Two of the little Rawnsley
children were bridesmaids. There were no literary celebrities from
London, not even their supporter Charles Kingsley.

Tennyson's own relatives were notable by their absence. His

mother had not been informed. Frederick was still in Florence, Septimus out of circulation, and his brother Charles announced he could not get down from his Lincolnshire estate; while his sisters Matilda and Mary remained in Cheltenham. Mary said afterwards that she thought Alfred's silence over the whole engagement was 'not fair towards his family'. She added gloomily: 'I hope they will be happy, but I feel very doubtful about it.'

Yet Tennyson obviously relished this lack of fuss or family, exactly what he had planned, and was in evident good humour. He made everyone laugh, when he looked round the vicarage dining room and innocently exclaimed that it was 'the nicest wedding he had ever been at'. Sophie Rawnsley, now Mrs Elmhirst, received a mischievous but good-natured note about the new Mrs Tennyson, on the following day. 'My dear Sophie. We seem to get on very well together. I have not beaten her yet. Ever yours affectionately AT.' The news spread slowly among his London friends, and though Emily was reluctantly approved of, there was of course the sense that the old original bachelor circle had broken up.

Yet the final decision to marry, and how Tennyson at long last made up his mind, always remained a mystery. Tennyson himself was aware of this, writing a solemn explanation to an old Cambridge friend, Robert Monteith. 'I have married a lady only four years younger than myself who has loved me and prayed for my earthly and spiritual welfare for fourteen years. God grant that we may get on well together ... My only fear is that I have lived so long unmarried that I may have crystallized into bachelorhood beyond redemption: then I hope that I have yet some plasticity left.'

Other bachelor friends would slowly be introduced, but FitzGerald was not included in their number for several awkward months. He wrote to Frederick Tennyson in December: 'As to Alfred, I have heard of his marriage, etc, from Spedding who also saw and was much pleased with her indeed.' FitzGerald's 'etc' was surely indicative of his reservations. He added sadly, but a touch defensively: 'But you know Alfred himself never writes, nor indeed cares a halfpenny about one; though he is very well satisfied to see one when one falls in his way. You will think I have a spite against him for some neglect, when I say this, and say besides that I cannot care for his In Memoriam. Not so, if I know myself . . .'

Carlyle, perhaps because he was married, and domestically established at his house in Chelsea with Jane, was introduced more promptly that autumn. He and Jane approved. But, even so, Carlyle could not avoid being faintly patronising about Emily herself. 'Mrs Alfred is a very nice creature, cheerful, good-manner, intelligent, sincere-looking: Alfred and she, I since hear, are in these parts, "looking for houses", but I have seen nothing of them since.'

In fact their first expedition together, immediately after the wedding, was a more private and heartfelt matter. They slipped away to the West Country and, for the very first time, visited Arthur Hallam's grave at Clevedon Court, on his grandfather's estate in Somerset. This, wrote Emily later in her new journal, was *her* 'special wish'. She must have understood quite how important – and symbolic – such a farewell was for her new husband. She obviously shared Tennyson's sense of a completion, a closure. 'It seemed a kind of consecration,' she noted with quiet satisfaction.

They went back via Cheltenham finally to pay their delayed respects to Tennyson's mother, and his remaining sisters. Old Mrs Tennyson had always liked Emily Sellwood, had wanted the marriage to take place for years, and was delighted to see her son settled. She gave them a wedding present of 'a tall crimson glass vase' perhaps as something that would always remind them of flowers, whatever the season. Meanwhile the faithful Aunt Russell produced another handsome and very practical gift of £50.

Duties done, they disappeared in mid-July for a real honeymoon in the Lake District, initially staying at various hotels around Ullswater, and then visiting Spedding's family house near Bassenthwaite. It was such a success that they remained in the Lake District for nearly four months, from July to mid-October, and at last felt like a truly married couple. They moved into the deep privacy of a lakeside house on Coniston, the dreamily named Tent Lodge, lent by one of Spedding's wealthy friends. Set in its own park, 'it looked as lovely as the Garden of Eden'.

Tennyson and Emily walked and talked through the whole summer in this idyllic setting, and rowed halfway round Lake Coniston in a little boat they found at the lodge moorings. It was a physical intimacy that had been delayed for so long, but they both felt health flooding back, and all the old anxieties fading away. One of the results was Tennyson felt able to confide in Emily his fears of depression, and the 'weird seizures' that had sometimes assailed him. He added a description of these to one of the Prince's speeches in the next 1851 edition of *The Princess*, moments of blankness and self-doubt when he 'seemed to move among a world of ghosts'. Another happier consequence was that they confidently expected a baby the following 'March or April': news which Tennyson, secretly bursting with pride, whimsically announced to the Rawnsleys – as 'an heir to nothing'.

Later they invited friends, and especially younger ones like Coventry Patmore and the starstruck poet William Allingham. Tennyson and Patmore climbed the Coniston Old Man, with plenty of brandy at the top. Patmore was impressed by Emily, 'perfectly simple and lady-like . . . her heart always deeper than her mind'. She was 'familiar with the best modern books' and would talk 'freely'. She had read Ruskin and F.D. Maurice, knew Charles Kingsley personally of course; and though evidently more pious than her husband, her religious views were both 'deep and wide'. She now followed Tennyson's work closely: in August she asked Patmore to obtain a recent review of *In Memoriam* in the summer issue of *Tait's Magazine*. Tennyson, he concluded, had made 'no hasty or ill-judged choice'.

Towards the end of their tour they met the young Matthew Arnold travelling in a coach, and visited Wordsworth's widow at Rydal Mount. At her request Tennyson inscribed a short poem in the visitors' album, which had been originally kept by Wordsworth's beloved daughter Dora. Written years earlier, it was in all the circumstances, a significant poem to have chosen to leave behind in the old Laureate's kingdom, and Tennyson published it the following year. 'The Eagle' is an extraordinary image of lonely, savage power. It has just two stanzas.

He clasps the crag with crooked hands;
Close to the sun in lonely lands,
Ring'd with the azure world, he stands.

The wrinkled sea beneath him crawls;
He watches from his mountain walls,
And like a thunderbolt he falls.

The poem may perhaps have been inspired by the eagles Tennyson
glimpsed long ago on the Pic du Midi, during his Pyrenean adventure with Hallam. Or perhaps by a native bird in his later solitary
visits to Scotland. Wonderfully compressed into bright, almost
photographic images, it is an exactly observed wildlife poem, with all
its poised violence brilliantly released in the last line. This vividly puts
into action Tennyson's heraldic phrase from *In Memoriam* No. 56,
'Nature, red in tooth and claw', and confirms the harsh view of relentless evolutionary forces he had struggled to accept. It incidentally
anticipates many such savage, wildlife pieces by the twentieth-century
Laureate Ted Hughes, such as 'Hawk Roosting'.

The poem can also be taken as something more symbolic: the
image of a native poet in absolute command of his world. Perhaps
Wordsworth in the Lake District, or perhaps even Tennyson himself?
Like his earlier iconic creature the slumbering Kraken, Tennyson's
awakened 'Eagle' may also have drawn on a biblical source from the
Book of Job.

Is it at your command that the eagle mounts up
    and makes his nest on high?
On the rock he dwells and makes his home,
    on the rocky crag and stronghold.
From there he spies out the prey;
    his eyes behold it from far away.
His young ones suck up blood,
    and where the slain are, there is he.

As autumn came on, Tennyson returned with Emily to his favoured retreat on Lushington's comfortable estate at Park House, and launched into the quiet domestic business of house-hunting around the Thames. Strangely he still had no further contact with FitzGerald. When Tennyson commissioned the painter Samuel Laurence to make a new official portrait in November, he wanted it based on a sketch that was still in FitzGerald's possession. But he left Laurence to chase it up. The painter agreed crisply: 'If FitzGerald is prompt in sending it, I will do it within a month.' FitzGerald evidently complied, but must have felt saddened by the impersonal nature of this 'prompt' request.

The need for a new portrait soon became clear. After the death of Wordsworth in April, there had been public debate and rumours about a successor all that summer. Many names were canvassed, but on 5 November Tennyson received news that Queen Victoria wished to appoint him Poet Laureate. It was not entirely a surprise. Coleridge was long dead; Leigh Hunt had not been considered suitable because of his scandalous private life; while Samuel Rogers had refused because of his classically advanced age. Rogers was eighty-seven, while Tennyson was just forty-one; in fact perilously young for a lifetime appointment. The final choice fell on Tennyson 'owing chiefly to Prince Albert's admiration for *In Memoriam*'.

He hesitated for a long weekend, aware of the momentous implications of becoming a public figure and part of the Victorian establishment. On Saturday night he wrote two letters, one accepting, one refusing, 'and determined to make up his mind after a consultation with his friend at dinner'. But it was Emily who finally convinced him, and on Sunday morning he sent his graceful acceptance. For the actual ceremony Rogers loaned Tennyson his expensive court dress, which he had previously loaned to Wordsworth for his own Laureateship. So the tradition was meant to pass on. But, to his disappointment, Tennyson found the trousers did not fit.

For the rest, the ceremony went smoothly, though Tennyson later confided to Sophie Elmhirst he had 'rather not been made Laureate', and anyway he would certainly not compose official Odes. But, he explained, he had been advised that, as the holder of a state pension, 'he could not *gracefully* decline the Queen's offer'. And of course both his wife Emily and his mother Elizabeth were decidedly of the same view. Indeed it was rumoured that old Mrs Tennyson had begun informing perfect strangers in the Cheltenham shops, and even on the Cheltenham omnibus, that the new Laureate would be her son.

Altogether this was an entirely new kind of fame. Tennyson began to see that 1850 was becoming his year of transformations. His early life of struggle and obscurity, his essential loneliness and homeless drifting, was suddenly and dramatically over. He was a published bestseller, he was married, he was Laureate, and from now on he was a national figure. Carlyle particularly observed an inner change: 'Alfred looks really improved, I should say; cheerful in what he talks, and looking forward to a future less *detached* than the past has been.'

His poetry had found a mass public, rather than a select reader-ship. Some of the initial consequences were unexpected. 'I get such shoals of poems that I am almost crazed with them; the two hundred million poets of Great Britain deluge me daily with poems: truly the Laureateship is no sinecure.'

He had so many congratulation letters to reply to that he wished he could 'hire the electric telegraph once a month, and so work off my scores with the wires at whatever the expense'. From this point on he would be besieged by the modern world. 'This old-world, slow pen and ink operation is behind the age . . .' He would also be increasingly besieged by visitors – at first his old friends like Carlyle and Brook-field; then new literary ones like Allingham, Patmore and Elizabeth Barrett Browning. But eventually there would be celebrities from every profession, the universities, the law and the Church. Yet at this date, there was still no visit from FitzGerald, who claimed to be dis-tracted all that summer by sailing boats at Woodbridge; though he would still frequently and tenderly quote from Tennyson's 'Old Poems' in letters to other friends.

Tennyson and Emily continued house-hunting at Twickenham in December, hoping to be settled in time for their expected baby in the spring. His sister Mary had written with an unlikely plan for them to move back to Cheltenham, and all share a large family house with their mother, his unmarried sisters and his younger brothers ('he would share the rent'). But clearly nothing was further from Tennyson's mind. He wanted complete independence for himself and Emily from the rest of the Tennyson family, and above all privacy. Ideally a large house with 'all the upper part to himself, for a study and smoking room etc' and no noise above his head.

The marriage was now accepted and even approved of by most of his friends. Aubrey de Vere, a recently converted Roman Catholic, wrote sententiously: 'I already observe a great improvement in Alfred. His nature is a religious one ... he has been surrounded from his youth up by young men, many of them with high aspirations, who believe no more in Christianity than the Feudal system ... his "wrath against the world" is mitigated ... He has unbounded respect for his wife, as well as strong affection, which has been growing stronger ever since his marriage.'

Others thought he was becoming tremendously respectable himself, and slipping – or rather shuffling – into his new Victorian role. A friend of the Rawnsleys noted with satisfaction: 'He realised my idea of a poet, the lines of his face more decided and deeper cut than in the published portraits. He is tall, perhaps six feet high, has a sort of shuffle in his walk, wears spectacles, and speaks with rather a full heavy voice.'

On 10 March 1851 the Tennysons moved from their country base with the Lushingtons, to their first family home at Chapel House, Montpelier Row, close by the Thames at Twickenham. It was a large detached Georgian building, 'a most lovely house' felt Tennyson, with high ceilings and four floors, with many panelled rooms, a fine oak staircase decorated with carved statues, and a big well-stocked garden: an ideal family residence. The upper rooms all had bright, stretching views of the river. Emily loved it, and Tennyson took it unusually on a long five-year annual lease of £50. It was their first proper home together, and this was a joyful time.

Yet the move may also have been physically disruptive for Emily, eight months pregnant and never strong in health. They were soon shadowed by a family tragedy. On 20 April the Tennysons' baby, their first son, was delivered stillborn. It was thought that a slight fall during the last weeks of Emily's pregnancy was responsible. Though the doctor also told Tennyson, perhaps in confidence, that his poor little boy 'got strangled' by the umbilical cord. For a grim moment he thought he might lose Emily as well. She recovered, but as expected, was deeply upset; yet showed the qualities that would sustain her for the rest of her life: patience, piety and immense stoicism. She said of her husband: 'We grew but the more closely together for it.'

More surprising was Tennyson's own shock and grief. This was perhaps unusual in a Victorian father at this date, when infant mortality was so high (one in six babies before their first birthday). Yet Tennyson certainly was greatly troubled. 'I have suffered more than ever I thought I could have done for a child still born,' he admitted. For the first few hours he comforted Emily, but could not bear even to look at the tiny body. Only gradually he nerved himself. The following day he found his son 'even majestic in his mysterious silence after all the turmoil of the night before'. He embarked on a series of agonised letters to friends, eventually sending out over sixty in all, but refused any idea of placing a notice in *The Times*. He repeatedly described the experience and tried to clarify it: 'It nearly broke my heart with going to look at him . . . He lay like a little warrior, having fought the fight . . .'

Finally he managed to contain his overwhelming feelings in a plain, rhythmically broken, but exquisitely rhymed and infinitely touching short poem. He never titled or published it. But it has an intimate and heart-stopping quality, in its own way as powerful as any of the more formal elegies from *In Memoriam*. It seems to mark a profound and permanent shift in Tennyson's sensibility.

Little bosom not yet cold,
Noble forehead made for thought,
Little hands of mighty mould
Clenched as in the fight which they had fought.

He had done battle to be born,
But some brute force of Nature had prevailed
And the little warrior failed.

Whate'er thou wert, whate'er thou art,
Whose life was ended ere thy breath begun,
Thou nine-months neighbour of my dear one's heart,

And howsoe'er thou liest blind and mute,
Thou lookest bold and resolute,
God bless thee dearest son.

This poem has a direct emotional charge like few others written by Tennyson after 1850, and indeed condenses so much that he had written before. 'Some brute force of Nature had prevailed / And the little warrior failed . . .' So perhaps this image of the little baby in the dark womb, fighting heroically to be born, can also be seen as some kind of redeemed version of his nightmare Kraken, fighting in the dark depths of the ocean, to surface into life? This might seem a strange connection to make. Yet the shared physicality of the images, in which the tiny lost child appears suddenly huge and powerful, seemed to haunt Tennyson and continued to bubble up like gasps of air in his letters. 'He was a grand, massive, manchild, noble brow, and hands, which he had clenched as in his determination to be born.'

The experience shook Tennyson profoundly, and he never forgot it. Forty years later he once burst into tears as he talked of his first born, and 'broke down describing the fists clenched as if in a struggle for life'. The bitter shock can be compared with Charles Darwin's reaction to the death of his own little daughter Annie, also in 1851. Only, unlike Darwin, Tennyson's religious faith – such as it was, with all his sceptical doubts and tender agnosticism – was not destroyed by his suffering. On the contrary, supported by Emily's unshaken faith, from this time onwards Tennyson experienced a steady revival in his religious beliefs and confidence in the idea of the immortal soul. Emily wrote a short story about it: 'We felt, as it were, lifted up somewhat higher above the earth by our trust in the Father who chastened us.'

Later Tennyson wrote what seems to be a short, follow-up poem, painfully meditating on the continuing memory of 'the child we lost', but now openly seen as a benign influence. The little warrior is here a healing and unifying force in their lives. The poem was slipped, without explanation, into a later edition of *The Princess* at the end of Section 1, but evidently kept a place in his heart, even if the child was never named and his actual grave is still unknown.

> As through the land at eve we went,
> And plucked the ripened ears,
> We fell out, my wife and I,
> O we fell out I know not why,
> And kissed again with tears.
>
> And blessings on the falling out
> That all the more endears,
> When we fall out with those we love
> And kiss again with tears!
>
> For when we came where lies the child
> We lost in other years,
> There above the little grave,
> O there above the little grave,
> We kissed again with tears.

Later the pain would also be eased by the birth of Tennyson's first surviving son the following summer of 1852. He would be christened Hallam, surely another kind of redemption.

Public life resumed in summer 1851, when Tennyson visited the Great Exhibition, at the Crystal Palace in Hyde Park, London. Energetically organised by Prince Albert, it was proclaimed as the Triumph of Arts, Sciences, Technology and Empire. It was a huge success, eventually visited by over six million people, a third of the British population. It was joyfully greeted by his old tutor William Whewell in a lecture: 'On the General Bearing of the Great Exhibition'. Charles Babbage also wrote a full but more critical commentary, 'On the Great Exhibition'. FitzGerald, dubious of imperial implications, refused to go, and would not even discuss it – 'The Great Exhibition I don't ask about! . . . One is tired of writing, and even seeing written, the word.'

Nonetheless the exhibition was generally welcomed as the moment when imperial England announced itself to the rest of the globe. Even so, Tennyson did not feel obliged to write a Laureate poem in celebration; instead he quietly admired the 'great Glass House', Joseph Paxton's huge glass and iron structure the Crystal Palace, and especially the fountains playing outside it. But his interest in the scientific exhibits was recognised by Prince Albert's administrative team, who organised him a place in the royal box at the grand opening. On a later visit, as he wandered through the art gallery section, he made a quite different personal discovery. He was dazzled by one of the earliest of the pictures by the Pre-Raphaelite Brotherhood, John Everett Millais's *Ophelia*, showing Shakespeare's beautiful mad heroine floating downriver to her death. The following year Millais painted the famous picture entitled 'Mariana' explicitly inspired by Tennyson's early poem. A quotation from the 'aweary, aweary' stanza, was engraved around the picture frame.

Tennyson's publisher Moxon, seeing how the decorative rather than the scientific aspect of Tennyson's poetry had caught the taste of the general public, proposed to commission a fine art de luxe edition, with a careful selection of his shorter poems. These turned out to be almost exactly the 'Old Poems' that were beloved by FitzGerald, though ironically he had no hand in their choice. Moxon thought this might enable Tennyson to purchase a bigger house than Twickenham. Such an edition, 'I am almost sure, within a very short time', would put in his pocket 'at least a couple of thousand pounds'. This Pre-Raphaelite edition, which became known as the *Moxon Illustrated Tennyson*, would appear five years later with limited commercial success in 1857, but immediately became a collector's item. It indicated the safe direction in which Tennyson's popularity as a Victorian poet would now expand. From then on Tennyson concentrated on his long-planned Arthurian epic poem, which over the next decade would become the *Idylls of the King*.

Meanwhile in July 1851 he had taken Emily off on a long, leisurely recovery tour of France and northern Italy. They started in Paris, where they met the Brownings; went over the Alpine passes to Bagni di Lucca (where the Shelleys had once stayed); and stood at dawn on the roof of Milan Cathedral, 'among the silent statues', watching the 'sun-smitten Alps'. For a period there was 'rain at Reggio, rain at Parma ... at Piacenza, rain'. But then they settled for several blissful weeks of sunlit sightseeing in Florence. There they had a surprisingly happy reunion with Frederick Tennyson and his Italian wife Maria, and were lent a four-room apartment with a garden court in their Villa Torrigiani.

Visiting the bustling cafés and packed Florentine galleries, especially the Uffizi, the tragic drama of the stillborn baby seemed to fade, though it would never be forgotten. The relaxed sunlit tone of this Continental holiday, and Tennyson's new domestic contentment, was later caught in a light-hearted travel poem he wrote especially for Emily, 'The Daisy'. It was inspired by a wild flower he picked for her on the snow-bound, zigzagging Splügen Pass in Switzerland, and later found she had tenderly pressed it between the pages of her guidebook. Its dancing metre and radiant imagery was announced from the very first stanza, an absolute break with what FitzGerald and others had criticised as the 'monotonous melancholy' of *In Memoriam*:

> O love, what hours were thine and mine,
> In lands of palm and southern pine;
> In lands of palm, of orange-blossom,
> Of olive, aloe, and maize and vine ...

At Florence, their easy, loving mood was captured in what were effectively a series of fond, dashed-off poetry postcards.

At Florence too what golden hours,
In those long galleries, were ours;
What drives about the fresh Casciné
Or walks in Boboli's ducal bowers.

In bright vignettes, and each complete,
Of tower or duomo, sunny-sweet,
Or palace, how the city glitter'd,
Thro' cypress avenues, at our feet.

When they returned to England in October, they found awaiting them at Twickenham a beautiful set of Italian landscape drawings sent by Edward Lear as a delayed wedding present. The first of many literary visitors began to call: including Francis Palgrave, Aubrey de Vere, William Allingham, Coventry Patmore, and the two young fashionable poets of the day, George Meredith (in his twenties) and soon Algernon Charles Swinburne (still a carrot-haired teenager). This was the first sign of a discrete new youthful fan club – if not a discipleship – forming round Tennyson. All of them started keeping extensive records of their meetings with him in their diaries, journals and letters.

Much of this was mere celebrity gossip and literary chitchat. But some of them, notably Allingham, wrote up long intelligent accounts of Tennyson's after-dinner conversation, and gave vivid impressions of him on long walks. There were many sudden and hitherto unexpected glimpses of Tennyson's views and character. 'To Farringford. After dinner T. spoke of boys catching butterflies. "Why cut short their lives? – What are we? We are the mere moths." Look at that hill (pointing to one before the large window) "It's four hundred millions of years old.; – think of that! Let the moths have their little lives!"'

There was the good news that Moxon had just issued another edition (the fourth) of *In Memoriam*, once again with a print run of five thousand copies. Altogether the Laureate's reputation was growing exponentially. Finally, before Christmas the Tennysons knew that another child was due the following summer.

It was at this point that FitzGerald, at long last, paid his first visit to Twickenham in December 1851, and was introduced to Emily. The visit passed off better than expected, given all FitzGerald's criticisms over the previous two years. Emily was gracious and hospitable, and though FitzGerald was guarded, he concluded Tennyson was 'I believe, happily married'. Having gone back to his Suffolk retreat for the New Year, FitzGerald then wrote a long letter to Frederick describing the demands of the Laureateship, and the effect of these on his old friend's future and poetical ambitions.

He took a long view. Given the continued political disruptions right across northern Europe, and especially in France where Louis-Napoleon was about to crown himself Napoleon the Third, he felt that England was 'the only spot in Europe where Freedom keeps her place'. Accordingly a new kind of national poet was needed to celebrate this unique situation. He wondered if Tennyson could fulfil this vital public role. Yet he feared his opportunity for greatness was slipping away, just at the very point he had become settled and established, his best work already done or dissipated. As usual, FitzGerald put this without mincing his words. 'Had I Alfred's voice, I would not have mumbled for years over In Memoriam and the Princess, but sung such strains as would have revived "the fighting men of Marathon" [*in Greek*] to guard the territory they had won. What can *In Memoriam* do but make us all sentimental?'

19

Empires

B ut fundamental changes were indeed occurring in both Tennyson's life and his writing. Rather as FitzGerald had hoped for a national poet, the imperial theme began to appear in his Laureate poetry throughout the 1850s, and this clearly confirmed him in his old dream of an Arthurian epic of mythic heroes and heroines. He had published 'To the Queen' (the 'revered, beloved' Victoria) in March 1851. In November 1852 he attended the Duke of Wellington's enormous and lavish state funeral in London, itself an imperial event, with ten thousand troops marching in procession down Pall Mall and Fleet Street, a military cortège including seventeen field guns and the duke's coffin in a heavily armoured gun carriage ('a monstrous bronze mass', according to Carlyle).

Tennyson's long 'Ode on the Death of the Duke of Wellington', strenuously stretched out to 281 lines, was published two days beforehand on 16 November. Private elegiacs for Arthur Hallam were now forcibly replaced by the gestures of public mourning. 'I wrote it because it was expected of me to write,' he explained to his Aunt Russell. The public Ode was a difficult form to master after the intimate voice of *In Memoriam*, and this showed not only in the flat formal rhymes that Tennyson summoned, but especially in his laboured attempts at funeral rhetoric. Here for example, the grand opening stanza, sets off in the muffled pace of a solemn slow march, only to stride out of step in its elongated fourth line:

> Bury the Great Duke
> With an empire's lamentation
> Let us bury the Great Duke
> To the noise of the mourning of a mighty nation,
> Mourning when their leaders fall,
> Warriors carry the warrior's pall,
> And sorrow darkens hamlet and hall.

Moxon, alive to the commercial possibilities of the Laureate market, swiftly published it in a pamphlet of ten thousand copies for a fee of £200, and offered Tennyson a huge advance of £1,000 on future work. This was doubly welcome since the birth of little Hallam Tennyson on 11 August that year. The baby was a week premature, but he flourished noisily, which delighted Emily even though he allowed her almost no sleep. 'The little sensual wretch roaring day and night for ailment lets her have no quiet,' wrote Tennyson, pretending he had rather wished for a little girl, but actually bursting with pride. Emily had difficulty nursing the child, and had to recuperate for several weeks on a sofa. But the whole household rejoiced, and Henry Hallam was invited to be the child's godfather.

Tennyson proudly announced his birth in *The Times*, and sent out innumerable letters. He did not write a poem, as he had so signally for his stillborn son. Instead he hosted a huge christening breakfast feast at Chapel House, on 5 October, with more than fifty guests. According to Emily, the writers and poets among them were 'as thick as blackberries', and they included Robert Browning (who bounced the baby on his knee), William Thackeray and F.D. Maurice. Jane Carlyle came unaccompanied by her husband, 'in consequence' was able to enliven the table with unexpected jokes. The Reverend Drummond Rawnsley provided a haunch of venison, beautiful peaches and 'the best champagne we could get'. In effect it was a long delayed house-warming party. But among notable absentees were his old bachelor friends, Spedding and FitzGerald.

Public affairs continued to interrupt domestic ones, and Tennyson began to read the newspapers with greater attention. Politics rather than science were now holding his thoughts, yet another change. Since Louis Bonaparte had seized the imperial crown and decreed the Second Empire in France, Tennyson was haunted by the idea of a future European war, and clearly felt his imperial duties as Laureate to warn and counsel. He first undertook the unlikely task of urging the formation of local militias. For this he drummed up a series of hearty patriotic songs, with percussive titles like 'Rifle-Clubs!!!' and 'Britons, Guard your Own!', each armed with battalions of exclamation marks. These he published anonymously in magazines, but never reprinted, though they hinted at a new kind of public poetry to come. They were not intended to be subtle. He sent one copy in confidence to his young supporter Coventry Patmore:

> We thought them friends, and we had them here,
> But now the traitor and tyrant rules!
> And Waterloo from year to year
> Has rankled in the heart of fools.
> We love peace, but the French love storm!
> Riflemen, form! Riflemen, form!
> Riflemen, riflemen, riflemen, form!

Tennyson added apologetically: 'Very wild but I think [not] too savage! Written in about 2 minutes. The authorship a most deep secret! ... My wife thinks it too insulting to the French, and too inflaming to the English.'

Amid these increasingly imperial diversions, FitzGerald did manage to make a second, follow-up visit to the Tennysons at the end of that year in December 1852. He diplomatically prepared the ground beforehand: 'Dear old Alfred ... I should like greatly to have a crack with you, and to see your dear old face again. As also Mrs Tennyson's *new* face. EFG.' He did not mention the missed christening feast for little Hallam in the summer.

When he arrived he behaved with all his tact and charm: 'I admired the Baby greatly and sincerely.' He was impressed to see that 'Alfred nurses him with humour and majesty.' Tennyson told him he had not seen the child 'in his full glory', however – which was 'sitting high and smiling'. Fitz was impressed by this unexpectedly doting father; but he wondered how much had changed. When they were finally alone, they talked of Tennyson's future poetry and his new Laureate and imperial duties. FitzGerald did not admit that he had secretly lost much faith in Tennyson's work, despite or perhaps because of Tennyson's public recognition, or even of his family happiness.

Yet he tried to be encouraging as in the old days. 'We had two long talks and smokes over the Ode [on Wellington]: which he has altered and enlarged quite successfully, I think. He is disappointed that people in general care so little for it – but I tell him they will learn to understand it by degrees: and that it will outlive all ignorance.' He was amused by Tennyson's anonymous 'Riflemen' songs, but secretly thought his old friend was rather 'full of *Invasion* . . . He also wrote some very fine Song on the Subject . . . but he says nobody listens or cares.' FitzGerald then made a casual suggestion that would eventually bear unexpected fruit. Despite his warlike concerns and new imperial anxieties, Tennyson should follow Fitz's own peaceful example, and start the study of medieval Persian poetry at Oxford.

The recurrent emphasis on 'dear old Alfred', as a term both of affection and of regret, was surely significant. Privately FitzGerald continued to prefer what he called 'Tennyson's *old* poems'. The previous spring, when walking up the hillside covered in cowslips above his lonely Suffolk cottage, he had found himself reciting 'The Lady of Shalott' and the 'Morte d'Arthur' out loud. 'As fresh as when I heard them nearly twenty years ago, perhaps *fresher* for being remembered so long.'

Now he went back to London to seek some sort of consolation from Carlyle, and found it one evening when he called unannounced at the Chelsea house. Carlyle was 'more than usually kind and subdued; wore spectacles to read with: his wife went out: and we lay down each at a side of the fire, and talked of Life slipping away!' It was evidently the kind of conversation – literally a fireside chat – that he longed to have with Tennyson, but which no longer seemed possible.

On reflection FitzGerald thought the domesticated Tennyson had looked rather 'shrivelled and dwindled', though in good spirits. 'He nurses his Child delightfully. But he will never write Poetry again, as I believe. I mean such Poetry as he was born to write.' This was an old refrain. There was also the question, once again, of 'old Alfred's port-drinking', obviously not entirely restrained by Emily. 'Some stimulus is necessary for him ... A Bottle of Port is really nothing to a man with such a Chest and Brain as his ...' FitzGerald said he did not wish to spread rumours or 'misunderstandings' in these reports about his brilliant old friend. But the sense of disappointment, and even a little jealous mischief-making, shadowed them.

The physical distance between them increased when Tennyson decided to move right away from London. He had not spoken about this with Fitz, but was anxious to settle into a true country home with Emily, and before their next child was born. He became 'so engaged in flying about the country in this wretched house hunting business – now in Kent – now in Surrey – now in Gloucestershire – now in Yorkshire – that I can never be sure of my whereabouts a day before-hand'. After further thoughts of the Cornish coast (with its Arthurian echoes), and even a snug seaside house at Lyme Regis, he and Emily took the ferry that summer from bustling Lymington across the Solent to sleepy Yarmouth. Here they went exploring right across the Isle of Wight. They soon found a large, rather rundown Georgian farmhouse at Farringford, with thirty acres of isolated gardens and grounds and most beautiful views. It was grandly positioned above Freshwater Bay on the south-western tip of the island.

It was a curious, fairy-tale-looking building. Outside it had a long, low brick frontage, with a pillared entry deeply shrouded in creeper. There were fifteen rooms, with windows wildly framed by imitation Gothic tracery. From a parapet above the third floor jutted a line of unlikely mock battlements. The homely farmhouse seemed to dream of becoming a lonely medieval castle at night, and this very much suited Tennyson's future literary plans. He reserved the single upstairs attic room as his library and study. To some extent it all felt like a return to the Gothic enchantments of the Somersby vicarage, but on a bigger scale, and now with Arthurian intimations.

Emily loved it too, but for different reasons, having exclaimed, 'I must have that view.' She was impressed by the grand reception hall with its high windows, the spacious farmhouse kitchen, and spare rooms for nurseries and servants quarters. She thought the little port of Yarmouth, three miles away down the hill, with its shops and postal service, was convenient enough. But she noted that two of their Twickenham servants 'burst into tears saying they could never live in such a lonely place'.

Tennyson had been looking for years for a lonely house by the sea, ever since the cottage at Mablethorpe, and now at last he had found it. He would stay there for the remaining forty years of his life (though he later added a fine country house at Aldworth on the North Downs). In November 1853 he rented – and finally purchased it – for the enormous sum of £4,350, to be paid over three years. Tennyson could never have remotely afforded anything like this before. But Moxon's promises of future royalties had transformed his financial situation, and much else besides. There was also a new set of shrewd capital investments advised by the faithful Lushington: no longer some doubtful artisanal wood-carving business, but thoroughly solid stock holdings in the East Lincolnshire railways. He was ringing down the grooves of change. This also enabled Tennyson to pass on the Twickenham house to his mother, so she could leave Cheltenham at last. She remained at Twickenham, together with her unmarried daughters, for the rest of her life; occasionally holidaying – by strict invitation – at the beautifully named Freshwater.

Farringford, withdrawn rather than commanding, seemed at last a suitable residence and retreat for a Poet Laureate. It offered enormous seascapes beyond its gardens, and long walks over the High Down to the west (soon to be known locally as Tennyson's Down). The resilient turf, alive with wild flowers in the spring, ran right along the steep chalk clifftops as far as the Needles. On stormy nights the sound of the sea roaring up the shingle beaches would reach the house, and struck Tennyson with familiar and reassuring music. It took him back to his earliest childhood memories of the North Sea in Lincolnshire, as if he had come full circle. It would sound again in his poem *Maud*, and mark a kind of reconciliation with the dark waters of the Kraken.

The big winter skies above Freshwater also took Tennyson back to his old love of astronomy, perhaps the most enduring (and uplifting) of all his scientific interests. His fascination with geology and evolution theory rather waned by comparison, though in May 1854 he would make a special visit to view the fine 'Iguanodons and Ichthyosaurs' models at the Crystal Palace, when they transferred from Hyde Park to the new site at Sydenham. He was 'much pleased by them', and shared the new enthusiasm for 'dinosaurs', and thought the whole site was 'certainly a marvellous place, but yet all in confusion'. But he now had the opportunity to use his new telescope, and even the time and leisure to draw complex star maps, some of which he shared with Emily. 'I hope we shall have another peep tonight,' he enthused to her in October.

That autumn he constructed a complete 'diagram of Orion', the constellation of the warlike Hunter with his Belt and glittering pendent Sword, hanging high overhead in the south. It seems that Tennyson's early fascination with Orion was still inspired by de Quincey's extraordinary essay on Lord Rosse's great telescope observations of the constellation. 'It is the famous nebula in the constellation of Orion: famous for the unexampled defiance with which it resisted all approaches from the most potent of former telescopes; famous for its frightful magnitude and for the frightful depth to which it is sunk in the abysses of the heavenly wilderness ...'

Tennyson observed two of the brightest stars in the night sky, yellow Rigel and red Betelgeuse, singular to his naked eye but revealed as a massive tumbling cluster of lights when viewed through his telescope. He carefully instructed Emily about this marvel: 'Look out at Orion, at a faintish star under the lowest star of the belt. That is really 8 stars all moving in connection with one another, a system by themselves, a most lovely object through the glass.' His own closer observation also disclosed the much smaller separate nebula (Messier No. 43) almost hidden halfway down the star line of Orion's Sword. He told Emily with mild triumph: 'I saw also the famous nebula which is in the 2nd star of the Sword and is amazing', and added in best pedagogic mode: 'Rigel one of the bright stars in Gemini is double two brilliant suns.'

To Moxon he wrote dreamily of buying new telescopes and having their 'night-glass' especially cleaned. 'In this lovely place one has nothing to look at but sea and stars.' But there was plenty to read about.

Over the coming years Tennyson would build up an extensive collection of astronomy books at Farringford, and establish a special cosmological section in his library there. These books would start with Whewell's recent *On the Plurality of the Worlds* (1853), and some basic star maps like Alexander Johnston's *School Atlas of Astronomy* (1856). They would expand to more advanced star studies such as Richard Proctor's *Half-Hours with the Telescope* (1868) and Norman Lockyer's *Spectroscopic Observations of the Sun* (1870). Later still Tennyson formed a lively friendship with Lockyer himself: one of the most famous of amateur Victorian astronomers, who established a public observatory in Devon, and became the founding editor of *Nature* in 1869, committed to explaining and popularising scientific ideas.

He and Lockyer exchanged letters, and Tennyson was invited several times to visit Lockyer's private observatory set up in his garden at West Hampstead. Lockyer – by then Sir Norman – also visited Farringford, and talked of the chemical composition of starlight, and the nature of meteorites. He quoted freely from Tennyson's poetry, and remarked wonderingly to Emily: 'His mind is *saturated* with astronomy.' Long after, Lockyer would compile an entire book around this idea, and his memories of Farringford: *Tennyson as a Student and Poet of Nature* (1910).

Stars were still much on his mind when, the next spring, their second son Lionel was born as he had hoped at Farringford, on 16 March 1854. Tennyson was especially triumphant, and named the child after the starry constellation of Leo, which he had been observing overhead that very night. 'A strong and stout young fellow came into the world, abusing it loudly, last night at nine . . . Mars was culminating in the Lion.' But here the Lionel name set off the new imperial note. 'Does that mean soldiership?' he joked with a new heartiness, while sending greetings to FitzGerald and other friends.

Yet the birth of his second son, and his growing awareness of what struggles and pain both children had caused Emily, brought home to Tennyson how lucky he was to be a father after all this time. The children's presence evidently transformed the whole household at Farringford, and released springs of feeling in their father that had never been expressed before, outside his poetry. He now watched them together with a tender doting eye, and put them into his letters. 'Little Hallam's behaviour to his small brother was very enchanting. He kissed him very reverently, then began to bleat in imitation of his cries; and once looking at him he began to weep, Heaven knows why: children are such mysterious things.'

This time FitzGerald was invited for the christening in June, and was apparently asked to be a godfather. But it turned out he was only to act as a proxy one, to stand in for Emily's old family friend Drummond Rawnsley who had married them at Shiplake. For FitzGerald it was a bittersweet occasion, his first visit to the Laureate's new kingdom at Farringford. As usual, he was careful to charm Emily, who spent most of her time on the sofa looking after the baby, 'very weak ... and not able to walk', according to Tennyson, but nonetheless commanding the household. FitzGerald fussed over the baby, and played Mozart on the piano in the drawing room to Emily ('one glorious air after another'), and tactfully absented himself for long afternoon walks along the Down, bringing back bunches of wild flowers for her.

This partially succeeded with Emily, who thought Fitz 'as amusing as man could be', and as a city man had actually 'learned to climb hills'. She liked the landscape sketches and beautiful bunches of 'horned poppies and yellow irises' he brought back for her. But perhaps these were rather too like peace offerings. She grew impatient when he kept Alfred too long in a corner in the evenings, peering at what would turn out to be Persian poetry, and hurting his eyes. It seemed a relief on both sides when FitzGerald left after a fortnight. This first visit was also his last ever to the Isle of Wight.

Subsequently he wrote regular letters, though it was Emily Tennyson who usually answered them. He took care to stay on good
terms with her, if only to observe his friend's imperial conversion at a
safe distance. As early as February 1852 he had noted, 'I had a letter
from Mrs Tennyson the other day, telling me Alfred is very warlike,
and has enrolled himself in a Rifle Corps. He now thinks himself that
an Invasion would make a Poet of him.'

FitzGerald was right to suspect that Tennyson's Laureateship had
subtly awakened his sense of patriotic duties. In the early 1852 poem
'To the Queen' he first wrote explicitly of the 'yokes' of empire. But
imperialism only came upon Tennyson slowly, and it was not until
twenty years later, and having nearly completed the whole of the
Idylls of the King, that he could confidently refer in a second poem
dedicated 'To the Queen' (1873) to unswerving imperial loyalties to
the Crown and all its stretching domains:

Our ocean-empire with her boundless homes
For ever-broadening England, and her throne
In our vast Orient . . .

It is true that there had been the military dreams, and foreign con-
quests (some erotic), which haunted the soldier-narrator of 'Locksley
Hall'. But what really first stirred Tennyson's imperial consciousness,
and roused his 'warlike' tendencies – though in a paradoxical way –
was his fascination with the unfolding of the Crimean conflict.

A vast naval expedition had been assembling just across the waters
from the Isle of Wight, in the great naval harbour of Portsmouth
throughout 1853. The plan was to sail down through the Dardanelles
and drive the Russians out of the whole of the Crimean peninsula, by
the simple device of capturing the port of Sevastopol. From the cliffs
above Freshwater, Tennyson could see the warships setting out down
the Channel all through the spring of 1854, and hear the boom of their
gunnery practice. He was also working on the final part of *Maud*, and
now decided to conclude it by having his madman, 'sane but shattered',
seek redemption by enlisting in the Crimean expeditionary force. 'I

Tennyson by James Mudd, c.1857. The first of many formal studio shots, featuring his new beard and Spanish hat, and celebrating not only his Laureateship (1850) but the birth of his two sons Hallam (1852) and Lionel (1854).

Emily Sellwood aged 22, while still living with her parents at Horncastle, Lincolnshire, a miniature on ivory, 1835. Tennyson got to know her at local dances and slowly fell in love; but after a broken courtship in 1840, mysteriously did not propose until 1849.

Rosa Baring, the beauty of Harrington Hall, near Spilsby, Lincolnshire, one of the inspirations for Tennyson's 'Locksley Hall'. She married a rich Lincolnshire landowner, Robert Shafto, in 1838, and this photograph dates from some twenty years later about the time of Tennyson's *Maud*, 1855, which revives his memories of her.

Charles Lyell in 1846, the revolutionary geologist whose great two-volume *The Principles of Geology* (1830–1833) transformed Tennyson's view of the natural world.

An unusual image of young Charles Darwin in 1840. His early *Beagle Journal* (1846) prepared Tennyson for the evolutionary ideas of his later work.

Robert Chambers, science journalist and author of the sensational *Vestiges of the Natural History of Creation* (1844) which popularised new concepts of universal Evolution before Darwin's *Origin of Species* (1859), and powerfully influenced Tennyson's *In Memoriam*.

Elizabeth Barrett Browning (1806–1861), leading Victorian poet and early fan of Tennyson's work, who often heard him give readings in London and Florence, criticised his *Princess* about an all-woman university, and later responded with her proto-feminist poem *Aurora Leigh* (1856).

Mary Somerville (1780–1872), Scottish mathematician and science writer, whose brilliant general survey *On the Connexion of the Physical Sciences* (1834) shaped much of Tennyson's thinking about the forces at work in nature. This portrait celebrates the many later editions of her work.

Charles Babbage (1791–1871), mathematician and computer pioneer, eccentric mentor of Ada Lovelace, and waspish critic of Natural Theology. He read Tennyson's poetry and sent him jokes about it.

A modern NASA image of the Whirlpool galaxy (M. 51), first identified by William Rosse, using his giant new telescope the Leviathan in 1845.

A contemporary drawing of the Leviathan, Rosse's 72-foot reflector with its 3-ton mirror, constructed at Parsonstown outside Dublin. The biggest telescope in the world for the rest of the century.

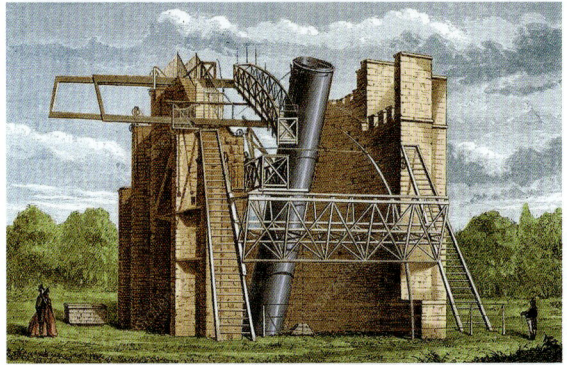

Sir John Frederick Herschel FRS (1792–1871), astronomer, cosmologist, the most celebrated public scientist of the early Victorian period. Like Tennyson, he became an iconic figure through photographs by Julia Margaret Cameron (1867). His books such as *A Treatise on Astronomy* (1833) deeply influenced Tennyson's view of the evolving star systems and a dynamic universe.

Danaë and the Shower of Gold, 1564, by Titian. A sensuous mixture of mythology and cosmology, this is the Vienna version, about which Arthur Hallam wrote ecstatically to Tennyson shortly before his death in 1833. It eventually inspired Tennyson's most beautiful love sonnet with its haunting line 'Now lies the Earth all Danaë to the stars' (1847).

Thomas Carlyle (1795–1881), Scottish historian and controversial essayist, author of *Chartism* (1839), photographed 1854. He befriended Tennyson, walked with him all over London at night, and held long metaphysical discussions and smoking sessions with him at his house in Chelsea. He kept a special pipe for Tennyson in a niche in his garden wall.

Jane Carlyle (1801–1866), witty and independent-minded wife, and brilliant letter writer, photographed 1855. She loved Tennyson's conversation, could keep him going all evening with whisky, and stoically ignored his appalling pipe.

John Ruskin (1819–1900), art historian and Victorian cultural critic, fascinated by geology, and so painted by a rocky stream, by J.E. Millais, 1854. He championed the work of Turner and the Pre-Raphaelites, and wrote personally to both Tennyson and FitzGerald, to praise *In Memoriam* and later *Omar Khayyam*.

A casual group of 13th Light Dragoons in Crimea before mounting up, a rare war photograph by Roger Fenton, 1855. Two thousand copies of Tennyson's 'Charge of the Light Brigade' were later sent to Sevastopol as printed leaflets, intended as part of a Christmas good-cheer package by Tennyson's publisher Moxon.

Alfred and Emily Tennyson peacefully with their children in the Farringford garden, 1864. Hallam on the right, aged 12, carries a hat in imitation of his father; Lionel, aged 10, firmly grasps his mother's hand.

Alfred Tennyson aged 56, a noble and defining photograph of the Victorian bard, but known in the family as 'The Dirty Monk'. One of a series of other celebrities, carefully composed by his neighbour Julia Margaret Cameron, 1865.

Emily Tennyson aged 52, a tender and expressive painting by their family friend, G.F. Watts, 1862. It rightly suggests both Emily's unworldly fragility and yet her great inner strength, both necessary for coping with the older Poet Laureate.

have felt with my native land, I am one with my kind.' It was a strange and controversial decision, which would shock many of his more liberal readers, FitzGerald among them.

Certainly as Laureate, Tennyson now identified increasingly with the Victorian establishment, and may have genuinely shared the patriotic mood of the moment. He explained that the final Part 3 of *Maud* was 'written when the cannon was heard booming from the battle-ships in the Solent before the Crimean War'.

But perhaps he felt, at the same time, that the early war-fever of 1854 was part of a general insanity. His madman finally rejects a 'hysterical' obsession with Maud, in favour of public duty and imperial warfare. The scene is both melodramatic and ambiguous. He imagines himself on the deck of one of those warships in the Solent and simultaneously among the jingoist crowd (the 'loyal people') cheering on its departure from the quayside. It is a patriotic fantasy which possibly Tennyson shared.

> 'It is time, O passionate heart and morbid eye,
> That old hysterical mock-disease should die.'
> And I stood on giant deck and mixt my breath
> With a loyal people shouting a battle cry . . .

The whole war divided the nation. It was regarded by some in Britain as a distant, heroic imperial duty; and by others as a foolish, expensive and unnecessary campaign. The British naval expedition had joined forces with the French to repel the Russian occupation. The Russians had seized Crimea from the Turks, whose crumbling Ottoman Empire was collapsing at various points around the Black Sea coastline. In particular the Russians had reinforced their naval base at the huge natural harbour of Sevastopol, on the south-western tip of the Crimea. From here they threatened to dominate all shipping across the Black Sea as far as the Bosporus and Constantinople. This menaced both British and French imperial interests, though the imperial cause – unlike the Greek War of Independence thirty years previously – was not universally popular with the British public.

Nevertheless the whole household at Farringford followed the fate
of the British naval expedition, involving over thirty thousand troops.
Even little Hallam, now aged two and a half, marched about the house
beating a toy drum, and flung himself on the carpet playing dead Russian soldiers, while baby Lionel squealed with delight. These childish
noises must also have been in Tennyson's ears, besides the naval guns.
In the autumn his involvement suddenly became more intense and
personal. He eagerly read newspaper accounts of the British landings
outside Sevastopol in September, and then the allied bridgehead
established on the plain above the port, secured by the Battle of Alma.
Then came the first confused account of the great Battle of Balaclava
in October 1854. He received a private letter describing the prelimin-
aries for the charge of the Heavy Brigade, which emphasised the
tremendous stunning and inescapable noise of battle. 'Our ears were
frenzied by the monotonous incessant cannonade going on for days
together.' This deadly inhuman noise, together with the cannonades he
had himself heard on the Solent, gave him the first slight suggestion
for what would become the thunderous refrain of his most famous,
headlong, galloping poem.

> Half a league, half a league,
> Half a league onward,
> All in the valley of Death
> Rode the six hundred.
> 'Forward, the Light Brigade!
> Charge for the guns!' he said.
> Into the valley of Death
> Rode the six hundred.

The plains of Balaclava lay immediately above Sevastopol, the allies'
strategic gateway to the besieged Russian city. The succeeding action,
soon to be made famous – or infamous – as the Charge of the Light
Brigade, was remarkable both for its reckless courage and for its violent, pointless slaughter. A brigade of British light cavalry, consisting
of nearly seven hundred troops carrying only lances and sabres, and

riding unarmoured horses, were mistakenly thrown into a suicidal cavalry attack. The vagueness of the final numbers involved, estimated between 607 and 669 troops, gives some indication of the lack of organisation. They were commanded to make a full-frontal galloping charge against a battery of fifty heavy Russian field guns, dug in on a hillside redoubt at the end of a narrow mile-long valley. This lethal approach, some 1,700 yards, became Tennyson's 'valley of Death', a term he originally took from Psalm 23: 'Even though I walk through the valley of the shadow of death I will fear no evil ...'

But there was plenty of evil. More than a third of the cavalry officers and troopers, 271 out of the 'noble six hundred', were killed or seriously injured, and very few returned unwounded. These numbers omitted the 335 horses killed. The overall casualty rate was estimated at more than 40 per cent. It may be noted in passing, its commander Lord Cardigan, who led the charge from the front, rode back unscathed, and dined that evening on champagne in his private yacht moored in Balaclava harbour.

The charge became celebrated – and then notorious – because it was accurately reported by the new generation of Victorian war correspondents. For the first time realistic on-the-spot battlefield accounts were sent uncensored to the London newspapers by independent witnesses. The first official report by Lord Raglan had been published in the *London Gazette* on 12 November, and briefly mentioned 'the brilliancy of the attack', and 'some misconception of the instruction to advance'.

But Tennyson was able to read a full account of the cavalry charge, vividly observed from the heights above Balaclava, by the young Irish reporter William Howard Russell, who was employed as special correspondent for *The Times*. Russell's first memorable piece (he despatched several) was published in the foreign news pages of the paper on 14 November. Already there was a feeling that something had gone terribly wrong:

'They swept proudly past, glittering in the morning sun in all the pride and splendour of war. We could hardly believe the evidence of our senses. Surely that handful of men were not going to charge an army position? But they were ... Through the clouds of smoke we could see their sabres flashing as they rode up to the guns and dashed between them. The blaze of their steel, an officer standing near me said, "was like the run of a shoal of mackerel".'

Tennyson picked up all these vivid images. But what really haunted him was a paragraph from *The Times* editorial of the previous morning, 13 November. It began by comparing the cavalry charge to some great 'heroic' sea disaster, but concluded with a bleak, accusatory phrase about an error. 'Causeless and fruitless, [the Charge] stands by itself as a grand heroic deed, surpassing even the spectacles of a shipwrecked regiment settling down into the waves, each man still in his rank. The British soldier will do his duty, even to certain death, and is not paralysed by feeling that he is the victim of some hideous blunder.'

The drowning sea image would have transfixed Tennyson, but it was that last phrase, 'some hideous blunder', which echoed in his thoughts, and subtly changed its rhythm. Tennyson later recalled: 'The Times account had "Someone had blundered", and the line kept running in my head, and I kept saying it over and over till it shaped itself into the burden of the poem.'

> 'Forward, the Light Brigade!'
> Was there a man dismayed?
> Not though the soldier knew
> Someone had blundered.
> Theirs not to make reply,
> Theirs not to reason why,
> Theirs but to do and die.
> Into the valley of Death
> Rode the six hundred.

It is clear that Tennyson initially regarded 'The Charge of the Light Brigade' as a kind of imperial protest poem. His first version, almost journalistic in presentation, appeared not in *The Times*, but in a special boxed column on the front page of the liberal – and sometimes radical – newspaper the *Examiner* on 9 December 1854. The original

MS was sent quite informally by Emily to John Forster, their friend and editor of the newspaper: 'Will you kindly put this into the Examiner for Alfred? It was written yesterday on a recollection of the first report in The Times which gave the number as 607. He prefers "six hundred" on account of the metre but if you think it should be altered to 700 which from later accounts seems to have been the number he says you are to alter it.'

Emily did not remark on what might be called Tennyson's poetic – or cavalier – attitude to exact military numbers, 'on account of the metre'. Instead she thought of the soldiers themselves, and added with a characteristic instinct of tenderness, that she wished she might be 'some bird of fabulous size and power to carry warm clothing and nourishing food to the poor soldiers'.

The poem was simply signed 'A.T', but contained seven stanzas, one more than the final popular version of six. This extra stanza, which appeared as the second verse, distinctly sharpened the tone of the poem, by emphasising the notion of the terrible mistake. In it Tennyson even explicitly named the officer, Captain Louis Nolan, who had carried the fatal 'instruction to advance' from Lord Raglan on the commanding heights to Lord Lucan down in the valley below, and then to Lord Cardigan at the head of the brigade itself.

> Into the valley of death
> Rode the six hundred,
> For up came an order which
> Someone had blunder'd.
> 'Forward the Light Brigade!
> 'Take the guns,' Nolan said
> Into the valley of Death
> Rode the six hundred.

The poem hugely increased Tennyson's public profile, partly because it was taken – and largely misunderstood – as a triumphant song of noble patriotism. In fact it tells the story of a grotesque double military blunder: the outward cavalry charge in the wrong direction,

followed by a horrific ride back under continuous enemy gunfire from
both flanks on return. The result was the appalling, unnecessary casualty count.

> Stormed at with shot and shell,
> While horse and hero fell,
> They that had fought so well
> Came through the jaws of Death
> Back from the mouth of Hell,
> All that was left of them,
> Left of six hundred.

Told with unflinching economy, it maintains a nightmare, drumming
beat across a very short dactylic line. Its headlong motion is constantly jerked back by the grim implication of its drum-roll rhymes:
'hundred – blundered – sundered – wondered – thundered'. It is now
difficult to read its rousing final stanza without irony, and wondering
if Tennyson – the master of doubt – partly intended this.

> When can their glory fade?
> Oh the wild charge they made!
> All the world wondered
> Honour the charge they made!
> Honour the Light Brigade,
> Noble six hundred!

Tennyson himself later reflected: 'My heart almost burst with indig-
nation at the accursed mismanagement of our noble little Army, the
flower of our men.' Ironically it became the most popular poem Tennyson ever wrote, even among the army. Among many unusual
tributes from the general public, he received a personal note from the
chaplain of the military hospital at Scutari, where Florence Nightingale would soon become the Lady with the Lamp. The chaplain told
him that some stanzas of the poem had spread among the troops.
'Half are singing it and all want to have it in black and white, so as to

read what has so taken them.' A similar letter came from one of the Scutari doctors, who said that the poem was read to convalescing cavalry men who had taken part in the actual charge. Here was more poetry to be learned by heart.

Accordingly, as Laureate, Tennyson approved of an astonishing form of publicity campaign proposed by the editor of the *Examiner* and his own publisher Moxon. Two thousand copies were printed as single-sheet leaflets, and shipped out directly to the British troops in Crimea, as part of a Christmas good-cheer package. The poem went on selling for the next forty years, and the total funds raised by its sale were eventually set aside in 1890 for the Society for the Well-Being of Crimean War Survivors. Tennyson was even persuaded, though with some reluctance, to record the poem on one of the new Thomas Edison phonograph devices. The equipment, with its waxed discs and varnished speaking tube, was carefully carried across to the Isle of Wight in May 1890. The recording session, overseen by an American Edison engineer Mr Stiegler, also included 'Blow, Bugle, blow', parts of *Maud*, and the song from *The Princess*, 'Ask Me No More'.

Remarkably, six minutes of his recording have survived, and can still be heard. Through the crackle and click of the wax cylinder, Tennyson's noble voice – though reduced to a metallic screech – quivers with heroic conviction, giving a passionate rendering which emphasises the thunderous beat and rhythm of the lines. At one moment there is a repeated banging sound that might almost be Tennyson striking the table with his hand to keep time. The other poem on the Edison tube is the lovelorn song, almost equally popular and frequently set to music, 'Come into the garden, Maud'. Both can now be heard together on YouTube, an extraordinary pairing of the apparently romantic and the apparently militaristic. Once recited out loud they are almost impossible to forget. They are two kinds of hypnotic Tennysonian music: so widely treasured, so frequently mocked, and so often misunderstood.

Moxon could see the commercial possibilities of both. The poems were published rapidly in his next collection, *Maud and Other Poems* (1855), together with pieces about other martial subjects, such as 'The

Charge of the Heavy Brigade' and the 'Ode' to the Duke of Wellington, which give the whole volume its decidedly warlike and inescapably
imperial air.

A group of survivors from the charge, with buttoned tunics but no
horses, were also photographed by Roger Fenton in Balaclava shortly
afterwards, one of the earliest of all war photographs. The debate
between Raglan and Lucan – who was to blame? – began and still
continues. Nolan himself did not survive to give his own explanation,
having had his head blown off in the first hundred yards of the charge.
All this added to the celebrity of the whole incident. Years later Tennyson was invited as guest of honour to the Light Brigade's reunion
at Alexandra Palace in 1875, though did not finally attend.

John Forster's highly emotional response, written for the *Guardian* at the end of 1854, gave a clear idea of the poem's national impact,
and how Tennyson – even withdrawn to the remote Isle of Wight –
had become identified with the heart of empire. 'How I value this
noble ballad, I need not say – how proud I am to print it first, and that
my old friend sent it to me, I *must* say. I hear little of you, but again
and again I think of you, and never have I done it so often as of late –
never, with such a throbbing heart, have read of those fights of heroes
at Alma, Balaklava, and Inkermann, that I have not been eager for *you*
to celebrate them – the only man that can do it up to their own
pitch – the only "muse of fire" now left to us that can of right ascend
to the level of such deeds.'

20

Stars

Tennyson's new fame as a poet, and growing popularity as a national figure, inevitably alienated some of his original friends, but most painfully FitzGerald. Once Tennyson's household had established itself on the Isle of Wight, their worlds and outlooks were increasingly and inevitably separated. Looking back, FitzGerald saw his old friend being gradually overwhelmed by, what seemed to him at least, self-important guests, doting admirers and celebrity hunters. 'Ah, if he [Tennyson] did not live on a somewhat large scale, with perpetual Visitors, I might go once more to visit him,' he lamented. Edward Lear experienced the same loss of intimacy, eventually complaining how 'an odious palaver and fuss succeeded to quiet home moments'. On his subsequent visits Lear had to stay at the tiny hotel in Freshwater village.

Yet Tennyson could still protect his family privacy, especially with Emily and his children. His own youth was coming back, even while he bought trumpets for little Lionel and kites for Hallam. With the huge open horizons, and unpolluted night skies above Freshwater, Tennyson turned his telescope away from naval shipping and war-like horizons, and back up again to the eternal stars. 'We looked at Orion last night . . . a most lovely object through the glass.' It was as if astronomy had become for him the alternative frontier of hope in the new Victorian world of military empire and colonial expansion. Instead of the old dark menace of the deep-sea Kraken, he turned to the radiant new light overhead. Moreover, it was a fascination he continued to share with Emily, and he sent her extensive notes about his star-gazing.

Their favourite constellation Orion was now augmented by the planet Saturn and its mysterious rings. In January 1856 he told her: 'I saw Saturn pretty well the night before last, but I could not see the division of the rings – it only seemed a big ring flaking off into thinness towards the edges, but I saw a bit of the dark ring at the base of the planet.' He then drew the planet with one large ring heavily shaded, and added carefully: 'The dark ring is the most interior, and I only saw it where it crossed the bright.' Yet the whole planet seemed to him endued with a strange marine life, and even its true colours were illusive. 'Others saw it slate-colour, pulsing within the other rings.' Sometimes he even believed the stars themselves had poured down his telescope, and somehow taken up occupation deep within his own eyes. These were probably just 'floaters', but he told his Aunt Russell that with his frequent night outings 'I have got some 15 *new* specks in my right eye; these all occur together, like a group of dark Pleiads.'

Despite all interruptions, the sea-girt Isle of Wight would frequently remain a place of magic, in which Tennyson, isolated in his top-floor study under his neo-Gothic battlements, could finally confront and appease many of his darker terrors. Here, 'morning and evening' throughout 1854, he composed many confessional lyrics, and earlier fragmented voices from his time at High Beech, to assemble the last poem of his youth, *Maud, or the Madness*. He presented it as a long, highly unconventional work, semi-dramatic in form, whose provocative subtitle Moxon eventually persuaded him to soften to the more explanatory: *Maud: A Monodrama*. Tennyson also later referred to it as his 'Drama of the Soul', and 'my little Hamlet'.

He filled its three unequal parts with the broken monologues of his mad and lovelorn narrator, pacing the cliffs or wandering through illusive landscapes of memory. Part 1, over nine hundred lines, in a bewildering variety of metrical forms, ends with the beautiful and sinister aria, 'Come into the garden, Maud'. Part 2, over three hundred lines, contains the narrator's terrible 'buried underground' monologues, with 'my heart is a handful of dust'. Their strangeness and broken interjections have no true parallels in Victorian poetry, and await the arrival of the broken voices of T.S. Eliot's *The Waste Land* in 1922. While the final Part 3 is delphically short at sixty lines, with its uncertain glimmerings of starlit hope and patriotic redemption. 'I have felt with my native land, I am one with my kind.'

Initially Tennyson placed his narrator on the Freshwater Down at night, listening to the sea in the shingle bay below, and watching the wheeling stars above. These passages are full of agonised memories, painfully vivid, yet now written with confident and steady power. Sometimes it is Maud's actual face he conjures up, with merciless clarity:

> Cold and clear-cut face, why come you so cruelly meek,
> Breaking a slumber in which all spleenful folly was drown'd,
> Pale with the golden beam of an eyelash dead on the cheek,
> Passionless, pale, cold face, star-sweet on a gloom profound;
> Womanlike, taking revenge too deep for a transient wrong
> Done but in thought to your beauty, and ever as pale as before
> Growing and fading and growing upon me without a sound,
> Luminous, gemlike, ghostlike, deathlike, half the night long
> Growing and fading and growing, till I could bear it no more . . .

Or sometimes it is only the memory of it, which tortures him at night, and drives him from his bed, to go outside and listen to the sea below, which itself seems 'maddened', or rather – in an extraordinary image – which has made the stones and shingle of the beach itself, scream in maddened pain:

> . . . But arose, and all by myself in my own dark garden ground,
> Listening now to the tide in its broad-flung shipwrecking roar,
> Now to the scream of a madden'd beach dragg'd down by the wave,
> Walk'd in a wintry wind by a ghastly glimmer, and found
> The shining daffodil dead, and Orion low in his grave.

The word 'Womanlike' gives pause for thought. Were these passages a final exorcism of the cold, beautiful Rosa Baring? Or was there still some faint, pale, infinitely distant, lurking ghost of the youthful Hallam? Was Tennyson's favourite constellation, Orion the Hunter, finally being used to bury more than one cruel layer of memory under the horizon of young Tennyson's past life? Certainly it was the first of many such burials, and voices from the 'grave' fill the whole poem in so many metrical forms.

These also now summoned the different asylum voices first heard at High Beech. Whatever their source, their violent images, and the whole obsessive love-story that *Maud* relates, seemed quite different from anything previously in *In Memoriam*. It also contained several long passages that appeared perilously close to confessional, such as the striking opening of Part 3.

> My life has crept so long on a broken wing
> Thro' cells of madness, haunts of horror and fear,
> That I come to be grateful at last for a little thing:
> My mood is changed, for it fell at a time of year
> When the face of night is fair on the dewy downs,
> And the shining daffodil dies, and the Charioteer
> And starry Gemini hang like glorious crowns
> Over Orion's grave low down in the west,
> That like a silent lightning under the stars
> She seem'd to divide in a dream from a band of the blest,
> And spoke of a hope for the world in the coming wars—
> 'And in that hope, dear soul, let trouble have rest.
> Knowing I tarry for thee,' and pointed to Mars
> As he glow'd like a ruddy shield on the Lion's breast.

For Tennyson's many new readers, these seemed strangely inappropriate revelations from the new Poet Laureate. As a result the new book, published as *Maud and Other Poems* on 28 July 1855, was much talked about, and something approaching a *succès de scandale*. It seemed extraordinary that the same collection could also contain such utterly different poems as the thunderous 'Charge of the Light Brigade', the charming traveller's keepsake 'The Daisy', and the solemn 'Ode to the Duke of

Wellington'. The contradictions became a great talking point, and a subject of intense literary speculation. Moxon was of course delighted with the large initial sales in consequence, and Tennyson's accumulated royalties leaped from £445 in 1855 to £2,058 in 1856. Yet he had to accept that the critical reception was very mixed, except for the faithful *Examiner*.

Several newspapers picked up the sense that Tennyson was struggling to deal with unfinished materials – and conflicts – from his youth. If so, most thought he had not succeeded. The *Morning Post* suggested cheerily that the true title was not Maud but frankly 'Mad' or indeed 'Mud'. *The Times* put this more loftily. 'Mr Tennyson has never yet presented to the public anything so crude, so shapeless, and so commonplace.' While the *Press* ridiculed what it identified as an atmosphere of reheated adolescent passions. 'Maud' expressed 'a strain of puling incoherent sentiment and disordered fantasy such as might flit through the brain of a love-sick youth in the measles'.

Other private critics included George Eliot, who criticised the poem's 'morbidity', and warmongering; while Matthew Arnold dismissed it in a letter to his friend Arthur Hugh Clough, as 'thoroughly and intensely *provincial*'. Although this did not stop Arnold sending Tennyson a presentation copy of his own 'The Scholar Gypsy' via a friend. Tennyson received it gracefully, considering it was a rival publication in 1855. Of Arnold he remarked, 'nobody can deny he is a poet'. But he could not forbear to add in the same letter that his own 'poor Maud' was about to be 'slashed all to pieces' in a Scottish paper, possibly the *Edinburgh Review*.

Evidently Tennyson felt all this deeply. 'My poor Maud,' he wrote with feeling after the first crop of bad reviews, 'she has been beaten as black and blue by the penny-a-liners as the "trampled wife" by the drunken ruffian in the opening poem.' He even got hostile anonymous letters addressed to him personally at Farringford, including one that signed off 'Yours in aversion, a former admirer'. His supporter John Forster in the *Examiner* was almost the only one to 'speak justly and honestly' about the collection as a whole. Consoled by Emily, he tried to be philosophical about all this. 'I always calculated on a certain quantity of anonymous insolence, but I have had more than my share of it ...'

Nevertheless surprising and very welcome support came from a quite different quarter: doctors and medical journalists. Dr Robert Mann, a retired physician living nearby at Bonchurch on the Isle of Wight, had invited Tennyson over to use his telescopes and microscopes, and then got him to discuss the medical aspects of *Maud*. As a result, Dr Mann published an entire pamphlet entitled *Maud Vindicated: An Explanatory Essay*, commenting in detail on the psychological development of its mad narrator. This was 'so finely conceived and so subtly enunciated', and expressed both in its imagery and its fantastic variety of metres and stanzaic forms.

Mann particularly dwelt at length on Tennyson's beautiful observation of stars, combined with his completely up-to-date use of modern astronomy – 'the science of the present age'. He saw that Tennyson responded to the philosophic terror of their newly discovered magnitudes. 'What a wonderful embodiment there is in this designation of the very essence, physically speaking, of star existence.' Here was another recognition of nature's alienating violence, not 'red in tooth and claw' as on earth, but 'stunning . . . with iron immensity' in the depth of space. Mann quoted with approval the fierce lines where the mad, lovestruck narrator of *Maud* gazes up into the night, and is

> . . . brought to understand
> A sad astrology, the boundless plan
> That make you tyrants in your iron skies,
> Innumerable, pitiless, passionless eyes,
> Cold fires, yet with power to burn and brand
> His nothingness into man . . .

Of the 'buried alive' passages such as 'my heart is a handful of dust', Mann wrote: 'The syllables and lines . . . actually trip and halt with abrupt fervour, tremble with passion, swell with emotion, and dance with joy, as each separate phase of mental experience comes on the scene.' Such critical understanding and appreciation Tennyson received gratefully, 'as true as it is full'.

A long essay also appeared in the October 1855 edition of the *Asylum Journal of Mental Science*. Written by the journal's editor John Charles Bucknill, *Maud* was reviewed seriously and favourably as a brilliantly dramatised 'case history' of insanity and its various phases, rather than as a purely literary work. Bucknill praised the Laureate's 'wonderful psychological insight' and stated that 'the writings of Tennyson are peculiarly metaphysical, or to use the new term *psychological*'.

Bucknill praised the technical or medical truth of Tennyson's imagined narrator, and 'vouched for the accuracy of the mental pathology of the protagonist'. Tennyson felt that 'the testimony' of such medical men was 'more valuable to me than that of the mere literary critic'. Concerning Bucknill's essay, Tennyson wrote proudly to Mann: 'I seem to have the doctors on my side if no one else. I have just received an article by a mad-house doctor giving his testimony as to the truth to nature in the delineation of the hero's madness. Valuable testimony it seems to me.'

Its value grew perhaps, when Bucknill was later knighted for his pioneering work in psychiatry and was elected a Fellow of the Royal Society.

Thus Tennyson's 'monodrama', with all its images of rejected love and psychic burial, could also be placed in a new and alternative scientific-literary tradition: that of psychiatric case studies. One might say that this would eventually lead by the end of the century to such works, part scientific analysis and part literary drama, as Sigmund Freud's famous studies like 'Dora' (1900) and 'The Wolf Man' (1914).

Nevertheless the popular criticism was a blow. Tennyson left the Isle of Wight for a few days, and went walking alone in the New Forest to recover. He got strangely lost, as he wrote to Emily from the Crown Hotel, Lyndhurst on 1 September 1855. In his distracted wandering he picked wild forget-me-nots along the banks of the River Itchen ('what beautiful water!'), and in his dreamy state left behind his precious tobacco pouch under a spreading beech tree: 'so like the colour of last year's beech leaves that I did not see it as I turned to leave'. He told Emily: 'I lost my way in the Forest today, and have walked I don't know how many miles. I found my way back to

Lyndhurst by *resolutely* following a track which brought me at last to a turnpike: on this I went a mile in the wrong direction . . .'

He had just been reading Dante's *Inferno*, so his account suggests the famous opening, in which Dante symbolises the creative midlife crisis as being plunged into a 'dark wood' in which the 'true direction' is lost: '*mi ritrovai per una selva oscura / ché la diritta via era smarrita*'. Unbeknown to him, FitzGerald had used exactly the same Dantesque image of his friend's 'lost' career. But for Tennyson the true direction would reveal itself clearly in the form of another kind of 'turnpike': the shaped and traditional narrative of the *Idylls of the King*, stretching ahead in twelve books of steadily pacing blank verse, from 1859 to 1885. Tennyson later said that these lonely wanderings through the New Forest directly inspired the first book of the series, *Merlin and Vivien* (1859), which recounts the unexpected seduction of the wise old wizard by the beautiful Vivien in an enchanted wood.

> For the pale blood of the wizard at her touch
> Took gayer colours, like an opal warmed . . .
> For Merlin, over talked and overworn,
> Had yielded, told her all the charm, and slept.

'My admiration for the Forest is great,' he told Emily on his return, evidently thinking of the Arthurian setting. 'It is true old wild English Nature – and the fresh-sweetened air is so delicious. The Forest is grand.'

His recovery from the attacks on *Maud* changed something else. Tennyson had always enjoyed reciting his own poetry in private, to his intimate friends, from the time he intoned 'The Lady of Shalott' at Trinity, and the 'Morte d'Arthur' to FitzGerald and Spedding long ago on Windermere. FitzGerald had even been persecuted by the frequency of these recitals during the difficult days at Camden Town. But now Tennyson wanted to *perform* his poetry *in public* for quite new supporters, especially the younger ones. A youthful American enthusiast reported back to Longfellow: 'One morning he read to us the whole of *Maud* in a style I cannot soon forget . . . His usual tone is a low unmelodious thunder-growl, but when he chooses he can melt, as well as rasp with his Lincolnshire tongue.'

In September 1855 the Brownings (temporarily back from Italy) invited him to a soirée with a few friends at their large rented London house in fashionable Dorset Street, Marylebone. Tennyson made the trip to London (now by ferry and railway), though again without Emily. It was something of a literary celebration. EBB, who had given birth to her own son 'Pen', and had recently published her immensely popular collection of forty-four love poems, *Sonnets from the Portuguese* (1850), was still working on her own highly experimental dramatic poem *Augusta Leigh* (1856). As Tennyson's supporters (despite their doubts over *The Princess*) the Brownings commiserated with him about the bad reviews, until he suddenly volunteered to give a private reading of the whole of *Maud* after dinner.

As they all determinedly settled into comfortable chairs, other friends came in until Tennyson found himself surrounded by half the Pre-Raphaelite Brotherhood. They included the poet Dante Gabriel Rossetti, the painters Ford Madox Brown and Holman Hunt, and the art critic William Rossetti. Rossetti remembered 'Tennyson seated on a sofa in a characteristic attitude, and holding the volume near his eyes . . . he read *Maud* right through.' Perhaps to while away the time, Dante Rossetti made surreptitious pen sketches, which he later gave to the Brownings as a souvenir of the occasion. They show Tennyson sitting sideways on a sofa, with one leg pulled up comfortably under the other, positioned for the duration. 'The Poet Laureate neither saw what Dante was doing, nor knew of it afterwards. His deep Grand voice, with slightly chanting intonation, was a noble vehicle for the perusal of mighty verse.'

But Rossetti also remembered, less nobly perhaps, Tennyson's repeated 'groanings and horrors' over the bad reviews of *Maud*. Despite this, or indeed because of it, his 'poor little Maud' remained Tennyson's favourite party piece for the next thirty years, and even Henry James remembered such an extraordinary performance in 1878. He was impressed, but strangely disconcerted by the difference beween the sublime poetry and the old poet actually 'spouting' it. It was

Jamesian paradox. As 'the organ roll of monotonous majesty' went on and on, he had to accept 'the full monsterous demonstration that Tennyson was not *Tennysonian*'.

At last in the summer of 1856, support suddenly came from FitzGerald. He had been away in Germany, looking at works by Rubens and Tennyson's favourite Titians. He was full of Continental enthusiasm, and even accepted the novel and fragmented form of Tennyson's latest poem. 'My dear Alfred . . . I like the *Drama* of Maud very much – the Characters each shadowed into his Proportional Place like an old Play – and all leaving something of [Walter Scott's] Bride of Lammermoor Gloom on one. I don't like the Lyrical Execution so well as I ought, I suppose. But it is doubtless an original form of Poem: with its Action etc altogether better to me than Princess and Memoriam – which is one of Fitz Crotchets.' This was an olive branch, even if spiked in the Fitz style; and perhaps there was more to come.

Meanwhile Tennyson found a better way of putting *Maud* in perspective by taking long walks over the wind-blasted Down above Freshwater Bay. He 'trudged out with the local geologist, Mr Keeping, on many a long expedition' and discussed the latest work of the controversial naturalist Richard Owen. Owen was the Hunterian Professor of Anatomy and Physiology at the Royal College of Surgeons, a brilliant lecturer and science writer, who had recently crowned his career by popularising the term 'dinosaur', meaning 'the terrible lizard'. This of course still fascinated Tennyson, with his own long-standing interest in those 'Dragons of the prime / that tare each other in their slime', as well as the symbolic life of the undersea monster of the Kraken.

Owen had reconstructed his own versions of these ancient land monsters from scientific drawings, and later presented them with huge success as life-size models. Tennyson had seen some thirty-three of these models at the Crystal Palace Exhibition in 1851, and met Owen himself the following August in London to discuss their evolution. Tennyson described the enthusiastic paleontologist as 'the Bone man, the Professor', while Owen observed the Poet Laureate with a diagnostic eye: 'He struck me as being ... care-marked, dark-eyed, rather bilious-looking ... with spectacles ...'

But it was the stars that still held Tennyson's attention. In the winter of 1855 he began reading his old Cambridge tutor, William Whewell's new astronomical study *On the Plurality of Worlds: An Essay* (1853). Despite its expansive title, he was puzzled to find it was a work of Christian revisionism. Whewell had originally proposed in his *Bridgewater Treatise* of 1829, a profoundly Romantic idea: that the creative power and bounty of God was 'boundless', and therefore the universe must surely contain numerous other civilisations on distant planets: a glittering 'plurality' of other earth-like worlds and intelligent life forms.

But the older Whewell now argued that neither John Herschel's recent *Outline of Astronomy* (1849), nor even the observations of Lord Rosse's telescope the 'Leviathan', could demonstrate that the most distant nebulae were anything more mysterious than 'cosmic gas clouds'. They were not resolvable into star clusters, and so they were not evidence of intelligent life forms, far, far beyond our own Milky Way. The 'universe' had thus shrunk back to a smaller, safer, emptier concept. 'Thus we appear to have good reason to believe nebulae are vast masses of incoherent or gaseous matter, of immense tenuity and thus to have made it certain that *these* celestial objects at least are not inhabited.'

Whewell fell back on narrower and more conservative Christian cosmology. God must have created mankind *unique* in the entire universe. His final reasoning was based on moral rather than scientific concepts. God's scheme of incarnation and resurrection must imply a *single* centre of intelligent life in the entire universe: that established by divine fiat on the planet earth. There was no need for anything further. As Whewell put it: 'One school of moral discipline, one theatre of moral action, one arena of moral contests for the highest prizes, is a sufficient centre for innumerable hosts of stars and planets, globes of fire and earth, water and air ...' In his own way, Whewell, now in his sixties, had become a Victorian.

Tennyson studied the book 'carefully', and at the same time reread

Dante and compared his astonishingly imaginative presentation of the medieval universe in the *Commedia*. By comparison he found Whewell's revisionist account dismayingly limited and earth-centred. Moreover it was arrogant in an imperial way, putting our planet ('a sufficient centre') in command of the entire universe ('innumerable hosts of stars'). Tennyson felt this was a conclusion that lacked both scientific, and poetic humility. He dismissed Whewell's revisions: 'I would not wish you to waste your eyes over it: it is to me anything but a satisfactory book ... It is inconceivable that the whole Universe was merely created for us who live on a third-rate planet of a third-rate sun.'

The growing evidence that the solar system occupied a marginal place in the Milky Way, and that the earth was so to speak at the edge of things, was being explored in new astronomical papers by John Herschel and Richard Proctor. Tennyson would keep up with these ideas in his continued scientific reading over the next thirty years. His modest reference to our 'third-rate' planet and sun was a deliberate rejection of religious hubris and imperial certainty, just as he had originally rejected it in the scientific agnosticism of *In Memoriam*. Tennyson, despite all imperial visions and Laureate grandeur, could still recognise that the true mystery of planet earth came down to its humblest and most familiar life forms. Walking about the walled section of garden at Farringford, he entered in a notebook:

> Flower in the crannied wall,
> I pluck you out of the crannies
> I hold you here, root and all, in my hand,
> Little flower – but *if* I could understand
> What you are, root and all, and all in all,
> I should know what God and man is.

21

Young Laureate

As Tennyson became famous nationally, more and more distinguished visitors took the ferry to the Isle of Wight, and flocked over the Down to Farringford. This was especially true after the controversial impact of the *Maud* volume in 1855. People were puzzled and intrigued: Who exactly was this young Poet Laureate, barely into his forties? Was he the secret, grieving tortured lover of *Maud*? Or was he the ebullient, patriotic bard of 'The Charge of the Light Brigade'? Or was he the subtle, tender agnostic of *In Memoriam*? The fascination with this enigma brought Tennyson the best book sales he had ever had, and considerable wealth. The slim but watershed publication of *Maud* was swiftly followed by the big de luxe edition of Moxon's *Illustrated Tennyson* in 1857, which gained him yet another kind of artistic readership and the fashionable support of the Pre-Raphaelite Brotherhood who contributed to it.

The handsome volume, with its scarlet binding, contained thirty beautiful woodcut illustrations by leading Pre-Raphaelite painters and another twenty by other more commercial artists. It contained especially fine illustrations – or, really, visual interpretations – of five of the poems: 'Mariana' (by Millais); 'The Lady of Shalott' (one by Dante Gabriel Rossetti, a second by Holman Hunt); 'Break, Break, Break' (a fine beach scene by Clarkson Stanfield); 'Godiva' (respectfully clothed, by Holman Hunt); and the 'Morte d'Arthur' (a sword in the lake, by Daniel Maclise). Hunt's vision of the Lady of Shalott, furiously tangled in her web, would later inspire a magnificent sequence of three full-scale paintings by John William Waterhouse between 1888 and 1915.

John Ruskin, who had 'greatly admired' *Maud*, wrote appreciatively from Denmark Hill, suggesting that such woodcuts were in reality 'always another poem' in parallel with the written one. If not aesthetically satisfactory, they would be 'of much use in making people think and puzzle a little'. He especially picked out Rossetti's pictures 'Sir Galahad' and 'The Lady of Shalott'. Moxon printed a first edition of ten thousand copies. The Pre-Raphaelites' overwhelming fascination with Tennyson's visual imagery dates from this time.

There was also growing interest in the *Idylls of the King*, which with the publication of the first four sections in 1859 ('Enid', 'Vivien', 'Elaine' and 'Guinevere'), began to be regarded as a kind of state of the nation saga; or even 'the state of its Soul' (if there was such a thing). Moxon confidently printed an astonishing first run for a single book of poetry: forty thousand copies. He also sold it to the American publishers Ticknor and Fields. Here was contemporary Victorian society apparently refracted through the coloured, medieval lens of the chivalric Arthurian legends: their loves, their battles, their hypocrisies and their ornate betrayals. Tennyson would work on it doggedly, some twelve thousand lines over the next thirty years, until 1885. But was such a refraction convincing or even relevant? Was its vision, and especially its science (Tennyson's beloved geology and astronomy), sacrificed for the decorative neo-Gothic manner from the moment it began?

> 'Thereafter – as he speaks who tells the tale
> When Arthur reached the field of battle bright
> With pitch'd pavilions of his foe, the world
> Was all so clear about him that he saw
> The smallest rock far on the faintest hill,
> And even in high day the morning star.'

Among those unconvinced was, of course, FitzGerald, though he eventually tried to tell Tennyson in a more tactful manner than usual. 'The whole Myth of Arthur's Round Table Dynasty in Britain presents itself before me with a sort of cloudy, Stonehenge grandeur.' Tennyson must have liked that monumental image. But FitzGerald could not restrain himself from pressing on with yet 'one more of Old Fitz's crotchets': his preference for the early work. 'But always excepting for the Morte d'Arthur ... I am not sure if the old Knight's adventures do not tell upon me better, touched in some lyrical way (like your own "Lady of Shalott") than when elaborated into Epic form.' Sadly FitzGerald would not live to see the paintings of the 'Lady of Shalott' by Waterhouse.

But by 1859 the *Idylls* had to contend with many other historic publications, mostly in prose, which redefined what Victorian society had become, and what it thought about itself. Just that same year saw the publication of Darwin's *On the Origin of Species*; as well as Mill's political classic *On Liberty*; Samuel Smiles's exemplary *Self Help*; George Eliot's first novel *Adam Bede*; Mary Ward's great illustrated scientific popularisers, *The Telescope* and *The Microscope*; and Dickens's *A Tale of Two Cities* (swiftly followed by the serialisation of *Great Expectations*). The same year also saw the fourth and revised edition of Elizabeth Barrett Browning's *Aurora Leigh*, which had now come to be seen as a controversial feminist novel in verse.

Yet the Farringford visitors were no longer strictly literary celebrities. There was the sculptor Thomas Woolner, who carved Tennyson's profile for a medallion, and arranged for his looming Christ-like portrait by G.F. Watts to be hung in the Royal Academy Summer show. There was the future Master of Balliol College, Oxford, Benjamin Jowett, anxious to discuss Plato and Florence Nightingale. Other visitors included the great physicist John Tyndall, keen to explain why the sky is blue; the eccentric Cambridge mathematician Charles Dodgson, just starting to photograph Alice Liddell, and soon a firm favourite of Tennyson's sons; and the controversial new chancellor of the exchequer – and one-time friend of Arthur Hallam at Oxford – William Ewart Gladstone, his mind set firmly against the Crimean War, being far too expensive for the Treasury, and making Tennyson suddenly worry if his shares in the East Lincolnshire Railway Company might collapse.

Adding social cachet to these gatherings were numerous self-improving members of the Great British aristocracy, such as the Duke and Duchess of Argyll. The 8th Duke, George John Douglas Campbell, was a Fellow of the Royal Society and revealed a passionate interest both in geology and in Tennyson's poetry. He was also the owner of a Highland castle and an expert in ornithology, both of which appealed to Emily. Exploring these subjects led to extensive walks along the Freshwater clifftops and long discussions about evolution, followed by even more extensive correspondence. The duke became president of the Geological Society, but like Lyell he argued against the ideas that eventually became fullblown Darwinian evolution, which he regarded as irreligious. He and the duchess were, however, progressive enough to encourage Tennyson to recite *Maud* in full, and at least excerpts of the *Idylls*, when he visited Argyll House in London.

In 1859 the duke gave a lecture at the Glasgow Athenaeum entitled 'Geology: Its Past and Present', which persisted in the traditional notion of a creative divinity at work, rather than Darwin's natural selection. It is notable that this lecture was specifically dedicated to Tennyson as Poet Laureate. The lecture concluded: 'It is thus that Science becomes the Vestibule of the Church, and Knowledge the handmaiden of Religion. All this, and more than this, is beautifully expressed in that great storehouse of poetic truth and feeling, Tennyson's *In Memoriam*.'

Some visitors like William Allingham were encouraged to stay for days, almost becoming members of the family, and tacitly allowed to keep extensive diaries of the Laureate's fine recitals, noble thoughts and terrible table manners. While other visitors were a demanding honour. In May 1856 Prince Albert himself dropped in, quite informally and without any warning. The ground-floor rooms were being refurnished, and the floors were covered with Hallam's toys. Emily recalled the dramatic moment. 'Two loud rings at the bell. The housemaid with a face that terrified me coming close whispered in a mysterious voice "His Royal Highness Prince Albert" ... He was looking out of the window when Alfred came in, and there he remained for [Alfred] never thought of asking him to sit down. He

offered him wine.' In fact Prince Albert became a great supporter of
Tennyson's poetry and passionately shared his scientific interests.

Among the more literary visitors was Edward Lear, briefly back
from his travels in Greece, half in love with Emily, but as it turned
out completely in love with Edmund Lushington's youngest brother
Franklin, whom he eventually followed to Corfu. While not provid-
ing him with rooms, Tennyson instead wrote him a tactful verse letter,
'To EL. On his Travels in Greece', with a glowing reference to his
travel journals and sketches:

> And trust me while I turn'd the page,
> And tracked you still on classic ground,
> I grew in gladness till I found
> My sprits on the Golden Age.

Other younger fans included a student rebel from Oxford, Algernon
Charles Swinburne, who worshipped the idea of Tennyson as the
Great Bard – or at least as the Bard in the masochistic agonies of his
Maud. While the seasoned feminist poet Caroline Norton, now cam-
paigning for divorce reform, was always ready to debate the virtues
and vices of *The Princess*. Allingham would eagerly note the most outlandish combination of such guests. On one occasion: 'Gladstone
dined here on Monday – Swinburne – Milnes – De Sade – Naked
model – the chastest thing I ever saw.' The naked model turned out to
be a painting by Rossetti, rather than an actual person. It was probably
the picture of Fanny Cornford, sitting modestly amid a tangled mass
of honeysuckle and roses.

All visitors, whether offered port or poetry, or the complete Farringford sit-down dinner (which always included both), tended to
remark on one striking physical transformation that had overtaken
the Laureate: the full Tennyson beard. This quickly became iconic
and its significance was variously debated among his friends. In general it reflected a decisive change in masculine styles and fashion at
this time, which heralded the Victorian frock coat, the ascending top
hat and the descending sideburns. These finally joined under the chin

as 'mutton chops', to become the full Victorian beard (though this was notably eschewed by Prince Albert). Tennyson stuck with his wide-brimmed Spanish hat and heavy cloak, as before. But the symbolism of his new beard – and the beard in general – led to wide speculation.

It was a mark of virility and male authority perhaps, the married man and his patriarchy within the family; but also of military prowess, political allegiance and expanding empire. Before 1850, there were striking pictures of clean-shaven Tennyson, Darwin, Dickens, Lyell and others. The Duke of Wellington had always remained clean-shaven until his death, and this certainly influenced most establishment men in public affairs until then. It was noted that in 1847 only a single member of the House of Commons wore a beard (George Muntz, MP for Birmingham); while by contrast many of the Chartist leaders (lobbying for universal male suffrage and annual Parliaments) sported 'radical' moustaches and beards. The male visitors to the Great Exhibition in 1851 were not usually bearded.

What changed the fashion in the midst of this decade was certainly the social impact of the Crimean War, as it changed much else in English sensibilities, including its poetry. Dickens's magazine *Household Words* first observed the shift in fashion with a light-hearted article by Henry Morley in summer 1853 entitled 'Why Shave?'; while the *Westminster Review* responded with a more serious general survey in 1854 raising the 'Great Beard Question'. Beards and moustaches were officially permitted in the British Army from 1850; by 1855 they had become virtually compulsory among the most hard-fighting Crimean foot regiments, as for example the 95th (Derbyshire) Regiment of Foot who were famously photographed on location.

Indeed photography became a decisive medium of this change. Freed from the primitive limitations of the 'French' daguerreotype in the 1840s, and through the English experiments of John Herschel, Fox Talbot and many others with glass plates and 'fixing' salts, photographic equipment became mobile in the 1850s. So it was through photography that the beard made a final visual break with the Romantic generation. There are no authentic photographs of Blake, Shelley,

Keats, Coleridge, Wordsworth or even the image-conscious Byron, all of whom remained clean-shaven. There is one possible studio portrait of Thomas de Quincey, who survived his opium addiction until 1859, which marks the exception. By contrast there are very extensive photographic records of the emergence of Dickens's new beard, representing Victorian fiction; Carlyle's beard, representing Victorian history; and Tennyson's beard, representing Victorian poetry and the coming of all things Arthurian.

As he adopted it, the Young Tennyson of Lincolnshire gradually began to disappear into Old Tennyson of the Isle of Wight. In fact he had first experimented with a small beard at Twickenham, to Emily's alarm in 1851. But increasingly painful teeth and then their replacement by a whole set of 'new teeth' during a session with his dentist in February 1853, encouraged him to adopt a much fuller beard, initially as a form of therapeutic disguise. As the years progressed it became fuller, wilder, and more demonstrative. One of the earliest public photographs of a dark, thoughtful, bearded Tennyson, with his now famous Spanish hat, was taken four years later by James Mudd during Tennyson's impromptu visit to the Manchester Art Exhibition in May 1857. Mudd somehow persuaded him to sit for a moment in a rather fine oak chair, possibly as one of the exhibits.

But Emily was never quite reconciled to the beard, and looked back fondly to the time Thomas Woolner cut the early Laureate medallion of a still beardless Tennyson in 1854, which she always treasured. She was also delighted when Woolner developed this into a full-size bust of Tennyson, still with his handsome unobscured jaw-line, which was purchased (with a private subscription organised through the Lushingtons) by his alma mater Trinity College in 1859. Whewell was urged to place this in the college's famous Wren Library, but since Tennyson was still a 'Living poet' – with or without the beard – Whewell cautiously consigned it to the library's 'Vestibule' until the necessary 'promotion' had been achieved. That is, until Tennyson was dead: so not yet.

Meanwhile Emily had complained to Woolner in March 1857: 'I wish the public could compel Alfred by act of Parliament to cut off his beard!' She even appealed to James Spedding in London: 'Pray do not defend the beard; it will surely fall.' At this time, among her other favoured Farringford visitors, Edward Lear had also taken to a beard, which he cultivated in exile in Corfu, until it eventually grew to tropical proportions. Looking around at his contemporaries Lear later produced one of his definitive limericks on the subject, which appeared in his third expanded and illustrated *Book of Nonsense*, in 1861. He may have had himself rather than Tennyson in mind, but it is notable that his nonsense songs became popular with Hallam and Lionel Tennyson as they grew up.

> There was an Old Man with a beard,
> Who said, 'It is just as I feared!—
> Two Owls and a Hen, four Larks and a Wren,
> Have all built their nests in my beard.'

Tennyson's image as Poet Laureate was finally defined by a series of black and white portrait photographs, taken with conscious artistic intent in the mid-1860s by Tennyson's new neighbour on the Isle of Wight – Julia Margaret Cameron. Charles and Margaret Cameron were wealthy heirs to imperial coffee plantations in Ceylon, who had downsized from their London family base at Little Holland House, to settle in a village near Freshwater. Here they restored a series of farm buildings, built a look-out tower, and constructed a photographic studio, christening the whole eccentric establishment 'Dimbula', after their original Ceylon plantation.

The rich crop they now harvested consisted of photographic portraits. What began as Julia's hobby soon became a professional craft. She swiftly mastered the latest camera techniques, especially methods of studio lighting and 'artistic' filtering, and the chemical development of high-quality black and white paper prints. She started by inviting first her family, and then all her holiday visitors to sit for her; and after many attempts finally got Tennyson – grumbling, and very reluctantly at first – to pose.

Several other photographers also tried to capture the elusive

Laureate, including Charles Dodgson, who produced a series of rather stiff portraits, with Tennyson apparently cornered against his will in a high-backed chair, but still defiantly sporting his black sombrero. It is interesting that in not one of these images is Tennyson shown wearing his usual spectacles or pince-nez. Was this forgivable authorial vanity, or perhaps on the presentational advice of Emily? In just one there is the hint of a monocle cord hastily tucked away behind a formal waistcoat.

But it was not until May 1863, that Julia Margaret Cameron finally fixed Tennyson's enduring image as the iconic Victorian poet of his generation, unmistakable in his wild black beard, caught in profile and clutching a book. The Laureate is intimate, collarless and disordered, and for once so hatless that he appears undressed – almost naked – under his long dark cloak. He has something of the sacred air of an actual Russian icon. He himself named this particular portrait, with a gleam of ill-disguised pleasure, 'The Dirty Monk'. His teenage son Hallam recalled: 'Mrs Cameron's wildly romantic ideas and performance used to call forth growls of amused satisfaction from him.' Emily reluctantly approved the image, though she still tried to increase the use of soap and water. The bearded, disordered and pungent Tennyson thus became the symbol of the British Poet, and soon of the Imperial Bard. From now on Alfred Tennyson was doomed to look ancient, patriarchal and prophetic. Many of these distinctive portraits are now held by the National Portrait Gallery, the National Gallery of Scotland, or in the Royal Collection. Margaret Cameron herself collected them in a special album presented to Sir John Herschel.

It was exactly at this time that a new professional readership, impressed by *In Memoriam*, emerged among the leading men of science. Since the death of the great Romantic chemist Sir Humphry Davy, scientists – however cultivated like Lyell or Darwin – were not normally expected to appreciate the latest poetry. But it was again his own son Hallam who later observed this cultural shift. 'Scientific leaders like Herschel, Owen, Sidgwick and Tyndall regarded him as a champion of Science, and cheered him with words of genuine admiration for his love of Nature, for the eagerness with which he welcomed all the latest scientific discoveries, and for his trust in truth.'

Another change to Tennyson's standing was the death of his faithful old publisher Edward Moxon in June 1858, and his replacement by his far more commercially minded brother William Moxon. From then on not only the twelve books of the *Idylls*, but numerous small poetry selections and special Tennyson editions poured from the Moxon press till the end of the century.

As other early friends and supporters dropped away, or fell out of Tennyson's ever-ascending social circle, Edward FitzGerald struggled to maintain the old emotional intimacy. After a gap of nearly two years, he was dismayed to encounter the transformed Tennyson in 1857, drifting through the stacks of the London Library, St James's Square. He found his old friend surprisingly aloof, and remote, 'staying at a great House in Kensington with which I won't meddle'. But in fact Tennyson, in turn, had noticed Fitz and had written anxiously to Emily. 'I called yesterday at the London Library and saw FitzGerald there looking thinnish and worn.'

Throughout the 1850s, FitzGerald had been turning his attention to less ephemeral matters than war and empire, at his Suffolk retreat in Boulge. There was sailing aboard his own yacht, the *Scandal*; romantic friendships with local fishermen; and several years of quiet labour, tutored in Persian linguistics by Edward Cowell, by now working at the Bodleian Library, Oxford. Yet FitzGerald had kept miraculously in tune with Tennyson through this new shared interest in the study and translation of Persian poetry. It was something entirely peaceful, unfashionable and bucolic, the very opposite of Tennyson's expanding public world of imperial duties, bardic readings and Arthurian chivalry. FitzGerald pursued this last outpost of their friendship with touching faithfulness, and it would finally produce a remarkable kind of reconciliation.

On that ultimate visit to Farringford of June 1854, they had sat down together in a corner, to discuss the challenge of translating Persian, until Emily had announced the tiny printed Persian grammar was bad for Tennyson's eyes, and hid the book away. Tennyson had joked to FitzGerald before he left that he would see 'Persian letters stalking like giants round the walls of his room'. But this saving extension of their friendship had been made.

In August FitzGerald had prompted again. 'My dear Alfred ... Have you done any Persian?' They later exchanged 'several versions' of the Odes of Hafiz, and FitzGerald continued to send copies of his own translations. Two years later in July 1856, FitzGerald's friend Cowell had made the discovery of an unknown Persian manuscript in the Bodleian. It was a long poetic fragment, not by the famous Hafiz, but by an unknown Persian mathematician whose name anglicised as Omar Khayyam – or, Omar the Tentmaker.

FitzGerald wrote to Tennyson in dreamy excitement: 'We read some curious Infidel and Epicurean Tetrastichs by a Persian of the 11th century – as savage against Destiny etc – as [Byron's] *Manfred* – but mostly of Epicurean Pathos of this kind: "Drink, for the Moon will often come round to look for us in this Garden, and find us not."' He clearly felt this had echoes of their old, lost intimacy.

Yet now he doubted if he would ever again visit Farringford, though Emily had once seemed to like his piano playing and tolerate his jokes. He observed wryly: 'I think I shall never again be other than a lumpy Companion.' But the old friendship was not forgotten in FitzGerald's heart. He wrote confidentially to Cowell: 'Tennyson's earlier poems bring back my own youth as well as that which must be personal in themselves. They make me cry like a fool – some of them.'

Cowell's Persian researches continued with a visit to Calcutta in 1857, and the library of the Bengal Asiatic Society. This produced a groundbreaking article entitled, 'Omar, the Astronomer Poet of Persia', which appeared in an obscure Anglo-Indian journal, the *Calcutta Review*, in March 1858. FitzGerald made his own selection from the sprawling mass of the newly discovered Omar manuscripts, work previously unknown in Europe. This was the first draft of a great long ballad-like poem that was to become *The Rubaiyat of Omar Khayyam*.

Omar's reputation in Persia had been largely that of a mathematician and astronomer. It has improved today, ironically due to the popularity of FitzGerald's own work in modern Iran, and a statue of

Omar has been installed in Laleh Park in Tehran. With Cowell's help, through extensive correspondence which continued while Cowell was away teaching in Calcutta, FitzGerald explored over 158 fragments from the Omar manuscript. 'What Scholarship it has is yours, my Master in Persian, as so much beside ...' he wrote gratefully to Cowell.

But from the beginning FitzGerald always seemed to associate the poem with Tennyson, having written to him in July 1857: 'I have really got hold of an old Epicurean so desperately impious in his recommendations to live only for *Today* ... he writes in little Quatrains ... He is very tender about his Roses and Wine, and making the most of this poor little Life ...' The grief of Omar for his lost drinking companion, and his speculations about cosmology and destiny, now echoed FitzGerald's own feelings about his lost Tennyson. He expressed these with increasingly open emotion. 'I don't know when I shall ever see you again; and yet you can't think how often I *wish* to do so, and never forget you, and never shall, my dear old Alfred, in spite of Epicurus.'

FitzGerald eventually selected from the original 158 Persian fragments to make an entirely new English poetic sequence, which he assembled in the form of 75 four-line stanzas, or repeated quatrains (*rubaiyat*). *The Rubaiyat of Omar Khayyam*, with its own distinctive mood and philosophic theme, was as much FitzGerald's own exotic creation, as any literal act of translation. In particular his imagery frequently moved very far from the Persian original, both simplifying its language and making its visual impact far bolder. The scientific and the literary were constantly combined. The opening quatrain, with its startling piece of astronomical magic, was particularly appreciated by Tennyson.

> Awake! for Morning in the Bowl of Night
> Has flung the Stone that puts the Stars to Flight:
> And Lo! the Hunter of the East has caught
> The Sultan's Turret in a Noose of Light.

As structured by FitzGerald, the poem was given the shape of a single day spent drinking wine and meditating on destiny. Omar, he wrote, 'begins with Dawn pretty sober and contemplative: then as he thinks and drinks, grows savage, blasphemous etc. and then sobers down into melancholy at nightfall'. Perhaps this also recalled many days – and nights – spent with Tennyson in the difficult decade of the 1840s.

The poem was first published privately by the firm of Bernard Quaritch in a tiny pamphlet edition of 250 copies in 1859. This was the same year as Tennyson's first instalment of the *Idylls of the King* from Moxon. Tennyson's book had instantly found a huge popular sale, while FitzGerald's was slow to find any general readership at all. Yet FitzGerald's Omar seemed to confront, in his own way, some of the same philosophical questions raised by *In Memoriam*: the possibilities of belief, the fragility of existence, the lessons of grief, and the final meaning of life. But Omar's (or FitzGerald's) first response was simply to deny the idea of any serious answers to such imponderables, and to dismiss them all as beautifully irrelevant.

> Oh, come with old Khayyam, and leave the Wise
> To talk; one thing is certain, that Life flies;
> One thing is certain, and the Rest is Lies;
> The Flower that once has blown for ever dies.

So the poem faced many of the profound issues as Tennyson had, but in a memorably different manner. Though cast in a similar, four-line repeating stanza form as *In Memoriam*, its tonal effect was something quite else: songlike and insouciant and irrepressible. The rhythm and scansion of FitzGerald's quatrains were also different. The two repeated rhymes in each of the *In Memoriam* stanzas perform a rocking, meditative movement across all four lines, holding everything in typically Tennysonian suspension.

> My own dim life should teach me this,
> That life shall live for evermore,

Else earth is darkness at the core,
And dust and ashes all that is . . .

The *Rubaiyat*, in contrast, normally has an unrhymed third line. This has the effect of releasing the sound and sense into the last, fourth, rhyming line, with tremendously emphatic force.

Then to the Lip of this poor earthen Urn
I leaned, the secret of the Well of life to learn:
And Lip to Lip it murmured – 'While you live,
Drink! – for once dead, you never shall return.'

His Omar elegantly – or languidly – dismisses any kind of certainty about belief itself – religious or otherwise. His Epicureanism is resigned and Oriental rather than rigorous and Classical. He answered Tennyson's haunted doubts and infinitely subtle agnosticism with exclamations of extreme scepticism, wild abandon and beautiful resignation.

Myself when young did eagerly frequent
Doctor and Saint, and heard great Argument
About it and about: but evermore
Came out by the same Door as in I went.

With them the Seed of Wisdom did I sow,
And with my own hand labour'd it to grow:
And this was all the Harvest that I reap'd—
'I came like Water, and like Wind I go.'

During the 1860s *Omar Khayyam* very gradually achieved a recognition as great as *In Memoriam*, though very slowly, and of a quite different order. It eventually ran to no fewer than four revised editions corrected by FitzGerald (1868, 1872, 1879, 1888) and expanded the original 75 to 101 stanzas or quatrains. By the end of the century it had become a favourite of book illustrators and luxury publishers. The first Omar Khayyam Dining Club was founded at a London restaurant in 1892, and an Omar Khayyam rose was propagated at Kew Gardens from 1884. But initially there was no reviewing, and indeed virtually no sales. Only when a copy of the first edition was found languishing on Quaritch's outside bookstall, was it taken up delightedly by the Pre-Raphaelite painters (notably William Morris, and Rossetti) just as they had earlier taken up Tennyson's poetry. Tennyson's wild admirer, the still rebellious poet Swinburne, enthused: 'I know none to be compared with it for power, pathos and beauty . . .', adding with a mischievous hint of blasphemy, '. . . except possibly *Ecclesiastes*.'

Four years after publication, in September 1863, the great critical champion of the Pre-Raphaelites and Tennyson, John Ruskin, also read *Omar Khayyam* and dashed off a fan letter to FitzGerald. 'I do with all my soul pray you to find and translate some more of Omar Khayyam for us: I never did – till this day – read anything so glorious, to my mind as this poem.' It was exactly the kind of letter that Fitz must have longed to receive from Tennyson, but never quite did.

FitzGerald's identity as the secret hermit of Suffolk would become merged with that of the Persian recluse Omar in his exotic Oriental paradise. Tennyson would recognise this transformation, and would refer to a 'Pagan Paradise' in England. That paradise – wherever located – became timeless in the famous lovers' picnic of No. 11, ending with its ecstatic exclamation mark:

Here with a Loaf of Bread beneath the Bough,
A Flask of Wine, a Book of Verse – and Thou
Beside me singing in the Wilderness
Ah Wilderness is Paradise enow!

The sense of elegy that permeates the whole of FitzGerald's *Rubaiyat* produces many specific echoes of particular scenes from *In Memoriam*. This is especially so of those set in the garden at Somersby, as No. 115, 'by night we lingered on the lawn'. Here Tennyson is left alone with his memories of Hallam, reading his letters, and feeling grief struck and isolated, as the rest of the company gently drift away:

> But when those others, one by one,
> Withdrew themselves from me and night,
> And in the house light after light
> Went out, and I was left all alone ...

Omar experiences a similar moment of intense abandonment in his garden, in the penultimate rubaiyat of FitzGerald's poem (No. 100):

> Yon rising Moon that looks for us again—
> How oft hereafter will she wax and wane;
> How oft hereafter rising look for us
> Through this same Garden – and for one in vain!

But the old friendship, like much else in the younger Tennyson's life, drifted away, overtaken by family affairs, or frankly overwhelmed by Tennyson's increasing wealth and celebrity. FitzGerald still wrote a letter faithfully every year to the Isle of Wight, but was invariably answered only by Emily or increasingly by his son Hallam. He lamented to a friend, 'I don't know what he thinks now, after so much Worship has been offered him.'

Tennyson's enthusiasm for his friend's poetry, which had suddenly blossomed so late at the age of fifty in his *Omar*, remained 'kindly' but restrained. FitzGerald wrote wistfully in 1872: 'You used to talk of *my* Crotchets: but I am quite sure you have one little Crotchet about this Omar ...' FitzGerald characteristically disguised his disappointment in a light piece of Laureate banter: 'I do not think it worthwhile to compete with you in your paltry poetical Capacity ...'

Tennyson once sent a note to FitzGerald asking for a replacement for the original presentation copy of *Omar*, which Tennyson had 'somehow' lost. FitzGerald sent a replacement, pointedly of the *third* edition, wondering if Tennyson thought it was improved? When he received no reply, FitzGerald contented himself by teasing the Laureate with mildly loaded questions, hoping to elicit some of the old playful response. 'Is Mr Rossetti a Great Poet, like Browning and

Morris? So the Athenaeum tells me. Dear me, how thick Great Poets do grow nowadays.' Tennyson did not rise to this either.

Behind Tennyson's back, FitzGerald still frequently repeated his old claim to be disillusioned with the development of the Great Poet's later career. He wrote to their mutual Cambridge friend Monckton Milnes, indulging that favourite pastime of late middle age, the comparison of worldly achievements (and failures). All three men were of roughly the same age, then in their mid-sixties. Milnes had been recently elevated to the peerage for his political work, as Baron Houghton. He was also recognised for having established the posthumous reputation of John Keats. Perhaps remembering all this, FitzGerald wrote ironically to Milnes in April 1874. 'I used to tell Tennyson thirty years ago that he should have been a Dragoon, or in some active Employment that would keep his Soul stirring, instead of revolving itself in idleness and Tobacco smoke.'

The outlandish Dragoon suggestion was an old joke between them, but perhaps freshly sharpened by the enduring popular success of 'The Charge of the Light Brigade'. It was clear that FitzGerald's complaint was as much about Tennyson's rich and comfortable domestic life at Farringford, as about his undoubted national standing as a poet. In addition there was now an unmistakable touch of sarcasm in his attitude to Emily. 'And now he has sunk into Coterie-worship, and (I tremble to say it) in the sympathy of his most Ladylike, gentle Wife. An old Housekeeper like Moliere's would have been far better for him, I think.' Here he was surely thinking about himself, and his own brief but disastrous marriage to a Suffolk neighbour, Lucy Barton, in 1856, which had lasted only a few months.

As to the later poetry, FitzGerald admitted to liking 'the dramatic element in *Maud*, and a few little bits in it', but not much else. 'I am told this is because I have shut up my mind, etc.' He was not impressed by the continuing *Idylls* either, thinking the Arthurian material, though well researched, never burst into new life. 'They seem to me to fail utterly in the one thing wanted – Invention, to make a new and better thing of old Legends . . .' The note of reluctant disappointment seemed only to deepen as the years passed.

'I mourn over him as a Great Man lost – that is, not risen to the Greatness that was in him – for he has done enough to out-last all others of his time, I think – up to 1842 ...' FitzGerald did not know that Tennyson had twice refused a peerage, offered once by Gladstone in 1873, and then again by Disraeli in 1874. Had he lived to see Tennyson finally accept a peerage in 1884 – Baron Tennyson of Aldworth and Freshwater – it would probably only have confirmed him in his opinions. Yet perhaps, beneath all this, FitzGerald was really mourning a Great Friend lost: his own after a lifetime of passionate loyalty, honest criticism, and deeply ironic affection. Perhaps only Omar's famous quatrain No. 51 could catch something of the profound, complex regrets of this situation:

> The Moving Finger writes; and, having writ,
> Moves on; nor all your Piety nor Wit
> Shall lure it back to cancel half a Line,
> Nor all your Tears wash out a Word of it.

In fact it was Tennyson who finally contrived a kind of poetic – and pastoral – reconciliation. In the autumn of 1876 he made one rare visit to FitzGerald's long established hideaway, the Little Grange, his small thatched cottage by the sea at Boulge. This time Tennyson was accompanied by his beloved son Hallam, now aged twenty-four and having just left Trinity College, Cambridge. Hallam was to be his father's amanuensis, and later faithful editor and memoir writer; and in him FitzGerald saw Tennyson's poetic legacy assured. He remarked, surely wistfully, that Hallam was a very nice fellow, 'who took all care of Papa'.

Tennyson had sent a warning postcard of their imminent arrival, in mid-September, simply scrawled in pencil. It contained only eleven words, but they said a great deal. 'Dear old Fitz, we are passing through, and will call again.' But the last three words were crossed out, and impatiently replaced with 'am here'. The familiar part of the greeting was obviously intended to reassure Fitz; but the speed of the visit seemed remarkably casual.

Despite this ill-omen, the visit lasted happily for two whole sunny days. Tennyson insisted on finding rooms at the local inn, the Bull, so he and Hallam would not impose on his old friend's domestic arrangements (or have to suffer them either). Tennyson never forgot the moment of meeting, after a gap of nearly twenty years. He and Hallam walked through the wicket gate, and suddenly came across FitzGerald reading peacefully on an iron bench in his back garden, under a trellis of late roses. To Tennyson's surprise FitzGerald was wearing a jaunty pair of blue-tinted sun-glasses (he explained that his eyes were now weak); and to Hallam's amusement, absolutely no shoes or socks. Evidently Old Fitz expected no formal guests. But he was not quite alone. FitzGerald was 'sitting among his doves, which perched about him on head and shoulder and knee, and cooed to him as he sat in the sunshine beneath his roses'. Indeed it was a scene almost directly out of *Omar Khayyam*, and possibly arranged deliberately by Fitz as such for his visitors.

Tennyson would soon respond accordingly in verse.

Old Fitz, who from your suburb grange,
Where once I tarried for a while,
Glance at the wheeling orb of change,
And greet it with a kindly smile;
Whom yet I see as there you sit
Beneath your sheltering garden-tree,
And watch your doves about you flit,
And plant on shoulder, hand, and knee,
Or on your head their rosy feet . . .

As FitzGerald remembered, they 'fell at once into the old humour, as if we had been parted twenty days instead of so many years'. He thought Tennyson looked much the same, 'except for his fallen locks', and of course the ever-maturing beard. They argued 'the same old grounds of debate' about poetry, and told 'some of the same old stories'. They talked of Cambridge and the Lakes; and Camden Town and Lincoln's Inn: and 'all was well'. But FitzGerald added, with a touch of his usual asperity, 'I suppose this is a sign of age – not altogether desirable.'

The following day they sailed down the River Orwell, not majestically in FitzGerald's own yacht the *Scandal* (long since given up) or in one of the hired fishing boats, but comfortably by tourist steamer from Ipswich quayside on a round trip to Harwich point and back. FitzGerald relaxed and was full of light-hearted banter as in the old days. He earnestly advised Tennyson to adopt vegetarianism for his health; and for his peace of mind to ignore all the endless bad poems and books sent for his approval as Laureate. He should imitate Charles Lamb and 'throw them into his neighbours cucumber frames'.

He was even 'impious enough' to advise Tennyson to stop writing any more poems altogether, especially the *Idylls*. This was perilous ground, even for a very old friend, and Hallam must have held his breath. FitzGerald frankly thought that Tennyson had enough 'Worshippers who tell him otherwise', though he did not say this explicitly. Instead he gestured to the river, and referred to the hardworking fishermen, and gently suggested Tennyson 'might as well ship his Oars now'.

They said gruff goodbyes the next morning, somewhat incongruously at Woodbridge railway station. 'The rail carried him off,' as FitzGerald put it. He accepted they would probably never meet again. But when he returned to Boulge, FitzGerald said his garden suddenly seemed 'to be haunted by that spirit which Tennyson says is heard talking to himself among the flower borders'. FitzGerald was thinking of that early 'Song', written at Somersby some fifty-five years ago:

A spirit haunts the year's last hours
Dwelling amid these yellowing bowers:
To himself he talks;
For at eventide, listening earnestly,
At his work you may hear him sob and sigh
In the walks . . .

Yet Tennyson never completely forgot their old comradeship, recalling: 'Dear old Fitz, – I had no truer friend – he was one of the kindliest of men and I have never known one so fine and delicate of a wit.' Over a decade later, in June 1883 while still deep in the final imperial fustian of the *Idylls of the King*, he was shaken into intimate poetry once again, by the news that FitzGerald had become ill. It produced one of his last verse letters, now returning to his magnificent first simplicity. It took the form of a birthday greetings, including his son Hallam's support and encouragment. He summoned that late pastoral and timeless visit to Suffolk among the doves and roses, and then went back to the ancient 'comradeship' of early London days. Finally, Tennyson looked back on that long friendship, tenderly and gratefully, yet spicing his recollections with a last touch of irony:

> ... And so I send a birthday line
> Of greeting; and my son, who dipt
> In some forgotten book of mine
> With sallow scraps of manuscript,
> And dating many a year ago,
> Has hit on this, which you will take,
> My Fitz, and welcome, as I know,
> Less for its own than for the sake
> Of one recalling gracious times,
> When, in our younger London days,
> You found some merit in my rhymes,
> And I more pleasure in your praise.

FitzGerald in turn had recalled their friendship, though also indirectly and unseen, in the course of a long late letter to Hallam. He had touched too, with his own irony, on the dark Kraken spirit that had so long shadowed his brilliant friend, invoking Alfred Tennyson as 'My dear old comrade, whom I should call "Master", and with whom (in spite, perhaps *because* of his being rather a "gloomy" soul sometimes, as Carlyle wrote of Emerson) I always did talk more nonsense than to anyone, I believe.'

And Tennyson did finally make a magnificent acknowledgment

of *Omar Khayyam* in his verse letter. He used one of his last and simplest astronomical metaphors:

> . . . your golden Eastern lay
> Than which I know no version done
> In English more divinely well;
> A planet equal to the sun
> Which cast it, that large infidel,
> Your Omar; and your Omar drew
> Full-handed plaudits from our best
> In modern letters . . .

In one sense this praise was all too late: Old Fitz died before the verse letter was delivered. 'I had written a poem to him within the last week,' lamented Tennyson, 'a dedication – which he will never see.' But in another sense it was perfectly timed. It was the Old Laureate keeping generous faith with the Young Tennyson.

END

Permissions and Acknowledgements

The source materials for any life of Alfred Tennyson are immense, glorious and daunting. Most are now gathered in manuscript archives at the Tennyson Research Centre, Lincoln; the Wren Library, Trinity College Cambridge; the Fitzwilliam Museum, Cambridge; and the Houghton Library, Harvard University. In consequence my book is indebted to the superb editorial work on these original Tennysonian materials, together with exceptionally detailed notes, made by the following scholars.

Christopher Ricks, editor, *Tennyson: A Selected Edition*, Pearson Longman 1969, 1989, revised edition 2007. His wonderful footnotes contain a rich and invaluable commentary on every aspect of Tennyson's work including his scientific reading.

Cecil Y. Lang and Edgar F. Shannon Jr, editors, *The Letters of Alfred Lord Tennyson*, 3 vols, Oxford, 1981, 1987, 1990. Again their footnotes supply a rich source of biographical materials, often including citations from Emily Sellwood, Edward FitzGerald, Thomas Carlyle, James Spedding and other members of Tennyson's early circle.

Alfred and Annabelle Terhune, editors, *The Letters of Edward Fitzgerald*, Vols 1, 2, 3, Princeton University Press, 1980. With additional materials concerning Tennyson's elder brother Frederick, his friends Thomas and Jane Carlyle, and the Persian scholar Edward Cowell.

Jack Kolb, editor, *Letters of Arthur Henry Hallam*, Ohio State University Press, 1981. Very revealing of the Hallam family background, so unlike Tennyson's.

I have also been able to draw on the work of four outstanding biographers, who rightly still dominate the world of Tennyson stud-ies. These are:

Hallam Tennyson (his eldest son, born in 1852), *Alfred Lord Tennyson: A Memoir*, 1897, 1899. The starting point for all Tennyson biography, a huge bran tub of historical information, family lore, manuscript quotation and vivid anecdote, all helpfully indexed.

Robert Bernard Martin, *Tennyson: The Unquiet Heart*, Oxford, 1980. The classic early biography of Tennyson, measured, tender and comprehensive.

Ann Thwaite, *Emily Tennyson: The Poet's Wife*, Faber, 1996. A big pioneering work, a true labour of love and scholarship, which places both Tennysons, and their family circle, in a new and vivid light. This wonderfully detailed and generous biography of Emily (earlier Emily Sellwood) has brought a whole new dimension to Tennyson's life, and most especially after his marriage in 1850.

John Batchelor, *Tennyson: To Strive, to Seek, to Find*, Pegasus Books, New York, 2013. A sensible, meticulous and scholarly account.

The first seeds for this book were sown (or scattered) long ago during two dream-like summers as Visiting Fellow Commoner at Trinity College, Cambridge (2001–2002), for which I express deep gratitude to the then Master Amartya Sen, and the then Senior Research Fellow, my old friend William St Clair. It was then that I first gazed up (blankly) at the forbidding bearded Tennyson statue.

After *The Age of Wonder* (2008) and various field trips ('The Lost Women of Nineteenth-Century Science', a series of articles in the *Guardian* and *Nature*, 2010–2015), the actual book was not begun until 2019–2020 during the Covid lockdown, when my usual bio-graphical 'footstepping' was sadly restricted. But this allowed me to write a different kind of biography, and I found instead the liberation of Tennyson's many wild beaches and empty seashores, which both begin and end this story. I had time to reconsider the whole impact of science on his imagination, and the very different friendships with Arthur Hallam, Edward FitzGerald and Emily Sellwood. The book was written slowly thereafter, as a kind of coming back to life. Instead of ageing, Alfred Tennyson seemed to grow younger; and his poetry stranger and more daring. By the end, I found a new Tennyson had emerged. And perhaps a new kind of national poet: the last of the Romantics, the first of the Moderns.

The need for persistence in such a long pursuit, goes way back to the example of Professor George Steiner, my original tutor at Churchill College, Cambridge in the 1960s; and to Sir Michael Holroyd, who I believe has been the benign and brilliant godfather of all literary biographers of my generation.

During this extended time of writing I owe special thanks for the encouragement and shrewd commentary on the many early drafts, patiently provided by my old and scholarly friend Professor Jon Cook; the publishing wizard Richard Cohen; my learned adviser Stuart Proffitt; and the novelist Rose Tremain. Not forgetting Dan Frank, my old friend and editor at Pantheon USA until his untimely death; and afterwards to his generous and understanding colleague at Pantheon, Edward Kastenmeier. For unfailing friendship and support, my thanks go to Alan Judd at the Reform (where Tennyson was once asked to take his feet off the table); to David and Heather Godwin at DGA (and sometimes the Ivy Norwich); and above all to my dauntless, peerless editor and patient friend at William Collins, Arabella Pike. Also to her resourceful team, Sam Harding, Lizzie Rowles and Hope Butler; my cool copy-editor Iain Hunt; and indexer Mark Wells. The book is dedicated to my beloved partner of more than thirty years, in fact and in fiction, Rose Tremain.

Notes

Abbreviations used in the notes

Batchelor – John Batchelor, *Tennyson: To Strive, to Seek, to Find*, Pegasus Books, New York, 2013

CT Alfred – Sir Charles Tennyson, *Alfred Tennyson*, Macmillan, 1949

CT The Tennysons – Sir Charles Tennyson and Hope Dyson, *The Tennysons: Background to Genius*, Macmillan, 1974

Fitz Letters – *The Letters of Edward Fitzgerald*, Vol. 1, 1830–1850; Vol. 2, 1851–1866; Vol. 3, 1867–1876; editors Alfred and Annabelle Terhune, Princeton University Press, 1980

Hallam – *Letters of Arthur Henry Hallam*, editor Jack Kolb, Ohio State University Press, 1981

HT Memoir – Hallam Tennyson, *Alfred Lord Tennyson: A Memoir*, 1 volume edition, 1899

Martin – Robert Bernard Martin, *Tennyson: The Unquiet Heart*, Oxford, 1980

Rader – Ralph Rader, *Tennyson's Maud: The Biographical Genesis*, California University Press, 1963

T Chrono – F.B. Pinion, *A Tennyson Chronology*, Macmillan, 1990

T Crit Heritage – *Tennyson: The Critical Heritage* (reviews 1831–1891), editor John D. Jump, Routledge, 1967

T Interviews – *Tennyson: Interviews and Recollections*, editor Norman Page, Macmillan, 1983

T Letters – *The Letters of Alfred Lord Tennyson*, editors Cecil Y. Lang and Edgar F. Shannon, 3 vols, Oxford, 1981, 1987, 1990

T Norton Memoriam – *Tennyson: In Memoriam*, editor Erik Gray, Norton Critical Edition, 2004

T Norton Poetry – *Tennyson's Poetry*, edited Robert W.E. Hill Jr, Norton Critical Edition, 1991

T Ricks Poems – *The Poems of Tennyson*, editor Christopher Ricks, Longman Annotated English Poets, 3 vols, 1969

T Selected Poems – *Tennyson: A Selected Edition*, editor Christopher Ricks, Longman, 1969, 1989

Thwaite – *Emily Tennyson: The Poet's Wife* by Ann Thwaite, Faber, 1996

The epigraph is drawn from Alfred Tennyson's 'last' poem, dated October 1889. Tennyson wrote: 'Mind you put "Crossing the Bar" at the end of all editions of my poems.' So here it is:

> Sunset and evening star,
>   And one clear call for me!
> And may there be no moaning of the bar,
>   When I put out to sea,
>
> But such a tide as moving seems asleep,
>   Too full for sound and foam,
> When that which drew from out the boundless deep
>   Turns again home.
>
> Twilight and evening bell,
>   And after that the dark!
> And may there be no sadness of farewell,
>   When I embark;
>
> For tho' from out our bourne of Time and Place
>   The flood may bear me far,
> I hope to see my Pilot face to face
>   When I have crost the bar.

1 Old Monster

2 *'As to the bearded Daguerreotypes'* Fitz Letters 2, 5 July 1860, p.362

3 *'Below the thunders'* T Selected Poems, pp.17–18

3 *'Black Tennyson'* W.H. Auden, *New Year Letter*, 1941

5 *'the spine-bone of the world'* HT Memoir, pp.16–17

5 *'See the account which Erik Pontoppidan'* Quoted T Selected Poems, p.17; Batchelor, p.118

6 'In these two miles' Richard Holmes, *Coleridge: Darker Reflections*, HarperCollins, 1998, p.497

7 *a long speculative article* Will Abberley, *Underwater Worlds*, Cambridge Scholars, 2018, pp.53–60; and James Wilson, 'The Kraken', *Blackwood's Edinburgh Magazine*, March 1818, online

7 'It was so enormous' Mary Somerville, *Personal Recollections*, 1873, p.21

8 'Mr Gosse has' Ann Thwaite, *Glimpses of the Wonderful: The Life of Philip Henry Gosse*, Faber, 2002, pp.180–3

9 'I found Him not in world or sun' T Selected Poems, p.470

10 'In the short piece' T Crit Heritage, p.423. See Peter Godfrey-Smith, *Other Minds: The Octopus and the Evolution of Intelligent Life*, William Collins, 2017

2 Somersby

11 *Outbuildings also housed* T Chrono, p.5

11 'overshadowed on one side' HT Memoir, p.2

11 *large family consisted of* T Letters p.199

12 *At the age of seven Alfred* T Chrono, p.3

12 'rude and ungovernable' Batchelor, p.13

12 *Elizabeth made a rich marriage* CT The Tennysons, pp.28–33

13 'We are three and twenty' Martin, p.22

13 'one of the most angelic natures' CT The Tennysons, p.95

13 *shot through with irony* Harold Nicolson, *Tennyson: Aspects of his Life, Character and Poetry*, 1923, p.40

14 'decidedly eccentric' CT The Tennysons, p.94

15 'Knights and heroes' HT Memoir, p.4

15 'Before I could read' Ibid., p.9

15 'thunderous roar' Ibid., pp.16–17

15 'Thou art the God' Ibid., p.13

15 'he had a passion for the sea' Hallam Tennyson in T Interviews, p.2

16 'Here often when a child' HT Memoir, p.135

16 'the cropped level of the marsh' Nicolson, 1923, p.132

16 'There is nothing here' T Letters 1, p.215

16 'Anything more unlike' HT Memoir, p.145

17 'The moanings of the homeless sea' In Memoriam, No. 35, T Selected Poems, p.378

17 *remembered sitting on the school steps* Batchelor, p.17

17 'A. Tennyson, Somersby' CT Alfred, p.39n

17 'I seemed to move' Princess, T Selected Poems, p.282

18 *'In my earliest teens'* HT Memoir, p.10, and T Chrono, p.4

18 *'scared by his father's fits'* HT Memoir, p.13; Martin, p.25; Nicolson, 1923, p.46

18 *terrifying bursts of ill-temper* Hallam Tennyson quoted in T Selected Poems, p.xl

18 *considered certifying him as insane* CT The Tennysons, p.62; Thwaite, p.69

18 *'Jackson, the saddler'* W.S. Rawnsley in T Interviews, p.21

18 *'When I was a lad'* Martin, p.36

20 *'I should* extremely *like to see'* Richard Holmes, 'Computer Science: Enchantress of Abstraction', *Nature*, 3 September 2015

20 *'I felt I had got hold of'* Richard Holmes, *The Age of Wonder*, HarperCollins, 2008, p.348

21 *His projected title was* Leonard G. Wilson, *Charles Lyell*, Yale University Press, 1972, p.169

21 *'The eye of a cow'* Tennyson Notes, in CT Alfred, p.36

21 *'share with his brothers and sisters'* Ibid.

21 *'A gate and field half ploughed'* Jenny Uglow, *Nature's Engraver: A Life of Thomas Bewick*, Faber, 2006

22 *'O suns and spheres'* CT Alfred, pp.41–2

22 *'a day when the whole world'* HT Memoir, p.3

23 *'always dawdlin'* 'Lincolnshire Memories' in T Interviews, p.3

23 *'my incomparable Dulcinea'* T Letters 1, p.3

23 *sometimes perched on his mother's head* HT Memoir, p.15

23 *'The Owl'* Ibid., p.19

23 *'When cats run home'* Christopher Ricks, *The Poems of Tennyson*, Longman Annotated English Poets, 1969, vol. 1, p.224

24 *'Deep glens I found'* HT Memoir, p.34n

24 *'Fred, think of Herschel's'* Ibid., p.16

25 *'You have probably received'* T Letters 1, p.7

25 *slim volume was issued* Ibid., pp.9–10

25 *'shared their triumph'* HT Memoir, p.19

25 *'I wander in darkness and sorrow'* T Norton Poetry, p.22

26 *'The glittering fly'* 'Love', *Poems by Two Brothers*, 1827, p.169; Harold Nicolson, *Tennyson's Two Brothers*, 1947, p.11

26 *'See Baker on Animalculae'* Nicolson, 1947, p.11

26 *'look like a drove of pigs'* Jane Marcet, *Conversations in Natural Philosophy*, 1824, p.47

27 *'Various philosophical doctrines'* Mary Shelley, *Frankenstein*, Intro, p.8

27 *'Rise, sons of Science and Invention'* Poems by Two Brothers, p.164

28 *'I wish he had something'* T Letters 1, p.14

28 *'dearest nephew'* CT The Tennysons, p.179

28 *destined for Royal Navy cadetships* T Letters 1, p.16

28 *'giddiness'* Ibid., p.21

29 *'Heavily hangs the broad sunflower'* T Selected Poems, p.13

29 *Malicious gossip flourished* T Letters 1, p.22n

3 Cambridge

30 *57 Trumpington Street* HT Memoir, p.29

31 *'I am an idle fellow'* Fitz Letters 1, p.153

31 *'something like the Hyperion'* Alfred Terhune, *The Life of Edward FitzGerald*, OUP, 1947, p.77

31 *'My dear Aunt'* T Letters 1, p.22

32 *'He was a thousand fathoms'* Falkland, 1827, p.269; T Selected Poems, p.17n

32 *'like Telescopes reversed'* T Letters 1, p.23

32 *almost his only income* T Chrono, p.39

32 *'the third of the Tennysons'* T Letters 1, p.39n

32 *'Six feet high'* Batchelor, p.29

33 *acted in a college production* Ibid., p.33

33 *'flying to Tennyson at Trinity'* T Letters 1, p.39n

33 *he had left behind* Batchelor, p.31

33 *'I kept a tame snake'* Tennyson quoted in HT Memoir, p.34n

34 *'Till that great sea-snake'* Poems of Tennyson 1, p.214

34 *previously identified in Lyme Regis by Mary Anning* For a romantic interpretation with fine locations, see the film *Ammonite*, directed by Francis Lee, 2020, with Kate Winslet (Mary Anning) and Saoirse Ronan as the geologist Charlotte Murchison (wife of the geologist Roderick Murchison, and friend of Mary Somerville)

35 *'A Monster then, a dream'* In Memoriam, No. 56, T Selected Poems, p.400

36 *'behind the veil'* Ibid.

36 *destructive impact* Ashley Dawson, *Extinction: A Radical History*, OR Books, 2022, *passim*

36 *'As when a man'* T Selected Poems, p.80, published 1832. Crewe MS/8/f. 30 Letter from R.M. Milnes to Arthur Hallam, written during an ascent in a balloon on 29 May 1829

37 *'I had many grapples'* AH to MM, 1 September 1829, note in T Selected Poems, p.344

38 *'And well I ween not Time'* CT Alfred, p.67

38 *'Rhyme has been said'* Gregory Tate, *Nineteenth-Century Poetry and the Physical Sciences*, 2020, p.156

38 'Oh might I be an arrow' 'To Poesy', lines 5–6, T Ricks Poems 1, p.186

38 'A still small voice spake unto me' 'The Two Voices', T Selected Poems, p.101

39 'it is not safe' T Letters 1, p.30

39 erased from family history CT Alfred, p.127; CT The Tennysons, pp.118–20; Thwaite, pp.86–7

40 'All day within the dreamy house' T Selected Poems, p.6

41 'that stupid Gallery of Beauties' Christopher Ricks, Tennyson, Macmillan, 1972, p.51

41 'Alfred enacted' Batchelor, p.74

41 'grotesque Grimness' Ibid., p.116

41 regathered that summer Fitz Letters 1, p.410n

42 'secretly wrote verses' Nicolson, 1923, p.85

42 Tennyson researched the subject See Aidan Day, Tennyson's Scepticism, 2004

42 Mungo Park had tried Holmes, The Age of Wonder, Chapter 5

43 'I had formed' René Caillié, Journal d'un voyage à Temboctou, 1830

44 'The clear galaxy' 'Timbuctoo' by Roger Ebbatson, in Valerie Purton, editor, Darwin, Tennyson and Their Readers, 2013, pp.9–10

45 'A Character' T Norton Poetry, p.30n

45 'With a half-glance upon the sky' T Selected Poems, p.14 and note

46 'promising fair to be' Batchelor, p.39; Martin, p.79

4 Apostles

47 'the Pope among us young men' HT Memoir, p.32

47 where Tennyson would often stay T Chrono, p.35

48 'This was one of the poems' HT Memoir, p.161fn

48 'He too was a friend to me' T Norton Poetry, p.75

48 'Sleep sweetly, tender heart, in peace' 'To J.S.', T Norton Poetry, p.76

48 hosting of a Science Fair Batchelor, pp.126–8

49 'pouring forth' HT Memoir, p.31

49 It took an oath of secrecy Frances M. Brookfield, The Apostles, 1907

50 'Saturday November 21st' T Letters 1, p.43

51 'He speaks of life and death' HT Memoir, p.862

51 'We soon grew' Nicolson, 1923, p.73

52 'the soul and foetal development' Rebecca Stott, 'Tennyson's Drift' in Purton, editor, Darwin, Tennyson and Their Readers, p.15

52 His polymathic reputation Tate, 2020, p.113

52 'the Lion-like man' HT Memoir, pp.32–3

52 *One by one these books* For Tennyson's science reading, see the list in my bibliography below. Also Gregory Tate, 2020, p.146; Anna Henchman, *The Starry Sky Within: Astronomy and the Reach of the Mind in Victorian Literature*, OUP, 2014, pp.89–91; p.250

53 *literary version of the Boat Race* Brookfield, *The Apostles*

53 *'a furious Shelleyist'* Seamus Perry, *Scepticism*, pp.74–5

53 *he admired his blank verse* 'Tennyson on the Romantic Poets', in T Interviews, p.176

54 *'keen delight'* Holmes, *Coleridge: Darker Reflections*, p.500; Samuel Taylor Coleridge, *Table Talk*, 1835

55 *spent a nostalgic evening* Martin, pp.173, 203

55 *they had apparently fled* Ibid., p.126

55 *'pleased with his hearers'* Robert Tennant in T Letters 1, p.46

55 *'a great deal of intellectual activity'* Martin, pp.125–6

56 *He was particularly intrigued* CT The Tennysons, p.210

56 *'a sort of seal on our friendship'* Batchelor, p.41

56 *'a true and thorough Poet'* Hallam, p.365

57 *'something of a tragedy queen'* CT The Tennysons, p.146

57 *'Are you a Dryad or a Naiad?'* Thwaite, p.76; Martin, p.82

57 *picked up the family donkey* HT Memoir p.66

57 *'O bliss, when all in circle drawn'* In Memoriam, No. 88, T Norton Poetry, p.259

57 *Effingham Wilson* Batchelor, p.40

58 *publicly dedicated to his old flame* Hallam, p.10

58 *a separate slim volume* Batchelor, p.40

5 Adventures

59 *Henry Hallam thought* T Chrono, p.12

60 *bundled illegally* Batchelor pp.42–5

61 *'Alfred went'* Hallam, p.387

61 *the friends wandered* Martin, p.119

61 *'and spend some weeks'* 26 July 1854, T Letters 2, p.94

61 *'Precipitous defiles'* Hallam, p.375; quoted Batchelor, p.43

61 *'There lies a vale in Ida'* T Selected Poems, p.36

62 *Tennyson also imagines* Nicolson, 1923, p.70; T Ricks Poems 1, p.386

62 *'One possible clue'* Martin, p.120n

63 *'It came into my head'* T Selected Poems, p.27n

63 *'With one black shadow at its feet'* Ibid., p.28

63 *'The river-bed was dusty white'* Ibid., p.31

64 *'All along the valley'* Ibid., p.590–1

64 *'It communicates its vibrations'* Mary Somerville, *On the Connexion of the Physical Sciences*, 1834, Section 16, 'Propagation of Sound', pp.148–9

64 *'The air itself'* Charles Babbage, *The Ninth Bridgewater Treatise: A Fragment*, 1837, p.112; and quoted in Gregory Tate, 2020, pp.154–5

65 *'During my last year'* Darwin's annotated copy of Herschel's *Introduction to the Study of Natural Philosophy* (1830) is held in the 'Darwin Library' section of the Darwin Archive, Cambridge University Library

66 *rigorous personal timetable* see Tennyson, *The Major Works*, edited Adam Roberts, Oxford World's Classics, 2009, p.528

66 *'PS. I have bought'* Fitz Letters 1, p.97

67 *'An artist'* Hallam, p.401; and quoted in T Selected Poems, p.63 footnote

68 *'We have spoken'* Hallam, 'On the Lyrical Poems of Alfred Tennyson', August 1831, in T Norton Poetry, pp.586–8

68 *'hoping to see his ghost'* T Chrono, p.15

68 *'Beautiful this harvest moon'* Hallam to Emily Tennyson, 28 August 1831, quoted in *Alfred Lord Tennyson, Selected Poems*, edited Christopher Ricks, Penguin Classics, 2007, Notes, p.344

68 *He drew on* T Selected Poems, p.18n

69 *'On either side the river lie'* Ibid., p.20

70 *'But in her web she still delights'* Ibid., p.25

70 *'All in the blue unclouded weather'* Ibid.

71 *'But Lancelot mused a little space'* Ibid., p.27

72 *Tennyson had essentially discovered* Batchelor, p.145

72 *'Call upon Mr Moxon'* AHH to Charles Merivale, 14 August 1831, in T Letters 1, p.66

73 *without completing his degree* T Chrono, p.16

73 *'more sociable'* Hallam, pp.508–9; and Thwaite, pp.82–3

6 Complications

74 *'There was a dark-haired'* Batchelor, p.120

74 *'musing and brooding'* T Letters 1, p.70

74 *'I built my soul a lordly pleasure-house'* T Selected Poems, p.52

75 *'on fire within'* Ibid., p.70

75 *'Alfred the Great will be in Town'* T Letters 1, p.73

75 *'The Gardener's Daughter'* Ibid., p.74

76 *'head and face almost too ponderous'* Ibid., p.75n

76 *a long, anxious letter* T Selected Poems, p.344n

76 *'Somersby looks glorious'* CT Alfred, p.125

76 *cancel his allowance* Ibid., p.124

76 *'dreadfully nervous'* AH in T Letters 1, p.74

76 *'I am now at Somersby'* Ricks, *Tennyson*, 1972, p.73

77 *'I have strange news'* T Letters 1, p.75

77 *'bug bitten, flybitten'* Ibid., p.77

77 *'Alfred is as sulky as possible'* CT Alfred, p.123

77 *But all was forgotten* Ibid., p.124

77 *long humorous accounts* AHH in T Letters 1, pp.75–79 notes

77 *'Alfred swears the Rhine'* Batchelor, p.70

77 *carry the date 1833* T Chrono, p.18

77 *'The title page must be'* CT Alfred, p.130

77 *did not reinstate the* Tale Ibid., p.129

78 *'Every shadow of not doubt'* AH to Brookfield, quoted by Garrett Jones, *Alfred and Arthur: An Historic Friendship*, 2017, p.76; and George P. Landow, 'Alfred and Arthur', Victorian Web, 2005

78 *'Alfred is, as I expected'* Ibid.

78 *'The Watcher on the column'* T Selected Poems, p.132

78 *'resting their weary limbs'* Ibid., p.70

78 *'my own pet'* Hallam, p.553

78 *'There is something horrible sad'* AH to ET, 11 April 1833, Hallam, p.747

79 *The woods of Holywell* AH to ET, 27 April 1833, Hallam, p.751

79 *'lions and tigers'* AH to ET, Hallam, p.748; Martin, p.177

79 *'as if we could destroy facts'* HT Memoir, p.232, and Henchman, p.93

79 *'Tuesday we went again'* AH to Emily Tennyson, T Letters 1, p.91n

80 *After the first version* T Norton Poetry, p.55, note 1

80 *'Tennyson, we cannot live in Art'* T Selected Poems, p.50n

81 *'If the Poem were not already'* Ibid., p.64n

82 *'Our stellar system'* Edwin Hubble, *The Realm of the Nebulae*, 1936, quoted Holmes, *The Age of Wonder*, p.205n

82 *'that brooding'* T Norton Poetry, p.55, note 3; Yeats, 'Poetry and Symbolism', *Ideas of Good and Evil*, 1903

82 *He would begin to reflect* T Selected Poems, p.243n

82 *'What be those two shapes'* Ibid., p.663

83 *'From shape to shape'* T Norton Poetry, p.60, note 9

84 *'we presume'* John Wilson Croker in T Crit Heritage, pp.73–83

84 *to visit the elderly Coleridge* Holmes, *Coleridge: Darker Reflections*, p.484

84 *'the art of painting a picture'* John Stuart Mill in T Crit Heritage, p.95

84 *'struggling upwards'* Ibid., p.93

84 *'whatever is comprehensive'* Ibid., T Crit Heritage, p.96

85 *'further effort on his part'* Ibid., T Crit Heritage, pp.95–7

85 *the first time he had* Martin, p.179

85 *'a sort of yearning'* Ibid., p.180

85 *The plan was to stay* T Chrono, p.19

85 *It kept them up till* T Letters 1, p.92n

86 *was he 'mad'* Batchelor, p.76; Hallam, pp.784–5

86 *'Adio, Carissima'* Hallam, p.553

86 *'Do you write'* Hallam, pp.784–5; Batchelor, p.77

86 *'Zeus came down'* T Selected Poems, p.310n

86 *'Now lies the Earth'* Ibid., p.319

7 Extinctions

87 *started out at four* Leonard G. Wilson, *Charles Lyell*, 1973, p.208

87 *'For if sedimentary'* Lyell note book, 1828, Ibid., p.903

88 *sulphurous activity* Charles Lyell, *Principles of Geology*, editor James Secord, Penguin Classics, 2005, p.118

88 *sinister erosions* Ibid., p.331

88 *riverbed rock carving* Ibid., pp.113–16

88 *'an hypothesis'* Ibid., pp.92–4

88 *'global' vision* Ibid., p.52

88 *'mighty waves'* Ibid., pp.431–5

88 *'Never, perhaps'* Ibid., pp.24–5

89 *avoided using the reassuring* Ibid., p.437

89 *to express so powerfully* E.E. Snyder, 'Tennyson's Progressive Geology', *Victorian Network*, Vol. 2, 2010

89 *doubts struck Tennyson* Ibid., and T Letters 1, p.145 (1836)

89 *'The existence of life millions'* Charles Lyell, *Journals*, no. IV, November 1859, p.297

90 *'The choice, the progress'* CT Alfred, p.249

90 *'Then might those genera'* Lyell, *Principles of Geology*, 'Climate', pp.66–7

90 *'standing by a railway'* HT Memoir, p.230, c.1846

91 *inspired Byron's poem* Lord Byron, 'Darkness', written Villa Diodati, Lake Geneva, July 1816

91 *'Have the changes'* Quarterly Review, November 1833; Stephen J. Gould, *Dinosaur in a Haystack*, Harmony, 1995, p.165. See William Glen, *The Mass Extinction Debates*, 1955

91 *'Earthquakes and Volcanoes'* Lyell, *Principles of Geology*, Volume 1, section 26, p.162

91 *'in as much as a round belly'* T Letters 1, p.145 and note

92 *'So dark a forethought'* T Selected Poems, p.308; *Life, Letters and Journals of Sir Charles Lyell*, 1881, 'Journey to Staffa', 1817, pp.53–6; Batchelor, p.202

92 *'A still small voice'* 'The Two Voices' T Selected Poems, p.191

92 *'"Yea!" said the voice'* 'The Two Voices', lines 157–65; T Norton Poetry, pp.88–9

93 *'some superb meditations'* T Selected Poems, p.101n

93 *'The design is so grand'* Spedding, 19 September 1834, in T Letters 1, p.118

93 *'the conflict in a soul'* Fitz in T Selected Poems, p.101n

93 *'To which the voice did urge reply'* Ibid., p.103

94 *other bleaker, more daring stanzas* Ibid., pp.103, 114, 115

94 *'No motion has she now, no force'* William Wordsworth: Selected Poetry edited Nicholas Roe, Penguin, 1992, p.89

94 *'When Mammoth, in the primal woods'* T Selected Poems, p.115n

95 *'When I wrote'* T Norton Poetry, p.84n

8 Memorials

96 *Matilda had claimed* Batchelor, p.78

96 *suggests how much Hallam* CT The Tennysons, pp.144–5

96 *hoping for some kind of obituary* T Letters 1, p.108

97 *'the earliest jottings'* HT Memoir, p.249

97 *'I seek the voice I loved'* Ricks, 1972, p.121

97 *'Dark house'* T Selected Poems, pp.351–2

98 *'there where the long street roars'* Ibid., pp.468–9

98 *twenty-nine other poems* 'In Memoriam: Image, Symbol, Motif: the Hand', Richard Schindler, Victorian Web, 20 February 2010

98 *eventually recovered from his Camden* T Letters 1, p.297 (1849)

99 *'It begins with death'* T Selected Poems, p.339n

99 *'a wonderful and deeply truthful'* Gould, *Dinosaur in a Haystack*, p.73

99 *'Old Yew, which graspest'* T Selected Poems, p.346

100 *'At our old pastimes in the hall'* Ibid., pp.373–4

101 *transformation is emphasised* T Norton Memoriam, p.ix

101   *allowed Tennyson to write his first drafts* 'Tennyson Mss Online', Trinity College Library, WordPress.com, 2017

101   *'affliction into which our family'* Frederick Tennyson, 18 December 1833, in T Letters 1, p.104

101   *'Hallam is dead!'* John Rashdall diary, 10 October 1833, quoted Ralph Rader, *Tennyson's Maud: The Biographical Genesis*, California University Press, 1963, p.12

101   *'left his heart a widowed one'* John Rashdall diary, 14 January 1834, Rader, p.16

102   *'She came down to us'* Hallam, p.795

102   *'In these overwhelming griefs'* CT The Tennysons, p.146

102   *'What hope is here for modern rhyme'* T Selected Poems, p.416

103   *'There is more about myself'* Ibid., p.138 notes

103   *'Tennyson has been in town'* T Letters 1, p.95

103   *'concentrated contemplation'* Ibid., p.103

103   *'till years after the burial'* T Norton Poetry, p.207n

104   *'till the silent midnight hour'* T Letters 1, pp.128–9

104   *'My Arthur, whom I shall not see'* T Selected Poems, p.354

104   *not publish certain of Hallam's own writings* Henry Hallam, 7 February 1834, in T Letters 1, pp.106–7

104   *practical list of shared costs* Jillian Hess, *How Romantics and Victorians Organized Information: Commonplace Books, Scrapbooks, and Albums*, OUP, 2022, p.239. She assumes it was a trip to France

105   *Though it is impossible to know* Martin, p.185

105   *'Break, break, break'* T Selected Poems, p.165

106   *'made one early summer morning'* Arthur Coleridge, 'Notes on Tennyson's Conversation' T Interviews, p.173

107   *'I sometimes hold it half a sin'* T Selected Poems, pp.348–9

107   *became 'Tithonus'* Ibid., p.992; p.583

107   *'I cannot rest from travel'* Ibid., p.141

108   *'Tho' much is taken'* Ibid., p.142

108   *'Newton had destroyed all the poetry'* Holmes, *The Age of Wonder*, p.319

108   *'The rings of Saturn'* Herschel, *A Treatise on Astronomy*, 1833, p.286

109   *'I am part of all that I have met'* T Selected Poems, p.142

109   *'originally a pendant to the Ulysses'* Ibid., p.992

109   *'The woods decay'* 'Tithonus', Ibid., pp.585 and 992

110   *'What is life to me!'* Ibid., p.583n

110   *'truly parental kindness'* Thwaite, p.104

110   *taking up singing* Ibid., p.105

110   *almost suicidally depressive* T Selected Poems, p.101

110   *also completed in draft* Ibid., p.124

110   *'Or in the night'* Ibid., p.130

111 *'Tears of the widower'* Ibid., p.358

112 *'A time to sicken and to swoon'* Ibid., p.366

9 Romances

114 *at the village of Spilsby* Rader, p.24

115 *'Alas for her that met me' Maud*, Part 2, Section IV, T Selected Poems, p.573

115 *'She would tell of how'* H.D. Rawnsley, *Memories of the Tennysons*, 1912, in Rader, p.27

116 *'peacock' feelings* Ibid., Rader, pp.27–8. Based on a conversation with Rosa Baring, about 1852–3, see Rader, p.123n

116 *'Come into the garden, Maud'* T Selected Poems, p.569; Batchelor, p.85

117 *'But all my blood in time to thine shall beat'* Rader, p.28

117 *'My dust would hear her and beat' Maud*, T Selected Poems, p.563

117 *'Lines written by'* Rader, p.28

117 *'To thee, with whom my best affections dwell'* Rader, p.62, from Harvard Mss

118 *'young gentleman of the ordinary type'* Rader, p.61; p.138n. From Sophie Rawnsley in H.D. Rawnsley, *Memories of the Tennysons*, 1912

118 *'Sometimes in the midst of'* Thwaite, p.96. From H.D. Rawnsley, *Memories of the Tennysons*, 1912, pp.67–8

118 *a survey of contemporary science* T Letters 1, p.128

118 *'organization, intelligence, life'* Whewell, *Astronomy and General Physics*, 1833, Book 3, p.270

118 *'leave our powers of conception far behind'* Ibid., p.278

119 *reaching an exhausted dawn* T Selected Poems, p.123

119 *'Thereto the silent voice replied'* Ibid., p.103; see W.B. Elliott, 'Tennyson and the Concept of Evolution before 1859', PhD, 1973, p.43

120 *'an inexhaustible field'* Michael J. Crowe, *Modern Theories of the Universe: From Herschel to Hubble*, Dover, 1994, p.162

120 *'Some vague emotion of delight'* T Selected Poems, pp.120–1; see W.B. Elliott, 1973, p.43

120 *'Alfred is delighted with her'* Emily Tennyson, T Letters 1, p.130n

120 *'a very good fellow but'* Fitz Letters 1, p.2

121 *almost comic physical contrast* Robert Bernard Martin, *With Friends Possessed: A Life of Edward FitzGerald*, Faber, 1985, p.30; p.79; p.99. Drawing on the only photograph of Fitz, much later in life, Martin suggests 'a round fleshy face, large startled eyebrows, and a cleft chin'

121 *more than enough to support* Fitz Letters 1, p.14

122 *'all the way to London'* Fitz Letters 2, p.32. Writing in June 1851, Fitz remembers first *hearing* of 'The Lady of Shalott' at Trinity College 'twenty years all but one since', that is in June 1832; but he cannot have actually read it until publication in 1833

122 *'King Arthur's sword, Excalibur'* T Selected Poems, pp.155–6

122 *'Not bad that, Fitz, is it'* FitzGerald 1835, in Interviews, pp.168–9

122 *'I met poor Hartley'* T Letters 1, pp.130–2 and note; T to Derwent Coleridge, 1850, T Letters 1, p.341

123 *Tennyson appears equally tousled* NPG No. 3940 'attributed to James Spedding, 1831'. Two similar informal sketches, but dated 1835, appear in Martin, Plates Vc and Vi, opp. p.117

123 *'but grumpy to you'* T Letters 1, p.132

123 *'So all day long the noise of battle rolled'* T Selected Poems, p.150

124 *'There was no declamatory'* HT Memoir, p.162

124 *'Dear Tennyson, though I am'* T Letters 1, p.134

124 *taking on protégés* Ibid., p.133n; Martin, *With Friends Possessed*, pp.110, 129, 138 and *passim*

125 *'Poets, as you say'* T Letters 3, 1835, p.451, quoted in *The Reception of Tennyson in Europe*, editor Leonee Ormond, Bloomsbury, 2017, p.26. Q.v. for other later continental contacts and translations in France, Russia etc., e.g., *Revue Britannique*, August 1842 including Godiva and Ulysses, p.27

125 *'the precise nature of the enchantment'* Mill in T Crit Heritage, p.88

125 *'states of emotion'* Ibid., p.95

125 *'philosophical speculation'* Ibid., p.96

126 *none of the Somersby Tennysons* T Chrono, p.24

126 *'I know no reason why'* Batchelor, p.100

126 *also an opium habit* Ibid., p.99

126 *promise of a capital investment* CT The Tennysons, p.121

126 *gloomy weeping fits* Ibid., p.120

126 *'You ask after Charles'* T Letters 1, p.131

127 *'For while the tender service'* Thwaite, p.116; from T Ricks Poems 2, p.90

127 *Louisa was making regular returns* Thwaite, p.116

127 *she owned her own copy* Ibid., p.91

127 *'You were in a silk pelisse'* T Letters 1, p.159

128 *'somewhat Eastern-looking as a girl'* CT Alfred, p.164; Thwaite, p.120

128 *summer of sonnet-writing* Batchelor, pp.84–90

128 *Much of this confusion* See Rader, 1963

128 *writing despairing love letters* T Letters 1, p.142; T Chrono, p.26

128 *'But what can T have'* Martin, p.228

129 *'By night we linger'd on the lawn'* T Selected Poems, pp.437–8

129 *immersed in the third volume* T Letters 1, p.145; and footnote
129 *He also read Babbage's* T Selected Poems, p.209n
130 *'arguments by design'* Babbage, 1837, Chapter 2
130 *'the administration of the universe'* Babbage, *Ninth Treatise*, 1837, title page
130 *'whilst the testimony of Moses'* Babbage, 1837, Chapter 5
130 *'objectionable passages'* Wilson, *Lyell*, p.461
130 *'Others there are'* Babbage, 1837, Chapter 7, p.91; T Selected Poems, p.209
130 *a vast calculating machine* Babbage, 1837, Chapter 2
131 *'The noiseless ether curdling'* T Selected Poems, p.209 n.
131 *would later be lost* See admirable research and discussion of surviving fragments and clues in Thwaite, pp.132–8
131 *'Yesterday I dined with Alfred'* T Letters 1, p.150
132 *'What a face'* Ibid., p.150
132 *'A dark Indian Taino maiden'* HT Memoir, p.47; Perry, p.13
132 *'It would be confuted by some Midshipman'* Martin, pp.124–5
132 *'I cannot have'* T Letters 1, January 1837, p.149
132 *'Oh! That 'twere possible'* T Selected Poems, p.989
133 *'We leave the well-beloved place'* Ibid., p.448. Somersby is frequently recalled, but especially in *In Memoriam* Nos 89, 100 and 101

10 Asylum

134 *Tennyson liked to mock* Martin, p.229
134 *'no sounds of Nature'* T Letters 1, p.159
134 *what would the sons do* CT The Tennysons, p.88
135 *the sole profession was poetry* Ibid., pp.88, 120
135 *able to afford servants* T Letters 1, p.177
135 *family was still large* CT The Tennysons, p.1
135 *'full of spirit'* T Letters 1, p.160
135 *'sufficiently hospitable'* Ibid., p.158
135 *'Has Charles been in any of his'* Ibid., p.160
135 *'with his long blue cloak'* CT Alfred, p.171
136 *'for months'* T Letters 1, p.171
136 *'this great black Babylon'* Ibid., p.157
136 *'I have been in this place'* Ibid.
136 *'the light of London flaring'* T Selected Poems, 'Locksley Hall', p.188, lines 114–15
136 *'the strongest most stinking tobacco'* T Letters 1, p.159

137 *different from the traditional lunatic asylum* Michel Foucault, *The History of Madness*, revised edition, 2006

137 *Born in 1783* Pamela Faithfull, *An Evaluation of an Eccentric: Matthew Allen*, PhD thesis, 2001, online

137 *appointed apothecary* T Letters 1, p.183

137 *'As early as 1807'* Matthew Allen, *Essay on the Classification of the Insane*, 1837, pp.viii–ix

138 *'not more than about 3 percent'* Ibid., p.33

138 *'detail to him a history'* Ibid., p.45

138 *sense of identity and achievement* Ibid., p.61

138 *doctor's assistant to Lincoln* CT The Tennysons, p.120

139 *'chair' the little boy* Ibid., p.120

139 *'some bustling active line'* T Letters 1, p.106

139 *'so miserable a termination'* CT The Tennysons, p.80

139 *'I have studied the minds'* T Letters 1, p.106; CT The Tennysons, p.80

140 *'Places of seclusion for exhausted Minds'* Allen, *Classification*, p.106

140 *'One day I went with Mrs. Montagu'* Online: Margaret C. Barnet, 'Matthew Allen MD (Aberdeen)', CUP, 1965, p.22

140 *healing comforts of Fair Mead* Martin, pp.236–7

141 *'which is absolutely necessary'* Elizabeth Tennyson to Charles Tennyson d'Eyncourt, 8 March 1843, T Letters 1, p.217

141 *in the wild hillsides of Tasmania* CT Alfred, p.168; T Letters 1, p.160n

141 *'Vanity, my dear Sir'* Martin, p.253

141 *'I love the forest and its airy bounds'* 'A Walk in the Forest', John Clare, *Major Works*, editor Eric Robinson, Oxford World's Classics, 1984, p.XXX

142 *'It is most singular'* Jonathan Bate, *John Clare*, Picador, 2003, p.430

142 *He never returned* Ibid., pp.430–48

142 *Tennyson grew increasingly fascinated* T Letters 1, p.183

142 *'has been on a visit'* Ibid., p.183n

143 *'None like her, none'* Maud, XVIII, T Selected Poems, p.551

143 *'whose mind was instantly wrecked'* T Selected Poems, p.576n

143 *'My heart would hear her and beat'* Maud, T Selected Poems, p.563; T Norton Poetry, p.10

144 *'O me, why have they not buried me deep enough?'* Maud, Part 2, Section V, stanza 11 in T Selected Poems, p.579

144 *'deeper, ever so little deeper'* T Norton Poetry, p.341n

144 *'It may be asserted'* Poe, 'The Premature Burial', originally in the *Philadelphia Dollar Newspaper*, July 1844; see Project Gutenberg, online

145 *extraordinary eleven-part monologue* Maud, Part 2, Section V, stanzas 1–11 in T Selected Poems, pp.574–9

145 *'Dead, long dead'* T Selected Poems, p.574

145 *entranced Tennyson* Martin, p.364

146 *'For a raven ever croaks' Maud*, T Selected Poems, p.534, line 246

146 *'the most original American genius'* HT Memoir, p.663

146 *'begged me to weave'* T Norton Poetry, p.339, and footnote

146 *'When I was wont to meet her' Maud*, Part 2, in T Selected Poems, p.569; dating also discussed in Batchelor, pp.105–8; and Ricks, 1972, p.148; pp.248–9

147 *'an* entirely new form' T Letters 2, pp.134–5

147 *'I am – yet what I am none cares or knows'* John Clare, 'I Am', 1844, Clare, *Major Works*, p.361

147 *'one of the best-known doctors'* HT Memoir, p.337; see also T Norton Poetry, p.341n

147 *'I took a man constitutionally diseased'* T Letters 2, p.138

147 *colonial life* CT Alfred, p.168

147 *'rather unused to the Planet'* Ibid., p.199

147 *'I am in the land of beauty'* CT The Tennysons, pp.121–2

147 *Allen's growing influence* Fitz Letters 1, p.259n

148 *'I shall be glad to see you again'* August 1840, T Letters 1, p.184

148 *draw the whole Tennyson family* Ibid., pp.189, 197

148 *convinced Tennyson to invest* This and John Clare is the subject of a contemporary novel, *The Quickening Maze*, by Adam Foulds, 2009

148 *'not so able as in old years'* T Letters 1, p.179

148 *transferred to the coast of northern France* T Norton Poetry, p.336, footnote composed '1830s'. Rader, p.8

148 *It was evidently inspired by* T Letters 1, p.179

148 *'See what a lovely shell'* T Selected Poems, pp.566–7 *Maud*, Part 2, Section II, 1–4

149 *'It sometimes appears extraordinary'* Lyell, *Geology*, 1833, quoted in T Selected Poems, pp.566–7, footnote

149 *'In Brittany'* T Norton Poetry, p.336n

11 London

151 *'We have had Alfred'* T Letters 1, p.161

151 *'A hand displayed with many a little art'* Rader, p.36

152 *'thou art mated with a clown'* T Selected Poems, 'Locksley Hall', p.185

152 *represents some continued exorcism* T Letters 1, p.168

152 *'coming down the hill over Torquay'* HT Memoir, pp.163–4

153 *'There on a slope of orchard'* T Selected Poems, p.195

153 *'Who'd serve the state?'* Ibid., p.196

153 *'old matters over'* Ibid., pp.196–7

154 *'So sang we each to either'* Ibid., p.197

154 *'A known landskip is to me'* T Letters 1, p.166

154 *inscribed 'Xmas Day, 1838'* Ibid., p.168n; Tate, 2020, p.147

154 *possibly a gift from Emily Sellwood* Thwaite, p.131

155 *'greatly augmented edition'* Richard Holmes, *This Long Pursuit: Reflections of a Romantic Biographer*, William Collins, 2016, p.207

155 *'The various musical instruments'* Somerville, *Connexion*, p.149

156 *'The splendour falls on castle walls'* T Selected Poems, p.265

156 *'Anyone who has observed'* Somerville, *Connexion*, pp.146–56

156 *university course in general science* Holmes, *This Long Pursuit*, pp.200–1

157 *'a masterly survey'* Ibid., p.202

157 *Samuel Johnson had proposed it* J. Killham, *Tennyson and The Princess: Reflections of an Age*, 1958, p.198

157 *talking it over with his sisters* Thwaite, p.131

157 *'The Husks'* CT Alfred, pp.145–6

158 *'Alfred wandering weirdly'* Ibid., p.175

158 *most of them have been lost* Thwaite, p.131

158 *entitled 'The New University'* HT Memoir quoting T Selected Poems, p.219n

159 *few fragments have survived* T Letters 1, pp.170–86

159 *'I murmured'* Ibid., p.168

159 *'Annihilate within yourself'* Ibid., p.174

159 *apologised for his 'preaching'* Ibid., p.182

159 *no record of any formal marriage proposal* Thwaite, p.131

160 *definitively forbid it* Martin, p.247

160 *curiously echoes* T Chrono, p.32

160 *definitely broken off* Martin, p.247

160 *'I fly thee for my good'* T Letters 1, p.182

160 *few letters are known between the two* Martin, p.245

160 *'I scarce expect thee'* T Letters 1, p.186

160 *'Thine dear, for ever and ever'* Ibid.

160 *'All send love'* Ibid., p.214

160 *'A true human soul'* Thomas Carlyle in T Interviews, p.16

161 *'His voice is musical metallic'* Ibid.

161 *'Is Alfred Tennyson among'* EB to Mitford, 1839, in Martin, p.241

162 *'I want AT to publish'* Fitz Letters 1, p.239

162 *'A feeling very generally exists'* Chartism, 1839, Chapter 1, p.2

163 *'I really do not know'* T Letters 1, p.177

163 *'in traveller's costume'* Ibid., p.184

163 *'Alfred Tennyson has reappeared'* Ibid., p.184n

163 *portrait of Tennyson* Ibid., p.235n

163 *'improved' to Pre-Raphaelite standards* Martin, p.240

163 *£900 in the therapeutic wood-carving business* Batchelor, p.100. These complicated transactions are explained in T Letters 1, footnote pp.183–4

163 *Tennyson may also have intended* T Letters 1, p.197; T Chrono, pp.32–3

164 *reported to Emily Sellwood* T Letters 1, p.214n

164 *'A fine, large-featured'* Ibid., p.214n

164 *Carlyle cut a special niche* Martin, p.243

164 *'preferred clubbing with his mother'* HT Memoir, p.156; T Interviews, p.16

165 *'Many a night from yonder ivied casement'* T Selected Poems, pp.183–4, 'Locksley Hall', lines 7–16

165 *'falser than all fancy fathoms'* T Selected Poems, p.185

166 *dreams of abandoning England* Ibid., p.188

167 *'For I dipt into the future'* Ibid., 'Locksley Hall', lines 119–28

167 *'Science moves, but slowly'* Ibid., p.189, 'Locksley Hall', line 134

167 *'Larger constellations burning'* Ibid., p.191

168 *'the trader' had never come* Ibid., 'Locksley Hall', line 161

168 *'There methinks would be enjoyment'* Ibid., lines 165–8

168 *'Mated with a squalid savage'* Ibid., p. 192, 'Locksley Hall', line 177

168 *'Could I wed a savage woman'* T Selected Poems, p.192n

169 *go abroad 'among savages'* Fitz Letters 1, November 1848, p.623

169 *'Fool, again the dream'* T Selected Poems, p.192, 'Locksley Hall', line 173

169 *South Seas languor* T Selected Poems, p.191

169 *'Not in vain the future beacons'* Ibid., p.192

170 *'The wheels in their resounding'* Blackwood's Magazine, August 1843

170 *'reverberating' throughout the poem* Martin, p.455

170 *'English people liked verse'* T Selected Poems, p.182n

170 *'I walk about the coast here'* T Letters 1, February 1841, p.188

170 *'got drunk'* Ibid., p.189

171 *'Dear old Fitz'* Ibid., p.188

171 *Sometimes they met at* Fitz Letters 1, March 1842, p.316

171 *'I would rather know'* Fitz Letters 1, p.57, quoted Martin, pp.262–3

171 *call 'a truce'* Fitz Letters 1, 17 February 1840, p.246

172 *'liker to flying than anything else'* HT Memoir, p.146; Fitz Letters 1, p.252n

172 *'hereditary tenderness of nerve'* Fitz Letters 1, April 1840, p.246

172 *'noble natured, with no meanness'* Ibid.

172 *several disappearances abroad* T Letters 1, p.192

173 *'I past beside the reverend walls'* No. 87, T Selected Poems, p.429

174 *'in great intimacy'* Fitz Letters 1, March 1841, p.272

174 *'I hope he will publish ere long'* Fitz Letters 1, November 1841, p.290

174 *have Mrs Tennyson sign* T Letters 1, p.213

174 *could somehow intervene* CT Alfred, pp.186–7

174 *'nothing here but myself and two starfish'* T Letters 1, p.215

175 *'with his short and keen sight'* CT Alfred, p.190

175 *young naval lieutenant she had met* T Letters 1, pp.194–5

175 *'I had such a romantic admiration'* Ibid., p.195

176 *Perhaps Jesse would prove* Ibid., p.194n

176 *'elflike'* CT Alfred, p.190

176 *continued his £300 annual allowance* T Letters 1, p.196

12 Publication

177 *New England Transcendentalists* HT Memoir, pp.135–6

177 *'a perfect music-box'* Emerson to Carlyle, Ibid., p.152

177 *'Yesterday, about 1 P.M.'* 'Dr Brydon's Report of the Khyber Disaster January 1842', by William Trousdale, 1983, www.Kyber.org/publications

178 *'So all day long the noise of battle roll'd'* T Selected Poems, p.150

178 *century later by Thomas Hardy* 'When I set out for Lyonnesse', 1870, Thomas Hardy, *Selected Poems*, edited by David Wright, Penguin, 1978, p.332. Accounts of the terrible retreat appeared throughout Europe, for example in the German novelist Theodor Fontane's ballad *The Tragedy of Afghanistan* (1858). See in a different style, George MacDonald Fraser's *Flashman in the Great Game* (1969), and with a full sense of imperial decline, William Dalrymple's *Return of a King* (2013)

179 *'What means this bitter discontent'* Chartism, Chapter 1, 'The Condition of England Question', 1839

179 *look 'detestable' in print* Fitz Letters 1, p.315

179 *'It was in 1842'* Fitz Letters 2, 4 December 1864, p.535

180 *'A love song I had somewhere read'* 'The Miller's Daughter', lines 65–72, in Gregory Tate, 2020, p.145

180 *admit just one short poem* T Selected Poems, p.165

181 *'With all his faults'* Fitz Letters 1, p.315

181 *'made a sensation'* T Letters 1, pp.209–10

181 *'No one but Coleridge among us'* T Crit Heritage, p.120

182 *'Heaven opens inward'* T Selected Poems, p.117, footnote 'Added to Heath MS which then ends'

182 *'so many doubts and hopes'* Richard Horne, 1844, T Crit Heritage, p.164

182 *'And forth into the fields I went'* T Selected Poems, p.123

183 *'Then fled she to her inmost bower' Alfred, Lord Tennyson: Poems Selected* by Mick Imlah, Faber, 2024, p.29

185 *'a noble poem'* Emerson in HT Memoir, p.152

185 *lines from it are carved beneath* 'Godiva' was not thought worthy of Christopher Ricks's *Tennyson: A Selected Edition* (2007), or the Norton Critical Edition of *Tennyson's Poetry* (1999), though included in the poet Mick Imlah's *Alfred, Lord Tennyson, Selected Poems*, Faber, 2004, pp.28–30

185 *'Sure never yet was Antelope'* T Ricks Poems 2, p.285

186 *favourably reviewed* Martin, p.266

186 *'admired it so much'* T Letters 1, p.212n

186 *'a kind of philosophical Keats'* T Crit Heritage, pp.134, 136, 152

186 *'It rests with Mr Tennyson'* Ibid., p.138

186 *'the direct outbirth'* Ibid., p.123

186 *'In the Two Voices'* HT Memoir p.161

187 *'Powers are displayed'* T Crit Heritage, p.152

187 *got to know Edmund Lushington* Martin, p.259; HT Memoir, p.152

187 *'There sinks the nebulous'* Ida in *The Princess*, T Selected Poems, p.265

188 *in the pages of the local press* 6 July 1842, Batchelor, p.128

188 *displayed clockwork boats* T Selected Poems, p.224

188 *rented a much larger* Martin, p.259

188 *'My head is yet vertiginous'* T Letters 1, p.212

188 *would form the opening* HT Memoir, p.169

188 *'We crossed into a land'* T Selected Poems, p.234n, from Harvard Ms copy book

189 *never met de Vere* T Letters 1, p.210

189 *'500 of my books are sold'* Ibid.

189 *'insane'* Martin, pp.266–7

189 *'the great Achilles'* T Selected Poems, p.145

190 *'greatest of poets'* Martin, p.267

190 *'clearness, solidity'* Ibid., p.266

190 *an inaccurate one* T Selected Poems, p.192

190 *'But it was a black night'* Ibid., p.192n

191 *'Fill the cup, and fill the can'* Ibid., p.215

191 *'In regard to metaphors'* Doron Swade, *The Difference Engine: Charles Babbage and the Quest to Build the First Computer*, Penguin, 2000

192 *'in the wail of the wires'* 'Poetry in the Age of New Sound Technology', Francis O'Gorman, *OpenEdition Journals*, 2009, online. See Seamus Perry, 2005, Chapter 4, 'Grieving'

13 Evolution

193 *first of his poems translated into Russian Revue Britannique*, August 1842, *The Reception of Alfred Tennyson in Europe*, Leonee Ormond, 2016, p.26; p.233

193 *'For Tennyson, as for a man'* Edgar Allan Poe in *Graham's Magazine*, Philadelphia, August, 1843

194 *'Goodbye old Fitz.'* T Letters 1, July 1842, p.205

194 *'Dear Dr Allen'* Ibid., p.205

194 *'proportion of anxiety'* Ibid., from Allen November 1841, p.197

194 *'Dr Allen has already'* Ibid., 15 October 1842, p.213

194 *'estranged himself from his family'* Ibid., p.214

195 *'I used to feel moods'* Tennyson, *Walks and Talks*, Tennyson Archive Lincoln, discovered by Ann Thwaite, 1996, p.74. She assigns it perhaps correctly to the 1830s

195 *at least four addresses* T Letters 1, January to July 1843, pp.215–19

196 *'Men to whom mental labour'* Edward Bulwer-Lytton, *Confessions of a Water Patient*, 1845, p.14

196 *an offshoot of Dr Gully's* T Letters 1, p.221

197 *'one was to see the West Indies'* HT Memoir, pp.219–20

197 *'No reading by candle light'* T Letters 1, pp.222–3 and notes

197 *continuing financial anxieties* HT Memoir, p.156n

197 *hideous sum of £8,000* T Letters 1, 15 October 1842, p.213

197 *'My dear Fitz – it is very kind'* Ibid., 2 February 1844, pp.222–3

198 *drifting back from Italy or France* CT The Tennysons, pp.125–6

198 *plenty of spare rooms* Batchelor, p.146

198 *intervention by Henry Hallam* Batchelor, p.134

198 *cash as much as water* T Letters 1, p.183n

199 *'the first attempt'* Examiner, 9 November 1844, in Killham, p.279

199 *'Is our race but'* Ibid., in Killham, p.285

200 *'I want you to get me'* T Letters 1, p.230. For a superb study of *Vestiges*, and the controversies surrounding it, see James A. Secord, *Victorian Sensation: The extraordinary Publication, Reception, and Secret Authorship of Vestiges of the Natural History of Creation*, University of Chicago Press, 2000

200 *'I trembled as I cut the leaves'* T Interviews, p.55

201 *'on the greatness of Turner'* T Letters 1, p.230

201 *'geology bad'* Darwin, Preface to 3rd edition of *Origins*, 1861; James A. Secord, *Victorian Sensation*, pp.430–3

201 *'How can we suppose'* Robert Chambers, *Vestiges of the Natural History of Creation*, 1844, p.154

201 *'Man is seen to be an enigma'* Ibid., p.345

201 *canine loyalty, photography and slavery* Ibid., pp.350, 356, 399

202 *not such a wild suggestion* Holmes, 'Computer Science: Enchantress of Abstraction', *Nature*, 3 September 2015

202 *'gives a remarkable suggestion' Examiner* quoted in Killham, Appendix, p.284

202 *found* Chambers's Encyclopaedia James A. Secord, *Victorian Sensation*, pp.290–6

203 *author was a woman* Ibid., pp.234–5

203 *'If the book be true'* Adam Sedgwick to Charles Lyell, 9 April 1845, in *Life and Letters of Adam Sedgwick*, editors John Willis Clark and Thomas McKenny Hughes, 1890, Vol. 2, pp.84–5

203 *'But why do not monkeys talk?'* Chambers, *Vestiges*, p.295; James A. Secord, *Victorian Sensation*, MS illustration on p.237

203 *'Most women have by nature'* James A. Secord, p.235

204 *'Be near me when my light is low'* In Memoriam No. 50, T Selected Poems, pp.392–3

204 *Tennyson later claimed* T Selected Poems, p.296n

205 *'It is clear from the whole'* Vestiges, quoted in T Selected Poems, p.397n

205 *'Oh, yet we trust that somehow good'* In Memoriam No. 54, T Selected Poems, p.396

206 *'so-called submarine forest of Happisburgh'* Charles Lyell, *Principles of Geology*, p.331

207 *'Amidst the vicissitude'* Charles Lyell, *Principles of Geology* quoted by John Carey, *Faber Book of Science*, p.77

207 *'None of the works'* Charles Lyell, *Principles of Geology*, p.333; in T Selected Poems, p.398n

207 *'"So careful of the type?" but no'* In Memoriam No. 56, T Selected Poems, pp.398–400, with an extensive footnote on the influence of Chalmers, Lyell and Babbage

210 *'Answer her riddle'* Thomas Carlyle, *Past and Present*, 1843, Chapter 3, 'The Sphinx'

211 *'Still am I sick of it'* 'To JH Reynolds', 25 March 1818: John Keats, *Letters*, edited Robert Gittings, OUP, 1970, pp.81–2

211 *Tennyson carefully omitted it* Batchelor, p.183

211 *'God cannot be cruel'* T Letters 1, p.175

14 Recognition

212 *'"The stars," she whispers'* T Selected Poems, p.347

212 *'the heat death of the sun'* William Thomson, later Lord Kelvin, 'On the Age of the Sun's Heat', *Macmillan's Magazine*, 1862, Vol. 5, pp.388–93. Expanded from an earlier article in the *Philosophical Magazine*, October 1852

213 *new kind of philosophical crisis* Batchelor, pp.177–84; Henchman, p.90

214 *'Behold a man raised up by Christ!' In Memoriam* No. 31, T Selected Poems, p.375

214 *'the cardinal point of Christianity'* T Norton Poetry, p.224n

214 *'It simply means that a man'* Thomas Huxley, 'Agnosticism', *Collected Essays V*, 1889

215 *'Tennyson is the only modern poet'* Martin, p.462; T.H. Huxley to Sir Michael Foster, January 1892, 'Astronomy and Geology: Terrible Muses' by A.J. Meadows, in *Notes and Records of the Royal Society*, 31 January 1992

215 *'Perplext in faith, but pure in deeds' In Memoriam* No. 96, T Selected Poems, p.441

215 *'No evolutionist is able'* Tennyson, in HT Memoir, p.271

215 *'Belfast Address'* Roland Jackson, *The Ascent of John Tyndall*, OUP, 2018, pp.217, 332

215 *'But the passage'* 'Scientific Materialism', by John Tyndall, in *Fragments of Science: Detached Essays, Addresses and Reviews*, 1872, quoted in HT Memoir, p.271. See also Tennyson's late poem 'By an Evolutionist', 1888, T Norton Poetry, p.570

217 *'Why this "Kraken" should'* Richard Horne, 'Alfred Tennyson', *New Spirit of the Age*, 1844, T Crit Heritage, pp.153–60

217 *did not speak of the elegies* T Letters 1, p.227

217 *He flitted between* Martin, p.257

217 *beds could be hired by the night* T Letters 1, p.227n

217 *'Don't you think the world'* Fitz Letters 1, p.478

218 *'After dinner'* T Interviews, p.9

219 *'he was very angry'* Ricks, 1972, p.182

219 *de Vere had become one* De Vere diary, in T Letters 1, p.237n

219 *'massive abundance'* Martin, pp.257–8

220 *'He and I made a plan'* Fitz Letters 1, June 1845, p.494

220 *'he had heard nothing before'* See T Selected Poems, p.219

220 *Mechanics' Institute's summer fair* Ibid., p.222

220 *'the much eulogised and calumniated'* Ibid., p.231

221 *'the fiery Son of Gloom'* Jane Welsh Carlyle's lively descriptive letter of 31 January 1845 is given in full in T Letters 1, p.233

221 *Lauterbrunnen valley* T Chrono, p.45

222 *'bad beer'* T Letters 1, p.260

222 *'Agreeable Swiss young lady'* Ibid., p.259

222 *'Their thousand wreathes'* T Norton Poetry, *The Princess*, Part 7, p.197

223 *'Myriads of rivulets'* T Selected Poems, pp.320, 322

223 *'Very feeble sunset'* T Letters 1, p.260; T Selected Poems, p.322n. Years later James Joyce took this joke about the smooth onomatopoeia of Tennyson's famous lines by referring to 'Alfred Lawn Tennyson' in *Ulysses* (1920)

223 *financing the entire trip* T Letters 1, p.260

223 *'that of the man who had been to Niagara'* Dickens to John Forster, 24 August 1846, quoted in T Letters 1, p.260

224 *coming round to Tennyson's work* HT Memoir, p.188

224 *'People fete and dine me'* T Letters 1, p.267

224 *'Ever thine'* Ibid., 12 November 1846, p.264

226 *catching the imagination* Kimberley Dimitriadis, 'Telescopes in the Drawing-Room: Geometry and Astronomy in George Eliot's *The Mill on the Floss'*, *Journal of Literature and Science*, 2018; Henchman, 2014, pp.158–60

226 *generally known by Tennyson* Henchman, 2014, pp.50, 90 and 94

226 *humorous but very odd* Fitz Letters 1, May 1847, p.559

226 *'a new era for the human intellect'* Henchman, 2014, pp.48, 52 and 92–3

227 *'The reader must look to'* Ibid., p.77

227 *What de Quincey had found* Ibid., Chapter 2, 'De Quincey's Disorientated Universe', especially pp.74–82

227 *'I met Carlyle last night'* Fitz Letters 1, p.534

227 *'Eh! old Jewish rags!'* Ibid., p.581n

228 *'You have him there'* Ibid.

228 *'to get a pint or two of fresh air'* Ibid., p.534

15 University

229 *'My dear Fitz'* Fitz Letters 1, December 1847, p.588; T Letters 1, p.281

230 *vehement in her feminist opinions* Thwaite, pp.131–2

230 *may be delusory* John Killham, *Tennyson and the Princess:Reflections of an Age*, The Athlone Press, University of London, 1958, p.197

230 *'It is in blank verse'* Ibid., p.12; T Selected Poems, p.219n

230 *'a great torment to Mr Moxon'* Killham, p.13

231 *'I was knocked up'* Fitz Letters 1, May 1847, p.559

231 *FitzGerald did not really approve* T Selected Poems, p.219

231 *'indecent … flippant'* Killham, p.103

232 *did not publish until 1870* Richard Holmes, *This Long Pursuit*, 2016, p.215

232 *'Quick answered Lilia'* Prologue, 128–37, T Norton Poetry, p.133

233 *'honest few'* T Selected Poems, p.505

233 *'They are not to study'* F.D. Maurice, Victorian Web, online. Rosalie Glynn Grylls, *Queen's College 1848–1948*, Routledge, 1948

233 *'The Objects and Methods'* Killham, p.132

233 *mathematician Ada Lovelace* Holmes, 'Computer Science: Enchantress of Abstraction', *Nature*, 3 September 2015

233 *'very* wonderful' T Letters 1, November 1844, p.23

233 *knew in considerable detail* T Selected Poems, *The Princess*, with notes, p.219

233 *'I believe the* Vindication' T Norton Poetry, p.129 Tennyson's note

234 *'Had fortune enabled me'* Holmes, *This Long Pursuit*, 2016, p.194

234 *'A man with knobs and wires and vials fired'* T Norton Poetry, pp.131–2

235 *'talk of College and of Ladies' rights'* T Selected Poems, p.230

235 *'LET NO MAN ENTER'* T Norton Poetry, p.145

235 *'Take Lilia, then for heroine'* Ibid., p.135

235 *'There sinks the nebulous star'* Ibid., p.160

236 *'Then we dipt in all'* Ibid., p.149, *The Princess*, Part 2, lines 357–63

236 *She is easily capable* T Norton Poetry, p.143

237 *'as formulated by Laplace'* Ibid., p.143 footnote 1, and Chambers, 1844, CUP, 2009, p.464

237 *'The world was once a fluid haze of light'* T Norton Poetry, p.143, *The Princess*, Part 2, lines 101ff

237 *'Poets, whose thoughts enrich'* Ibid., p.145, *The Princess*, Part 2, lines 109–64

238 *'"Why Sirs"'* Ibid., p.250, *The Princess*, Part 2, lines 346–55

238 *'that gives the manners of your country women'* Ibid., p.163, *The Princess*, Part 4, line 133

238 *The men gallantly plunge in* Ibid., pp.163–4, *The Princess*, Part 4, lines 134–70

238 *'mock-heroic gigantesque'* Ibid., p.201

238 *'Man is the hunter'* T Selected Poems, p.288, *The Princess*, Part 5

238 *'For woman is not undeveloped man'* Ibid., p.324, *The Princess*, Part 7

239 *various disputants and rivals* T Norton Poetry, p.183, *The Princess*, Part 5

239 *university is dissolved* Ibid., p.193, *The Princess*, Part 7; T Selected Poems, p.313

239 *'So was their sanctuary violated'* T Norton Poetry, p.193, *The Princess*, Part 6

239 *'The grand error'* T Crit Heritage, p.167

239 *'we wound'* T Norton Poetry, p.159

240 *'I would the old God of war himself'* Ibid., p.176, *The Princess*, Part 5, 139–43

240 *Similar images* Michelle Geric, *Tennyson and Geology: Poetry and Poetics*, Macmillan, 2017, p.43

240 *'crammed with theories out of books'* T Norton Poetry, p.201, *The Princess*, Conclusion, lines 29–35

240 *first produced in 1870* At the Olympic Theatre in London on 8 January 1870. Gilbert called the piece 'a whimsical allegory'

241 *seventeen editions* Killham, p.5

241 *Tennyson responded immediately* T Letters 1, p.282n

241 *'Why should Mr Tennyson'* Killham, p.14

241 *'stately Amazon'* T Crit Heritage, pp.166–7

242 *'I know nothing which'* 13 January 1848, Fitz Letters 1, p.592

242 *set it to music* Thwaite, p.186

242 *'Sweet and low'* T Norton Poetry, p.151

243 *'Now sleeps the crimson petal'* Ibid., p.196, *The Princess*, Part 7, lines 158–74

244 *very fully for himself* The mysterious glimmering, sexual quality of the whole poem inspired by Titian would later appear in the work of the Viennese painter Gustav Klimt, and his famous picture *Danaë* of 1907. The poem was, in its own mysterious way, a life-giving answer to that earlier death-giving sonnet 'The Kraken'

244 *'elaborate trifling'* T Selected Poems, p.188

244 *'Tennyson is now in Ireland'* Fitz Letters 1, May 1848, p.604

245 *only 'great poet' alive* Ibid., June 1848, p.607

245 *did not seem to include* Ibid., September 1847, p.278

245 *'You will understand'* T Letters 1, p.273

245 *'Long conversation with him'* F.T. Palgrave, Journal, in T Letters 1, p.298n

246 *'eight distinct echoes'* T Norton Poetry, p. 159n, *The Princess*, Part 3; and Memoir

246 *Holst's settings of 1905* Gustav Holst, 'Songs from The Princess Op 20a', Holst Singers, available on Hyperion Records 1994, and YouTube

16 Epic

247 *Here he was lapped* T Letters 1, p.283

248 *'I do not see that'* De Vere Journal, in HT Memoir, p.240

248 *'an excellent dancer'* Ibid.

248 *'a life-guardsman spoilt by making poetry'* HT Memoir, p.157

248 *'a Poet to whom Nature'* Martin, p.242

248 *'A truly interesting Son of Earth'* T Letters 1, p.281 and footnotes

249 *'I never in my whole life'* HT Memoir, p.222

250 *'Perhaps, however'* William Howitt, 'Tennyson', *Homes and Haunts of the Most Eminent British Poets*, 1847, Vol. 2, pp.529–30. See Project Gutenberg, online

250 *no sign of Edward FitzGerald's name* T Letters 1, p.286

251 *'Provisions are growing very scarce'* Emily T letters, in HT Memoir, pp.227–8. Online source: https://kimberlyevemusings.blogspot.com/2015/05/meet-tennyson-sisters-of-somersby.html

251 *'pile her barricades with dead'* In Memoriam No. 127 in T Selected Poems, p.473

252 *powers on this prophecy* Thomas Carlyle, 'Signs of the Times' (*Edinburgh Review*, 1829; revised 1845, 1858), Victorian Web, paragraph 31

252 *'I hear that there are larger waves'* T Letters 1, p.288n

253 *'fanged with cobbles'* Ibid., p.290

253 *'Coast looked gray'* Ibid., p.291

253 *'Large rich crimson clover'* 'Tennyson's Cornwall Journal' in T Letters 1, p.288. Fuller version HT Memoir, pp.228–9

254 *'zoophytes, corallines and a spider'* Journal, T Letters 1, p.290

254 *'Turf – fires on the hills'* Ibid.

254 *'pretty railway by the sea'* T Letters 1, pp.288–92

255 *'And shall Trelawny die?'* 'Trelawny' by Robert Stephen Hawker, 1826; Dickens, *Household Words*, 30 October 1852

255 *'Seated on the brow of the cliff'* T Letters 1, p.290n; Martin, p.320

255 *clamped on his head* See Batchelor, pp.153–5

255 *'they put me into the Cornish newspapers'* T Letters 1, p.290

256 *'many weeks of privation and penance'* Fitz Letters 1, November 1848, p.295

256 *'go among Savages'* Ibid., p.623

256 *'The Noble Savage'* Dickens, *Household Words*, June 1853

256 *'some savage woman'* T Selected Poems, p.191

256 *It was to join his brother* Fitz Letters 1, p.622

257 *'And be sure'* Charles Kingsley, *Alton Locke*, 1850, Chapter 15

257 *ran to a second edition* Gregory Tate, 'The Poetry of Victorian Science, 1842', *The Public Domain Review*, July 2018, online

257 *'In science we find'* Robert Hunt, *The Poetry of Science*, 1844, Project Gutenberg, online

258 *'painted, with considerable correctness'* Hunt, *The Poetry of Science*, p.318

258 *as Mary Somerville had observed* Gregory Tate, 'Tennyson's Sounds', *Nineteenth-Century Poetry and the Physical Sciences*, 2020, pp.145–60

258 *'To rest content'* Hunt, *The Poetry of Science*, Introduction, p.xiv

258 *'The task of wielding'* Ibid., p.411

259 *'To show that Science'* Dickens in the *Examiner* 9 December 1848, quoted in Gregory Tate, 'Robert Hunt, *The Poetry of Science: Studies of the Physical Phenomena of Nature, 1844*', Project Gutenberg, 2016, online

259 *'I often think it is not'* Fitz Letters 1, p.566

260 *'The falls of Niagara afford'* Lyell, *Principles of Geology*, pp.113–15

260 *'It is not only that this vision of Time'* Fitz Letters 1, p.566; Martin, p.140; T Selected Poems, 'Parnassus', p.662n

260 *'As to my Epic theory'* Fitz Letters 1, p.569

260 *'The Facts of Man's history'* Ibid.; Martin, p.140

261 *'There rolls the deep where grew the tree'* T Selected Poems, pp.468–9

262 *'It required little geological practice'* Darwin, Journal, 1 April 1835, *The Voyage of the Beagle: Journal of Researches into Geology and Natural History*, 1845, p.332

17 Private

264 'this desolate sea-coast' T Letters 1, p.309

264 several days in November Ibid., pp.310–12

265 a more painful symbolism Ibid., p.313n

265 'I was amused' The Cracroft Diary, T Letters 1, p.311n

265 'Tennyson, it seems, has returned' T Letters 1, p.297n

265 'still the same noble and droll fellow' Fitz Letters 1, p.627

265 'Contemplate all this work of Time' No. 118, T Selected Poems, p.464

266 'island universes' Henchman, pp.90–4; Holmes, The Age of Wonder, Chapter 4

266 'The solid earth whereon we tread' In Memoriam, No. 118, T Selected Poems, pp.464–5

267 'higher race' Matthew Rowlinson, 'History, Materiality and Type' in Purton, editor, Darwin, Tennyson and Their Readers, pp.35, 43, 51

267 divine purpose guiding the universe CT Alfred, p.250

267 'I trust I have not wasted breath' No. 120, T Selected Poems, p.466

268 work of Herbert Spencer, See Herbert Spencer, A System of Synthetic Philosophy, 1864; Alfred Russel Wallace, 'The Origin of Human Races and the Antiquity of Man Deduced from the Theory of "Natural Selection"', 1864; T.H. Huxley, Evidence as to Man's Place in Nature, 1863

268 'The exertions of the present race' Chambers, Vestiges, p.405. See also Batchelor, pp.177–84

268 'higher type' See Rebecca Stott, 'Tennyson's Drift', in Valerie Purton, 2013; and William Brent Elliott, 'Tennyson and the Concept of Evolution in Victorian Poetry before 1859', University of British Columbia, PhD thesis, 1973, online

268 'Whereof the man, that with me trod' In Memoriam: Epilogue, T Selected Poems p.484

268 'You speak of the Flimsiness' Ruskin, Letter to Henry Acland, 24 May 1851, Complete Works, vol. 36, p.115. See also George P. Landow, 'John Ruskin: Loss of Belief', Victorian Web

269 'poems on A. Hallam, some exquisite' T Letters 1, p.298n

269 'I heard Alfred had been seen' Fitz Letters 1, p.641

269 'I only wonder that' Ibid.

269 effectively establish Moxon Matthew Sangster, Living as an Author in the Romantic Period, Palgrave Macmillan, 2021

270 remains obscure This is despite the brilliant and dedicated research of Ann Thwaite, Emily Tennyson: The Poet's Wife, 1996. See especially her Chapter 3, 'Learning to Love and Weep' and Chapter 4 'Living so long unmarried …'

270 'unannounced before breakfast' Thwaite, p.171

270   *'had even definitely refused him'* Rader, p.79

270   *'She had grown to feel'* Ibid., pp.78–9

270   *'make one music'* T Selected Poems, p.343

271   *begun to find oppressive* Thwaite, p.173

271   *'Strong Son of God, Immortal Love'* T Selected Poems, p.341

272   *'Our little systems have their day'* T Selected Poems, p.343, *In Memoriam*, lines 17–24

272   *'Let knowledge grow from more to more'* Ibid., lines 25–8

273   *'Forgive my grief for one removed'* Ibid., p. 344, *In Memoriam*, lines 37–40

273   *'Forgive the wild and wondering cries'* Ibid., lines 41–4

274   *'the willing and deliberate champion'* Charles Kingsley, *Fraser's Magazine*, September 1850, in T Crit Heritage, p.173; Thwaite, pp.182–3

274   *Instead he drew attention to* Fitz Letters 1, December 1849, p.657

274   *'I have made up my mind'* T Letters 1, p.316

274   *'I found A. Tennyson in chambers'* 17 January 1850, Fitz Letters 1, p.661

275   *'A.T. is gone'* Ibid.

276   *suggested a quite new role* T Letters 1, pp.307–9n; Thwaite, p.187

276   *'Give none away'* T Letters 1, p.322

277   *still exist in her handwriting* Thwaite, p.182

277   *reply is strange and awkward* T Letters 1, p.322n

277   *may have been written earlier* Thwaite, p.182

277   *passage of unrestrained praise* T Letters 1, p.323; CT Alfred, p.242

278   *'I have read the poems'* T Letters 1, p.323n; CT Alfred, p.242; Thwaite, pp.182–3

278   *'After such big words'* Emily Sellwood in T Letters 1, p.323n

18 Public

279   *dedicated with due formality* Batchelor, p.177

279   *fifty thousand sales by mid-autumn* CT Alfred, p.248

279   *'varied and profound reflections'* T Norton Memoriam, pp.111–12

280   *nineteen-page eulogy to Tennyson* Ibid., p.115

280   *'the great epochs in the history of poetry'* Ibid., p.113

280   *'stood at last on a pedestal'* CT Alfred, p.248

280   *no less than sixty thousand copies* Batchelor, p.189

280   *'The Way of the Soul'* CT Alfred, p.248

280   *whose 'bounty' had never failed* T Letters 1, p.331; HT Memoir, p.279

281   *'Blessed it is to find'* Charles Kingsley in T Crit Heritage, pp.184–5; T Letters 1, p.323n; Batchelor, p.160

281 *'fell on Alfred Tennyson's poetry'* Charles Kingsley, *Alton Locke: Tailor Poet: An Autobiography*, 1850, Chapter 9

281 *which Ada Lovelace had claimed* Holmes, 'Computer Science: Enchantress of Abstraction', *Nature*, 3 June 2015

282 *'It is the cursed inactivity'* Fitz Letters 1, p.696

282 *Had Tennyson refused his?* Thomas Carlyle, *On Heroes, Hero-Worship, and the Heroic in History*, 1841

282 *'The measure is of too obvious facility'* T Norton Memoriam, p.116

283 *'The sad mechanic exercise'* T Norton Memoriam, p. 9, *In Memoriam*, No. 5

283 'In Memoriam *can, I think'* T.S. Eliot, 'In Memoriam', *Essays Ancient and Modern*, 1936, T Norton Memoriam, p.626

284 *'an almost total absence'* T Norton Memoriam, p.117

284 *'undecided as to Mr Tennyson's faith'* Ibid., p.118

284 *'the physical world is always'* Henry Sidgwick, Letter to Hallam Tennyson, 1897, HT Memoir, pp.253–4

284 *'I remember being struck'* Henry Sidgwick quoted at length in HT Memoir, pp.252–6

285 *echo Tennyson's own sense of writing* Batchelor, p.278

285 *under the Tennysonian strapline* Bernard Lightman, *Victorian Popularizers of Science: Designing Nature for New Audiences*, Chicago University Press, 2007, pp.329, 335

285 *'It must be remembered'* Tennyson quoted in HT Memoir, p.255

286 *he assured Knowles* Knowles, 'Aspects of Tennyson', 1893, reprinted in T Crit Heritage, p.172n

286 *arrived at the last moment* Thwaite, pp.192–6, gives a witty account of the wedding

286 *'I told nobody'* CT Alfred, p.244

286 *'Alfred Tennyson of Lincoln Inn Fields'* Batchelor, p.160

286 *no literary celebrities from London* T Letters 1, p.327; Thwaite, pp.194–5; T Chrono, p.56

287 *'I hope they will be happy'* T Letters 1, p.325n

287 *'My dear Sophie'* Ibid., p.328

287 *'I have married a lady'* Ibid., p.330

287 *'But you know Alfred'* Fitz Letters 1, p.696

288 *'Mrs Alfred is a very nice creature'* Ibid., p.691

288 *'It seemed a kind of consecration'* CT Alfred, p.244; Emily Sellwood's Journal, T Letters 1, p.330 note

288 *practical gift of £50* Thwaite, p.195; T Letters 1, p.325n

288 *'it looked as lovely'* T Letters 1, p.334

289 *'seemed to move among a world of ghosts'* T Selected Poems, p.282

289   '*an heir to nothing*' T Letters 1, p.344

289   '*no hasty or ill-judged choice*' Ibid., p.335; Thwaite, p.208

289   *Tennyson published it the following year* CT Alfred, p.205

290   '*He clasps the crag with crooked hands*' T Norton Poetry, p.81; T Selected Poems, p.96

290   *Pyrenean adventure with Hallam* Martin, p.119

290   '*Hawk Roosting*' From Ted Hughes's second collection *Lupercal* (Faber, 1960), which also includes the famous 'Pike'

290   '*Is it at your command*' Book of Job, 39:27–30 RSV

291   '*If FitzGerald is prompt*' T Letters 1, p.344

291   '*owing chiefly to Prince Albert's admiration*' HT Memoir, p.280

291   *trousers did not fit* Ibid.

292   '*he could not* gracefully *decline*' T Letters 1, p.343

292   *new Laureate would be her son* CT Alfred, p.256

292   '*Alfred looks really improved*' Carlyle to Jane, 3 October 1850, in T Letters 1, p.339

292   '*I get such shoals of poems*' 28 September 1852, T Letters 2, p.45

292   '*This old-world*' HT Memoir, pp.281–2

292   *new literary ones like* T Letters 2, p.36

293   *nothing was further from Tennyson's mind* T Letters 1, p.340n

293   '*all the upper part to himself*' Ibid.

293   '*I already observe*' Aubrey de Vere, in T Letters 1, p.340

293   '*He realised my idea of a poet*' Walter White, *Journals*, 23 October 1850, in T Letters 1, p.341n

293   *five-year annual lease* T Letters 2, p.3; Martin, p.357

294   '*We grew but the more closely together*' Thwaite, p.227

294   *refused any idea* T Letters 2, p.15

294   '*It nearly broke my heart*' Ibid., pp.13–16; Thwaite, p.227

294   '*Little bosom not yet cold*' Christopher Ricks, 1972, p.234; T Ricks Poems 2, p.465

295   '*He was a grand, massive, manchild*' T Letters 2, p.14

295   '*broke down describing*' Martin, p.359

295   '*We felt, as it were*' Thwaite, p.227

296   *grave is still unknown* T Selected Poems, p.239

296   '*As through the land at eve we went*' *The Princess*, Part 1, altered 1850, 1862 editions, T Selected Poems, p.239

297   '*The Great Exhibition I don't ask*' Fitz Letters 2, pp.30, 38

297   '*great Glass House*' T Letters 2, p.17

297   *place in the royal box* Martin, p.358

297   '*aweary, aweary*' stanza HT Memoir, p.320

297   '*I am almost sure*' T Letters 2, p.77n

298 *their Villa Torrigiani* Thwaite, p.236

299 '*At Florence too what golden hours*' T Selected Poems, p.501

299 '*To Farringford*' Allingham, *Diary*, 29 July 1865, T Interviews, p.53

300 '*I believe, happily married*' Fitz Letters 2, p.46

300 '*Had I Alfred's voice*' Ibid., pp.45–6

19 Empires

301 '*I wrote it because it was expected*' T Letters 2, p.50

302 '*Bury the Great Duke*' T Selected Poems, p.489

302 *advance of £1,000 on future work* T Letters 2, p.48

302 *recuperate for several weeks* Ibid., p.41; Martin, p.366

302 *notable absentees* T Letters 2, p.48

303 *battalions of exclamation marks* Martin, p.365

303 '*Very wild but I think*' T Letters 2, p.21

303 '*Dear old Alfred*' Fitz Letters 2, p.82

304 '*sitting high and smiling*' Ibid., p.83

304 *made a casual suggestion* Ibid., pp.56, 83, 131

304 '*As fresh as when I heard them*' 27 May 1851, Fitz Letters 2, p.28

304 *evidently the kind of conversation* 24 January 1853, ibid., p.83

305 '*He nurses his Child delightfully*' 19 October 1853, ibid., p.112

305 '*so engaged in flying about*' September 1853, T Letters 2, p.72; Batchelor, p.203

306 '*burst into tears*' Ibid., p.204

306 *enormous sum of £4,350* 9 January 1854, T Letters 2, p.77

307 '*certainly a marvellous place*' Ibid., p.90

307 '*diagram of Orion*' Ibid., p.73

307 '*It is the famous nebula*' Thomas de Quincey, 1846, in Henchman, 2014, pp.73–5

308 '*Rigel one of the bright stars*' Ibid.

308 '*In this lovely place*' 28 February 1854, ibid., p.80

308 *expand to more advanced* Tennyson's book lists in Henchman, 2014, p.90; and her notes, pp.250–1

308 *garden at West Hampstead* Batchelor, p.278

308 '*His mind is* saturated' HT Memoir, p.739

308 '*A strong and stout young fellow*' T Letters 2, p.81

309 '*Little Hallam's behaviour*' 20 March 1854, ibid., p.82

309 *married them at Shiplake* Martin, p.179

309 '*very weak*' June 1854, T Letters 2, p.92

309 *last ever to the Isle of Wight* 15 June 1854, Emily T, in Fitz Letters 2, p.132

310   *'I had a letter'* Fitz Letters 2, p.48

310   *'yokes' of empire* T Selected Poems, p.485

310   *'Our ocean-empire'* Ibid., p.974

310   *'I have felt with my native land'* T Norton Poetry, p.346

311   *'written when the cannon'* Ibid., p.344n, from HT Memoir

311   *'It is time, O passionate heart'* *Maud*, Part 3, Section III, T Norton Poetry, p.345

312   *besides the naval guns* Thwaite, pp.293–4

312   *'Our ears were frenzied'* HT Memoir, p.320

312   *'Half a league'* T Selected Poems, pp.509–10

313   *the lack of organisation* Military background from Christopher Hibbert, *The Destruction of Lord Raglan: A Tragedy of the Crimean War, 1854–55*, Viking, 1984

313   *'Victorian war correspondents'* John Simpson, *We Chose to Speak of War and Strife: The World of the Foreign Correspondent*, Bloomsbury, 2016, 'Palaeojournalism', pp.39–47

313   *'They swept proudly past'* The Times, 14 November 1854

314   *'Causeless and fruitless'* The Times, 13 November 1854

314   *'The Times account'* T Interviews, p.24

314   *'Forward, the Light Brigade!'* T Selected Poems, pp.509–10

315   *'Will you kindly put'* T Letters 2, p.100

315   *'some bird of fabulous size'* Ibid.

315   *This extra stanza* Corrected proof copy illustrated in Thwaite, p.296; *Examiner* December 1854, original text from British Library Research online

315   *'Into the valley of death'* T Selected Poems, p.509, footnote 1854 not 1855

316   *'Stormed at with shot and shell'* Ibid., p.510

316   *'When can their glory fade?'* Ibid., p.511

316   *'My heart almost burst'* CT Alfred, p.288

317   *poetry to be learned by heart* Ibid.

317   *good-cheer package* August 1855, T Letters 2, pp.117, 120

317   *'Blow, Bugle, blow'* CT Alfred, p.519

317   *'on YouTube'* See for example the extraordinary *poetryreincarnations* 2011, online

318   *photographed by Roger Fenton* See online at *allworldwars.com/Crimean-War*

318   *'How I value this'* John Forster to Tennyson, 9 December 1854, in T Letters 2, p.102

20 Stars

319   *'go once more to visit him'* Terhune, *Life of Edward FitzGerald*, p.186

319   *Lear had to stay* Martin, p.431; Batchelor, p.277

319   *'We looked at Orion'* T Letters 2, p.141

320 *'Others saw it slate-colour'* January 1856, ibid., pp.141, 143

320 *'I have got some 15 new specks'* 1854, T Letters 2, p.84

320 *'morning and evening'* T Selected Poems, pp.511–12

320 *'my little Hamlet'* Ibid., p.515n

321 *'I have felt with my native land'* Maud, Part 3, T Norton Poetry, p.146

321 *'But arose, and all by myself'* Maud, Part 1, ibid., p.313

322 *'My life has crept so long'* Maud, Part 3, ibid., pp.344–5

323 *Tennyson's accumulated royalties leaped* T Letters 2, p.144n

323 *'strain of puling incoherent sentiment'* Batchelor, p.216

323 *'thoroughly and intensely provincial'* T Chrono, p.72; Batchelor, p.218

323 *'slashed all to pieces'* T Letters 2, p.124

323 *'I always calculated'* Ibid., p.127

324 *'What a wonderful embodiment'* R.J. Mann, 1856, in T Crit Heritage, pp.204–5

324 *'with iron immensity'* Ibid., p.205

324 *'brought to understand'* T Norton Poetry, p.328

324 *'my heart is a handful of dust'* T Selected Poems, p.574

324 *'The syllables and lines'* R.J. Mann in T Critical Heritage, p.199

324 *'as true as it is full'* HT Memoir, pp.332, 342

325 *'the writings of Tennyson are peculiarly metaphysical'* John Bucknill, *Asylum Journal of Mental Science*, October 1855

325 *'more valuable to me'* T Letters 2, p.131, and footnote on Bucknill and the *Asylum Journal*

325 *'I seem to have the doctors'* T Letters 2, pp.145,147, 161; Edgar Shannon Jr, 'The Critical Reception of Tennyson's "Maud"', *PMLA*, 24 March 2021, p.405, note 38

325 *'I lost my way'* T Letters 2, p.126

326 *'For the pale blood of the wizard'* 'Merlin and Vivien', T Norton Poetry, p.413

326 *'It is true old wild English Nature'* T Letters 2, p.126

326 *'One morning he read to us'* James Field, *Atlantic Monthly*, 1859; T Letters 2, p.235n

327 *'The Poet Laureate neither'* D.G. Rossetti, *A Memoir,* September 1855, T Letters 2, p.128

327 *'groanings and horrors'* Batchelor, p.221

327 *Henry James remembered* H. James, *Autobiographies,* edited by Philip Horne, Library of America, 2016, pp.620–5

328 *'I like the Drama of Maud'* 15 July 1856, Fitz Letters 2, p.234

328 *'trudged out'* Michelle Geric, *Tennyson and Geology*, 2017, pp.151–3

328 *'Dragons of the prime'* In Memoriam, No. 56, T Norton Poetry, p.238

329 *'He struck me as being'* T Letters 2, p.32, and footnote on Richard Owen

329 *reading his old Cambridge* T Letters 2, p.84

329 *intelligent life forms* Michael J. Crowe, *Modern Theories of the Universe*, 1994, pp.164, 168, 175

329 *'Thus we appear to'* William Whewell, *On the Plurality of Worlds*, 1853, in Crowe, *Modern Theories of the Universe*, p.175

329 *'One school of moral discipline'* William Whewell in Crowe, 1994, p.379

329 *'carefully'* T Letters 2, p.84

330 *'I would not wish'* Ibid.; HT Memoir, p.319

330 *and Richard Proctor* Crowe, 1994, p.196

330 *'Flower in the crannied wall'* 1869, T Norton Poetry, pp.372–3

21 Young Laureate

331 *contributed to it* On Millais see T Letters 2, p.79n; p.89

331 *The handsome volume* A copy in the British Library catalogue 11647.e.59

331 *especially fine illustrations* Ibid., p.89

332 *picked out Rossetti's pictures* HT Memoir, p.354

332 *Ticknor and Fields* T Chrono, p.86; T Letters 2, p.146

332 *'Thereafter – as he speaks'* 'The Coming of Arthur', 1859, T Norton Poetry, p.381

333 *'But always excepting'* January 1870, Fitz Letters 3, p.183

333 *Royal Academy Summer show* T Chrono, p.84

333 *Railway Company might collapse* T Letters 2, p.135

334 *'beautifully expressed in'* George Douglas Campbell, Duke of Argyll, 'Geology: A Lecture', Glasgow 1859, pp.49–50; Batchelor, p.213; T Letters 2, p.562 and Appendix C on correspondence with the Duke of Argyll

334 *'Two loud rings at the bell'* A rapturous account of the royal surprise visit by Emily in T Letters 2, pp.150–1

335 *youngest brother* Franklin Lushington, appointed Judge in the Supreme Court of the Ionian Islands, 1855, Batchelor, pp.199–200

335 *'And trust me while I turn'd the page'* T Norton Poetry, p.293

335 *Swinburne, who worshipped* first visit in spring 1858, when Tennyson thought him 'very modest and intelligent', Batchelor, p.236

335 *probably of Fanny Cornford* Allingham *Diary*, June 1864, in T Letters 2, p.377n

336 *'radical' moustaches and beards* Kathryn Hughes, *Victorians Undone*, 2017, p.100

336 *not usually bearded* Christopher Lawrence and Steven Shapin, editors, *Science Incarnate: Historical Embodiments of Natural Knowledge*, Chicago University Press, 1998, pp.274–5; Thomas Maguire's 'Sixty Lithographs of Scientific Men', Ipswich Museum, 1847–51

337 *therapeutic disguise* T Letters 2, p.58; p.160

337 *taken four years later* NPG, one of twelve taken or printed by James Mudd 1857–61

337 *until Tennyson was dead* Batchelor, p.228

337 *'I wish the public'* Thwaite, p.331

338 *'Pray do not defend the beard'* Kathryn Hughes, *Victorians Undone*, p.124

338 *finally got Tennyson* March 1854, T Letters 2, p.82; Batchelor, pp.254, Thwaite, p.331

339 *hint of a monocle Victorian Giants: The Birth of Art Photography*, NPG, 2022, pp.111–13

339 *'Mrs Cameron's wildly romantic'* HT Memoir, p.487; p.490

339 *album presented to Sir John Herschel* 'The Herschel Album', Science Museum Archive

339 *'Scientific leaders'* This was his son Hallam Tennyson, HT Memoir, p.250

340 *'staying at a great House'* Fitz Letters 2, p.272 and Batchelor, p.242

340 *'I called yesterday'* T Letters 2, p.171

341 *'Persian letters stalking'* HT Memoir, p.315

341 *'Have you done any Persian?'* Fitz Letters 2, p.135

341 *'We read some curious'* 15 July 1856, ibid., pp.233–4

341 *'I think I shall never again'* Ibid., p.166

341 *'Tennyson's earlier poems'* Ibid., p.31

341 *'Omar, the Astronomer Poet of Persia'* January 1859, ibid., pp.325–6n

342 *'What Scholarship it has is yours'* Terhune, *Life of Edward FitzGerald*, p.174

342 *'I don't know when I shall'* Fitz Letters 2, 18 July 1857, pp.291–2

342 *particularly appreciated by Tennyson* See FitzGerald's exchange with Tennyson about 'the Dawn of Nothing', 'Nothingness', and 'the French *Néant*', April 1872, Fitz Letters 3, p.342; p.345

342 *'Awake! for Morning in the Bowl of Night'* Edward FitzGerald, *Rubaiyat of Omar Khayyam*, No. 1, First edition 1859, edited Laurence Housman, Collins 1928; Edward FitzGerald, *Rubáiyát of Omar Khayyam*, editor Daniel Karlin, Oxford World's Classics, 2009, p.16

343 *'begins with Dawn'* Letter to Quaritch, quoted Terhune, *Life of Edward FitzGerald*, p.227

343 *'Oh, come with old Khayyam'* Omar, No. 26, Housman, p.72; *Omar*, Karlin, p.29

343 *'My own dim life should teach me this'* In Memoriam, No. 34; T Norton Memoriam, p.28

344 *'Then to the Lip of this poor earthen Urn'* Omar, Housman, No. 39; *Omar*, Karlin, p.33

344 *'Myself when young did eagerly frequent'* Omar, No. 27–8, Housman; *Omar*, Karlin, pp.29–30

345   '*I know none*' *Omar*, Karlin, pp.l–li; *Omar*, edited by Tony Briggs, Phoenix, 2009, p.xxiii

345   '*I do with all my soul*' Terhune, *Life of Edward FitzGerald*, p.212

345   '*Pagan Paradise*' To E FitzGerald, T Selected Poems, p.634

345   '*Here with a Loaf of Bread*' *Omar*, No. 11, Housman, p.69; *Omar*, No. 12, editor Tony Briggs, p.7; *Omar*, No. 11, Karlin, p.21

346   '*But when those others, one by one*' T Selected Poems, p.438; T Norton Memoriam, p.69

346   '*Yon rising Moon*' *Omar*, No. 100, Housman; *Omar*, No. 104, Tony Briggs, p.25; *Omar*, No. 74, Karlin, p.53

346   *Tennyson's increasing wealth and celebrity* Batchelor, p.338

346   '*I don't know what he thinks now*' 8 April 1872 to Milnes, Fitz Letters 3, p.344; Batchelor, p.304

346   '*I do not think it worthwhile*' 12 April 1872, Fitz Letters 3, p.346

346   '*somehow*' lost April 1872, Fitz Letters 3, p.346

346   '*Is Mr Rossetti a Great Poet*' Ibid., p.243. On Omar's 'Nothingness' p.342 and on Keats p.344

347   '*I used to tell Tennyson*' 12 April 1874, Fitz Letters 3, p.487

347   *lasted only a few months* Terhune, *Life of Edward FitzGerald*, pp.192, 201

347   '*They seem to me*' 1874, Fitz Letters 3, p.487; Batchelor, p.214

348   '*I mourn over him*' April 1874, Fitz Letters 3, p.487

348   '*The Moving Finger writes*' *Omar*, No. 51, Karlin, p.41; *Omar* No. 76, Tony Briggs, p.19

348   '*who took all care of Papa*' Terhune, *Life of Edward FitzGerald*, p.321

348   '*am here*' Ibid.

349   '*sitting among his doves*' HT Memoir, p.683

349   '*Old Fitz, who from your suburb grange*' T Selected Poems, p.631

350   '*I suppose this is a sign of age*' HT Memoir, p.683

350   *to Harwich point and back* Martin, p.515

350   '*might as well ship his Oars now*' Terhune, *Life of Edward FitzGerald*, p.321

350   '*to be haunted by that spirit*' T Interviews, p.126

350   '*A spirit haunts the year's last hours*' T Selected Poems, p.13

350   '*Dear old Fitz*' Terhune, *Life of Edward FitzGerald*, and Batchelor, p.339

351   '*And so I send a birthday line*' T Selected Poems, p.631

351   '*My dear old comrade*' CT Alfred, p.467

352   '*your golden Eastern lay*' T Selected Poems, p.632

352   '*I had written a poem to him*' Ibid., p.369

Bibliography

Early published works by Young Tennyson

Poems by Two Brothers (privately published, Louth), 1827

Poems Chiefly Lyrical (Effingham Wilson, London), 1830

Poems (Moxon, London), 1833 (1832)

Poems in Two Volumes (Moxon, London), 1842

The Princess: A Medley (Moxon, London), 1847

The Princess: A Medley (Moxon, London), 1850 (revised, with additional lyrics)

In Memoriam (Moxon, London), 1850 (composed 1833–1849; revised editions 1851, 1864)

Maud and Other Poems, 1855 (including 'The Charge of the Light Brigade'; revised editions, 1859, 1865)

The Illustrated Tennyson (Moxon, London), 1857

Some popular science books read by Young Tennyson
(in chronological order of publication)

The Microscope Made Easy, Henry Baker, 1742

Dialogues Concerning Natural Religion, David Hume, 1779

History of British Birds, Thomas Bewick, 1797

Natural Theology, William Paley, 1802

Conversations in Chemistry, Jane Marcet, 1805

Conversations in Natural Philosophy, Jane Marcet, 1819

Journal d'un voyage à Temboctou, René Caillié, 1828

On the Study of Natural Philosophy, John Herschel, 1830

Astronomy and General Physics, The Third Bridgewater Treatise, William Whewell, 1833

A Treatise on Astronomy, John Herschel, 1833

The Principles of Geology, Charles Lyell, Vol. 1, 1830; Vol. 2 1833

On the Connexion of the Physical Sciences, Mary Somerville, 1834

The Architecture of the Heavens, John Pringle Nichol, 1837

The Ninth Bridgewater Treatise: A Fragment, Charles Babbage, 1837

A History of the Inductive Sciences, William Whewell, 1837

Hydropathy; or The Cold Water Cure, as practiced by Vincent Priessnitz, at Graefenberg, Silesia, Austria, Vincent Priessnitz, 1842

Modern Painters, John Ruskin, Vol. 1, 1843

Vestiges of the Natural History of Creation, Robert Chambers, 1844

The Poetry of Science: Studies of the Physical Phenomena of Nature, Robert Hunt, 1844

The Premature Burial, Edgar Allan Poe, 1844

Physical Geography, Mary Somerville, 1845

The Voyage of the Beagle: Journal of Researches into Geology and Natural History, Charles Darwin, 1845

The System of the World, John Pringle Nichol, 1846

System of the Heavens as Revealed by Lord Rosse's Telescope, Thomas de Quincey, *Tait's Magazine*, 1846

On the Plurality of Worlds, William Whewell, 1853

On the Origin of Species by Means of Natural Selection, Charles Darwin, 1859

Select Bibliography

Will Abberley, *Underwater Worlds*, Cambridge Scholars, 2018

Matthew Allen, *Essay on the Classification of the Insane*, 1837

William Allingham, *A Diary, 1824–1889*, edited by John Julius Norwich, Penguin Lives and Letters, 1985

Isobel Armstrong, *Victorian Poetry: Poetry, Poets and Politics*, Routledge, 1993

W.H. Auden, *Collected Poems*, edited by Edward Mendleson, Faber, 2004

Jane Austen, *Emma*, 1815

Charles Babbage, *Passages from the Life of a Philosopher*, 1864, CUP, 2011

Samuel Bamford, *Passages in the Life of a Radical*, 1844

Julian Barnes, *Levels of Life*, Cape, 2013

Elizabeth Barrett, *Poems*, 1844

Elizabeth Barrett-Browning, *Sonnets from the Portuguese*, 1850

Elizabeth Barrett- Browning, *Aurora Leigh*, 1856

John Batchelor, *Tennyson: To Strive, to Seek, to Find*, Pegasus Books, New York, 2013

Jonathan Bate, *John Clare: A Biography*, Picador, 2003

Frances M. Brookfield, *The Apostles*, 1907

'Dr Brydon's Report of the Khyber Disaster January 1842', by William Trousdale, 1983, www.Kyber.org/publications

Edward Bulwer-Lytton, *Falkland*, 1827

Edward Bulwer-Lytton, *Confessions of a Water Patient*, 1845

John Carey, *Faber Book of Science*, Faber, 2012

Thomas Carlyle, *The French Revolution*, 1837

Thomas Carlyle, *Miscellaneous Essays*, 1839

Thomas Carlyle, *Chartism*, 1839, 1845

Thomas Carlyle, *On Heroes, Hero-Worship, and the Heroic in History*, 1841

Thomas Carlyle, *Past and Present*, 1843

Thomas Carlyle, *Life of John Sterling*, 1851

Robert Chambers, *Vestiges of the Natural History of Creation, Together with Explanations: A Sequel*, 1844, 1846, Cambridge Library Collection, CUP, 2009

John Clare, *The Rural Muse*, 1835

John Clare, *The Shepherd's Calendar*, 1827

John Clare, *Major Works*, editor Eric Robinson, Oxford World's Classics, 1984

John Clare, *Selected Poetry*, edited by Jonathan Bate, Faber, 2004

S.T. Coleridge, *Aids to Reflection*, 1825

S.T. Coleridge, *Biographia Literaria*, 1817

S.T. Coleridge, *Poetical Works*, 1829

S.T. Coleridge, *Table Talk*, edited by Henry Nelson Coleridge, 1835

S.T. Coleridge and W. Wordsworth, *Lyrical Ballads*, 1798

Joseph Conrad, *An Outcast of the Islands*, 1896

Wendy Cope, *Collected Poems*, Faber, 2024

Michael J. Crowe, *Modern Theories of the Universe: From Herschel to Hubble*, Dover, 1994

M.J. Crowe, *The Extra-terrestrial Debate: A Source Book*, University of Notre Dame Press, 2008

William Dalrymple, *Return of a King*, Bloomsbury, 2013

Richard Dawkins, *The Blind Watchmaker: Why the Evidence of Evolution Reveals a Universe without Design*, Norton, 1986

Ashley Dawson, *Extinction: A Radical History*, OR Books, 2022

Aidan Day, *Tennyson's Scepticism*, Palgrave Macmillan, 2005

Charles Dickens, *Dombey and Son*, 1846

Charles Dickens, *David Copperfield*, 1850

Charles Dickens, *Hard Times*, 1854

Charles Dickens, *A Tale of Two Cities*, 1859

Joan Didion, *The Year of Magical Thinking*, 4th Estate, 2012

Kimberley Dimitriadis, 'Telescopes in the Drawing-Room: Geometry and Astronomy in George Eliot's *The Mill on the Floss*', *Journal of Literature and Science*, 2018

George Eliot, *Adam Bede*, 1859

T.S Eliot, *The Wasteland*, 1922

T.S. Eliot, 'In Memoriam', *Selected Essays*, 1936, Faber, 1955

William Brent Elliott, 'Tennyson and the Concept of Evolution in Victorian Poetry before 1859', University of British Columbia, PhD thesis, 1973

Kimberly Eve, 'Meet the Tennyson Sisters of Somersby', *Victorian Musings*, 2015, online

Pamela Faithfull, *An Evaluation of an Eccentric: Matthew Allen*, University of Sheffield, PhD thesis, 2001, online

James Fenton, *Selected Poems*, Penguin, 2006

Edward FitzGerald, *Rubáiyát of Omar Khayyám*, editor Daniel Karlin, Oxford World's Classics, 2009

Edward FitzGerald, *Rubaiyat of Omar Khayyam*, First edition 1859, edited Laurence Housman, Collins, 1928

Edward FitzGerald, *Rubáiyát of Omar Khayyám*, edited by Tony Briggs, Phoenix, 2009

Theodor Fontane, *The Tragedy of Afghanistan*, 1858

Michel Foucault, *The History of Madness*, revised edition, Routledge, 2006

Adam Foulds, *The Quickening Maze*, Jonathan Cape, 2009

J.A. Froude, *The Nemesis of Faith*, 1849

Elizabeth Gaskell, *Mary Barton*, 1848

Michelle Geric, *Tennyson and Geology: Poetry and Poetics*, Palgrave Macmillan, 2017

William Glen, editor, *The Mass-Extinction Debates: How Science Works in a Crisis*, Stanford University Press, 1994

Rosalie Glynn Grylls, *Queen's College 1848–1948*, Routledge, 1948

Peter Godfrey-Smith, *Other Minds: The Octopus and the Evolution of Intelligent Life*, William Collins, 2017

P.H. Gosse, *The Aquarium: An Unveiling of the Wonders of the Deep Sea*, 1854

P.H. Gosse, *A Handbook to the Marine Aquarium*, 1856

Stephen J. Gould, *Dinosaur in a Haystack: Reflections in Natural History*, Crown Trade Paperbacks, New York, 1995

Henry Hallam, *Remains in Verse and Prose of Arthur Henry Hallam*, 1834

Robert Stephen Hawker, *Echoes from Old Cornwall*, 1846

Robert Stephen Hawker, *The Quest of the Sangraal*, 1864

Alethea Hayter, *FitzGerald to his Friends*, Scolar Press, 1979

Anna Henchman, *The Starry Sky Within: Astronomy and the Reach of the Mind in Victorian Literature*, OUP, 2014

John Herschel, *Outlines of Astronomy*, 1849

Jillian Hess, *How Romantics and Victorians Organized Information: Commonplace Books, Scrapbooks, and Albums*, OUP, 2022

Christopher Hibbert, *The Destruction of Lord Raglan: A Tragedy of the Crimean War, 1854–55*, Viking, 1984

Richard Holmes, *Coleridge: Darker Reflections*, HarperCollins, 1998

Richard Holmes, *The Age of Wonder*, HarperCollins, 2008

Richard Holmes, 'Computer Science: Enchantress of Abstraction', *Nature*, 3 September 2015

Richard Holmes, *This Long Pursuit: Reflections of a Romantic Biographer*, William Collins, 2016

Homer, *Iliad*

Homer, *Odyssey*

Richard Horne and Elizabeth Barrett, *The New Spirit of the Age*, 1844

William Howitt, *Homes and Haunts of the Most Eminent British Poets*, Vol. 2, 1847

Edwin Hubble, *The Realm of the Nebulae*, 1936

Kathryn Hughes, *Victorians Undone: Tales of the Flesh in the Age of Decorum*, 4th Estate, 2017

Ted Hughes, *Hawk in the Rain*, Faber, 1957; *Lupercal*, Faber, 1960

Alexander Humboldt, *Cosmos*, 1845

Robert Hunt, *The Mount's Bay: a Descriptive Poem*, 1829

Robert Hunt, *The Poetry of Science*, 1844

James Hutton, *The Theory of the Earth*, 1788

T.H. Huxley, *Evidence as to Man's Place in Nature*, 1863

Thomas Huxley, 'Agnosticism', *Collected Essays V*, 1889

Mick Imlah, *Alfred, Lord Tennyson: Poems Selected*, Faber, 2024

Roland Jackson, *The Ascent of John Tyndall: Victorian Scientist, Mountaineer and Public Intellectual*, OUP, 2018

Samuel Johnson, *The History of Rasselas, Prince of Abyssinia*, 1755

Alexander Johnston, *School Atlas of Astronomy*, 1856

Garrett Jones, *Alfred and Arthur: An Historic Friendship*, independently published, 2017

John Keats, *Letters of John Keats*, edited by Robert Gittings, OUP, 1970

John Keats, *Life, Letters and Literary Remains*, 2 vols, edited by Richard Monckton Milnes, Moxon, 1848, CUP, 2013

John Killham, *Tennyson and the Princess: Reflections of an Age*, Athlone Press, University of London, 1958

Charles Kingsley, *Alton Locke*, 1850

Charles Kingsley, *The Water Babies*, 1863

George P. Landow, 'John Ruskin: Loss of Belief', Victorian Web, online

Pierre-Simon Laplace, *Mécanique Céleste*, 1799

Philip Larkin, *Collected Poems*, edited by Anthony Thwaite, Faber, 2003

Christopher Lawrence and Steven Shapin, editors, *Science Incarnate: Historical Embodiments of Natural Knowledge*, Chicago University Press, 1998

Edward Lear, *Book of Nonsense*, 1861

C.S. Lewis, *A Grief Observed*, Faber, 1961

Bernard Lightman, *Victorian Popularizers of Science: Designing Nature for New Audiences*, Chicago University Press, 2007

Carl Linnaeus, *Systema Naturae*, 1735

Norman Lockyer, *Spectroscopic Observations of the Sun*, 1870

Norman Lockyer, *Tennyson as a Student and Poet of Nature*, 1910

Ada Lovelace, *Notes*, 1843, in *Ada, The Enchantress of Numbers: Poetical Science* by Betty Alexandra Toole, 2010

Charles Lyell, *Principles of Geology*, editor James A. Secord, Penguin Classics, 1997

George MacDonald Fraser, *Flashman in the Great Game*, 1969

Dr Robert Mann, *Maud Vindicated: An Explanatory Essay*, 1856

Robert Bernard Martin, *Tennyson: The Unquiet Heart*, Faber, 1983

Robert Bernard Martin, *With Friends Possessed: A Life of Edward FitzGerald*, Faber, 1985

Frederick Denison Maurice, *Theological* Essays, 1853

Herman Melville, *Moby Dick*, 1850

John Stuart Mill, *On the Subjection of Women*, 1869

John Stuart Mill, *On Liberty*, 1859

John Stuart Mill, *Mill on Bentham and Coleridge*, CUP, 2010

Hugh Miller, *Old Red Sandstone*, 1845

John Milton, *Lycidas*, 1638

Harold Nicolson, *Tennyson: Aspects of his Life, Character and Poetry*, Constable, 1923

Harold Nicolson, *Tennyson's Two Brothers*, CUP, 1947

Leonee Ormond, editor, *The Reception of Tennyson in Europe*, Bloomsbury, 2017

Laura Otis, editor, *Literature and Science in the Nineteenth Century*, Oxford World's Classics, 2009

Ruth Padel, 'Introduction', *The Folio Tennyson*, Folio Society, 2009

Francis Turner Palgrave, *Golden Treasury: English Songs and Lyrics*, 1861

Coventry Patmore, *The Angel in the House*, 1854

Seamus Perry, *Alfred Tennyson*, Northcote House, 2005

Pliny the Elder, *Naturalis Historia*, AD 77

Edgar Allan Poe, *The Raven and Other Poems*, 1845

Edgar Allan Poe, 'The Premature Burial', 1844, *Works Volume 2*, Raven Edition, Project Gutenberg

Erik Pontoppidan, *The Natural History of Norway*, 1755

Richard Proctor, *Half-Hours with the Telescope*, 1868

Valerie Purton, editor and 'Introduction', *Darwin, Tennyson and Their Readers: Explorations in Victorian Literature and Science*, Anthem Press, 2013

Thomas de Quincey, *System of the Heavens as Revealed by Lord Rosse's Telescope*, *Tait's Magazine*, 1846

Ralph Rader, *Tennyson's Maud: The Biographical Genesis*, California University Press, 1963

Christopher Ricks, *Tennyson*, Macmillan, 1972

Christopher Ricks, editor, *The Poems of Tennyson*, Longman Annotated English Poets, 3 vols, 1969

Christopher Ricks, editor, *Tennyson: A Selected Edition*, Longman, 1969; revised edition 1989, 2007

Mary Robinson, *Letter to the Women of England*, 1799

John Ruskin, *The Stones of Venice*, 1851

Alfred Russel Wallace, 'The Origin of Human Races and the Antiquity of Man Deduced from the Theory of "Natural Selection"', 1864

Matthew Sangster, *Living as an Author in the Romantic Period*, Palgrave Macmillan, 2021

James A. Secord, *Victorian Sensation: The Extraordinary Publication, Reception, and Secret Authorship of Vestiges of the Natural History of Creation*, University of Chicago Press, 2000

James A. Secord, *Visions of Science: Books and Readers at the Dawn of the Victorian Age*, OUP, 2014

William Shakespeare, *Love's Labour's Lost*, 1597

William Shakespeare, *Twelfth Night*, 1601

William Shakespeare, *Measure for Measure*, 1604

Edgar Shannon Jr, 'The Critical Reception of Tennyson's "Maud"', *PMLA*, 24 March 2021

Mary Shelley, *Frankenstein, or the Modern Prometheus*, 1818, revised 1831

Percy Bysshe Shelley, *Queen Mab*, 1813

Percy Bysshe Shelley, *Adonais*, 1821

Percy Bysshe Shelley, *Epipsychidion*, 1821

John Simpson, *We Chose to Speak of War and Strife: The World of the Foreign Correspondent*, Bloomsbury, 2016

Samuel Smiles, *Self Help*, 1859

E.E. Snyder, 'Tennyson's Progressive Geology', *Victorian Network*, Vol. 2, No. 1, 2010, online

Mary Somerville, *Personal Recollections of Mary Somerville*, 1873

Rebecca Stott, *Tennyson*, essays edited and introduced, Longman, 1996

Doron Swade, *The Difference Engine: Charles Babbage and the Quest to Build the First Computer*, Penguin, 2000

Gregory Tate, *Nineteenth-Century Poetry and the Physical Sciences*, Palgrave Macmillan, 2020

Gregory Tate, 'Robert Hunt, *The Poetry of Science: Studies of the Physical Phenomena of Nature, 1844*', Project Gutenberg, 2016, online

Charles Tennyson, *Sonnets and other Fugitive Pieces*, 1830

Sir Charles Tennyson, *Alfred Tennyson*, Macmillan, 1949

Sir Charles Tennyson and Hope Dyson, *The Tennysons: Background to Genius*, Macmillan, 1974

Hallam Tennyson, *Alfred Lord Tennyson: A Memoir*, 1 volume edition, 1899

Alfred Terhune, *The Life of Edward FitzGerald*, OUP, 1947

Ann Thwaite, *Emily Tennyson: The Poet's Wife*, Faber, 1996

Ann Thwaite, *Glimpses of the Wonderful: The Life of Philip Henry Gosse*, Faber, 2002

Jenny Uglow, *Nature's Engraver: A Life of Thomas Bewick*, Faber, 2006

Aubrey de Vere, *The Search after Proserpine: Memories of Greece*, 1843

Jules Verne, *Twenty Thousand Leagues Under the Sea*, 1872

Mary Ward, *The Telescope*, 1859

Mary Ward, *The Microscope*, 1859

H.G. Wells, *New World Order*, 1940

Basil Willey, 'Tennyson's Honest Doubts', *More Nineteenth Century Studies*, 1956

John Willis Clark and Thomas McKenny Hughes, editors, *Life and Letters of Adam Sedgwick*, 1890

Leonard G. Wilson, *Charles Lyell: The Years to 1841*, Yale University Press, 1972

Mary Wollstonecraft, *Thoughts on the Education of Daughters*, 1787

Mary Wollstonecraft, *Vindication of the Rights of Woman*, 1792

William Wordsworth: Selected Poetry, editor Nicholas Roe, Penguin, 1992

W.B. Yeats, 'Poetry and Symbolism', *Ideas of Good and Evil*, 1903

List of Illustrations

FIRST PLATE SECTION

Alfred Tennyson, by Samuel Laurence, and Sir Edward Burne-Jones, c.1841. © National Portrait Gallery, London

The Old Rectory, Somersby. Postcard, early 20th century. © Look and Learn/Bridgeman Images

Mablethorpe Beach, Lincolnshire. Photograph by R. Kawka, 1894. © Alamy Stock Photo

Great Court, Trinity College, Cambridge University, c.1870

Alfred Tennyson as an undergraduate, by James Spedding, 1831. © National Portrait Gallery, London

Edward FitzGerald as an undergraduate, by James Spedding, c.1832. © Fitzwilliam Museum

Arthur Hallam, bust by Sir Francis Chantrey. Photogravure from *Alfred Lord Tennyson and his Friends*, printed in 1893. © Heritage Image Partnership Ltd/Alamy Stock Photo

William Whewell, by James Lonsdale, 1825. © Art Collection 3/Alamy Stock Photo

The Lady of Shalott by William Holman Hunt, from *Poems by Alfred Tennyson, D.C.L., Poet Laureate*, London, 1857. © Chronicle/Alamy Stock Photo

The Lady of Shalott by John William Waterhouse, 1888. © Tate, Photo: Tate

The Lady of Shalott by John William Waterhouse, 1894. © Leeds Museums and Galleries, UK/Bridgeman Images

Edward FitzGerald in middle age, by or after Cade & White of Ipswich, 1873. © Bridgeman Images

SECOND PLATE SECTION

Tennyson, by James Mudd, 1857. © National Portrait Gallery, London

Emily Sellwood. Miniature on ivory, 1835. In the collection of the Beinecke Rare Book & Manuscript Library, Yale

Rosa Baring, by Camille Silvy, 1861. © National Portrait Gallery, London

Charles Lyell, taken from a daguerreotype by J.E. Mayall, 1846

Young Charles Darwin, by George Richmond, 1840. © Historic England/Bridgeman Images

Robert Chambers, by Daniel John Pound, after John Jabez Edwin Mayall's engraving, 1860. © The History Collection/Alamy

Portrait of Elizabeth Barrett Browning, from *The poetical works of Elizabeth Barrett Browning* (London, 1889–90). © UniversalImagesGroup/Getty

Mary Somerville, by James Rannie Swinton, 1844

Charles Babbage, by Samuel Laurence, c.1844–45. © National Portrait Gallery, London

The Whirlpool Galaxy M51, from the Hubble Space Telescope. © NASA

The Leviathan Telescope, illustration. © Science History Images/Alamy

Sir John Frederick Herschel FRS, by Julia Margaret Cameron, 1867. © The Rubel Collection, Gift of William Rubel

Danaë and the Shower of Gold, 1564, by Titian. © Kunsthistorisches Museum, Vienna

Thomas Carlyle, by Robert Scott Tait, 1854. © Princeton University Art Museum. Robert O. Dougan Collection, Gift of Warner Communications, Inc.

Jane Carlyle, by Robert Scott Tait, 1855. © National Portrait Gallery, London

John Ruskin, by Sir John Everett Millais, 1854. © The Print Collector/
    Alamy
Officers and men of the 13th Light Dragoons. Photograph by Roger
    Fenton, 1855. © National Army Museum
Tennyson with his wife and children at Farringford, c.1862–64. Pho-
    tographed by Oscar Gustav Rejandler. © Granger/Bridgeman
Alfred Tennyson, by Julia Margaret Cameron, 1865. © Alto Vintage
    Images/Alamy
Emily Tennyson, by George Frederick Watts, 1862. © Painters/Alamy

Index

records/accounts of Tennyson in, 219–20, 265; 22 St James's Place, 220; British Museum, 227, 250; Ebury Street, Belgravia, 245; Tennyson's new social life (1848), 245, 249, 250; Chartist gatherings (spring 1848), 251; Great Exhibition, Hyde Park (1851), 296–7; Dorset Street, Marylebone, 327; Argyll House, 334; London Library, St James's Square, 340

London Committee of Spanish exiles, 59–60

London Gazette, 313

London Review, 84–5, 122, 125

Longfellow, Henry Wadsworth, 193, 218, 326–7

Louis-Philippe, King of France, 251

Louth grammar school, 17

Lovelace, Ada, 157, 201–2, 233

Lowell, James Russell, 193

Lucan, Lord, 315, 318

Lucas, Hippolyte, 125

Lucretius, *De Rerum Natura*, 244–5

Lushington, Edmund: support and sustenance of Tennyson, 3, 48, 162–3, 175, 187–8, 198, 220; in Tennyson's friendship group at Cambridge, 3, 47, 48; background and character, 48; hospitality of, 48, 162–3, 175, 187–8; marriage to Cecilia Tennyson, 48, 175, 176, 188, 270, 286; Park House, near Maidstone, 48, 152, 162–3, 166, 175, 187–8, 220, 231, 234, 270, 291; Professor of Greek at Glasgow, 48, 187, 226; and Tennyson's mourning, 111–12; chambers at 2 Mitre Court Buildings, 195, 217, 220, 245; visits Samuel Rogers, 220; and 'Woman's University' idea, 230; in *The Princess: A Medley*, 235; *In Memoriam* dedicated to, 279; advises on investments, 306

Lushington, Franklin, 335

Lushington, Henry, 48

Lyell, Charles, 4, 113, 203, 272, 336, 339; and Jane Marcet, 19, 20–1; background of, 20–1, 87–8; and extinction theory, 36, 87, 88–90, 91, 149–50, 205–6, 207; and deep geological time, 87, 88–90, 91, 207, 252, 259–61; and Natural Theology, 88, 89, 130, 206; rejects Darwinian evolution, 88, 334; images from in Tennyson's poetry, 91–2, 149–50, 205–6, 208; and the relentlessly changing earth, 262; *The Principles of Geology* (1830/1832), 21, 36, 88, 91, 129, 149–50, 205–6, 207, 208, 223, 240, 259–61

Lyonnesse, 123, 178, 255

Lyrical Ballads (Coleridge and Wordsworth, 1798), 56

Mablethorpe beach, 25, 105–6, 148–50, 160, 165, 264; Tennyson's childhood escapes to, 4–5, 15–16; Tennyson family cottage at, 15, 16, 170–1, 174, 195

Macaulay, Thomas Babington, 1, 42

Maclise, Daniel, 331

The Maidstone and Kentish Advertiser, 187–8

Maidstone Mechanics' Institute festival, 48, 187–8, 220, 231, 234

Malory, Sir Thomas, 68

Malvern spa, 195, 198, 254, 255–6

mammoths, 35, 240

Manchester, 275–6; Art Exhibition, 337

Mann, Dr Robert, 324, 325

Mansfield, Katherine, 233

Marcet, Alexander, 19

Marcet, Jane, 19–21, 26

Martin, Robert Bernard, 62

Martineau, Harriet, 19, 216, 233

materialism, 45, 80, 215

mathematics, 7, 64, 130, 155, 191, 201–2, 208

Maud: A Monodrama (Alfred Lord Tennyson, 1855): Harrington Hall in, 115, 143; 'Come into the garden, Maud' lyric, 116–17, 317, 320; and Tennyson's women friends, 116–17, 128, 146–7, 152, 321; love lyric in, 132–3, 146; madness theme, 139, 143–4, 146, 147, 150, 152, 320–2, 324–5; 'buried alive' passages, 144, 145, 320–1, 324;